A TOP-DOWN, CONSTRAINT-DRIVEN DESIGN METHODOLOGY FOR ANALOG INTEGRATED CIRCUITS

A TOP-DOWN, CONSTRAINT-DRIVEN DESIGN METHODOLOGY FOR ANALOG INTEGRATED CIRCUITS

Henry CHANG
Edoardo CHARBON
Umakanta CHOUDHURY
Alper DEMIR
Eric FELT
Edward LIU
Enrico MALAVASI
Alberto SANGIOVANNI-VINCENTELLI
Iasson VASSILIOU
University of California
Berkeley, California, USA

KLUWER ACADEMIC PUBLISHERS
Boston/London/Dordrecht

Distributors for North America:
Kluwer Academic Publishers
101 Philip Drive
Assinippi Park
Norwell, Massachusetts 02061 USA

Distributors for all other countries:
Kluwer Academic Publishers Group
Distribution Centre
Post Office Box 322
3300 AH Dordrecht, THE NETHERLANDS

Library of Congress Cataloging-in-Publication Data

A C.I.P. Catalogue record for this book is available
from the Library of Congress.

Printed on acid-free paper.

Printed in the United States of America

CONTENTS

Contents vii

1

INTRODUCTION

The complexity of electronic systems being designed today is increasing in many dimensions: on one hand the number of components is growing constantly, on the other several radically different functions must be integrated. For example, in the exploding personal communications market, a product is the combination of wireless transmission, analog and digital signal processing, and digital computing. Antennas, radio-frequency components, and analog and digital sub-systems have to be designed in a unified way to meet the performance, power, and size constraints imposed by the application.

Designing integrated circuits is inherently a complex task involving human expertise as well as aids intented to accelerate the process. The objective is the implementation of a system from specifications to a marketable product. A fundamental requirement for success is a clear strategy that coordinates the entire design process. In large projects, non-systematic or inconsistent design methodologies often result in limiting the efficiency of design teams.

Traditionally, integrated circuit design has been classified into two categories—digital and analog. While microprocessors and microcontrollers are inherently digital components, certain functions can be realized in analog or digital form. A typical example is signal processing. Signals can be manipulated as waveforms, or they can be first encoded and then manipulated in the digital domain. Noise is often the limiting factor in the quality of analog signal processing. In the digital domain, noise has much less influence. Hence there has been a strong trend towards moving computation carried out in the analog domain to its digital counterpart. However, there are still functions that have to be carried out in the analog domain. An example is signal conversion (the "real" world is analog!). In addition, there are computations that are still much more efficient in the analog domain. An example is filtering in an analog signal path (the signal conversion overhead is too expensive). Because of the continuous quest towards smaller and smaller electronic systems, many integrated circuits being designed today are mixed analog-digital. Hence, more and more chips depend upon the ability to design effectively analog components. Because of its noise sensitivity and its critical dependence on parasitics, analog design is inherently time consuming. In mixed digital-analog systems, noise injected from digital circuits further complicates analog design. Thus, in mixed-signal systems where the analog circuits are small, analog design is often a bottleneck.

1.1 ROLE OF COMPUTER-AIDED-DESIGN

The main objective of computer-aided-design (CAD) is the creation of *methodologies* and *tools* for the design of engineering systems, helping human designers build functionality while satisfying intended performance specifications. Over the past three decades, the development of computer aids for the design of electronic systems has been one of the fastest growing areas of activity. In particular, CAD for the physical assembly of electronic systems, either in the form of an integrated circuit (IC) or of a printed circuit board (PCB), has become one of the largest research areas in the field.

Electronic ICs have rapidly evolved from the relatively low complexity of the early days to the high sophistication of today. The task of circuit designers has become increasingly difficult, hence the need for more advanced design aids. In particular, the study of effective methodologies for the design of high-speed analog and mixed-signal ICs and of tools supporting it has been a very active topic of research in the past decade. This subject is the central topic of this book.

1.2 COMPUTER-AIDED-DESIGN OF ANALOG AND MIXED-SIGNAL ICS

In digital systems signals are represented by sequences of binary digits, so these signals can assume discrete values only. Due to the binary nature of these signals, digital circuits are realized using gates with only two states, with each state defined as a specific range of the continuous signal. This makes digital circuits, to a large degree, immune to various noise and parasitic sources inherent in ICs. Hence the design effort can be directed mainly towards trade-offs between power consumption, speed, and area.

In analog and mixed-signal systems, however, signals are continuous, and the design of these circuits exploits more degrees of freedom than the design of digital circuits. Analog circuits often utilize the full spectrum of capabilities exhibited by individual devices. In analog circuits the individual devices often have substantially different sizes and electrical characteristics. These circuits require optimization of various performance measures. As an example, among the performance measures for operational amplifiers are gain, bandwidth, noise, power supply rejection, dynamic range, offset voltage, and so on. The importance of each performance measure depends upon the circuit application. For this reason, fine tuning plays a crucial role in the design of analog circuits.

Because of the rather wide range of parameter spreads in ICs, analog designers have developed circuits which cancel out the first-order effects caused by variations in key parameters. Second-order effects dominate performance. Typical examples are the matching of input devices in differential pairs, or capacitor matching in switched capacitor filters. Second-order effects become especially critical during the circuit's physical assembly because of the numerous non-idealities and parasitics introduced.

For these reasons designing CAD tools for analog applications is, in general, a difficult task. Consequently, while it is sometimes possible to share CAD tools between the digital and analog portions of a circuit, such as design rule checkers, extractors, and databases, there are many tools that must be designed for use primarily on analog circuits. A general and consistent methodology is required to properly guide the tools towards the satisfaction of all specifications at the system level. In addition, design failures must be interpreted effectively so as to organize appropriate redesign schemes.

Research on analog CAD systems has progressed at a considerably slower pace than research on digital CAD. Part of the reason has been the intrinsic difficulty of defining and controlling performance in analog circuits. High performance can be achieved by taking advantage of the physical characteristics of integrated devices and of the correlation between electrical parameters and their variations due to statistical fluctuations of the manufacturing process. Device matchings, parasitics, thermal effects, and substrate effects must all be taken into account. The nominal values of performance functions are subject to degradation due to a large number of parasitics which are generally difficult to estimate accurately before the actual layout is completed. Another reason might be the present difficulty in identifying a level of abstraction where generic models, such as the ones developed for digital synthesis, can be derived.

All these concerns need to be addressed in each phase of the design with equal care. Severe performance degradation, even if localized only in some components, often jeopardizes the functionality of the whole system of which the component is a relevant part.

1.3 ORIGINS OF COMPUTER-AIDED-DESIGN FOR ELECTRONIC SYSTEMS

1.3.1 Circuit Simulation

The last three decades have seen a tremendous increase in the complexity and sophistication of electronic systems. Designing to realize functionality while meeting a set of performance specifications soon required the need for tools capable of overcoming relatively inaccurate and lengthy hand analysis. Not surprisingly, the first developed computer aids addressed the problem of circuit simulation and verification.

In the early 1950s digital computers began to be actively utilized in electrical engineering for the solution of simultaneous algebraic equilibrium-condition equations of linear electrical networks in the sinusoidal steady state [104]. Only a decade later, the first viable programs were developed for the simulation of circuits in the time domain. NET1 [195] and SCEPTRE [197] used explicit-integration and predictor-corrector techniques in the solution of integral-differential equations of nonlinear systems. To maintain stability, however, very small time steps were needed, which significantly increased the time needed to converge to a solution. It was only in the mid 1960s, with the introduction of the implicit integration scheme in combination with the backward-Euler method, that superior convergence performance could be achieved. In implicit integration, the set of integral-differential equations turns to a set of static algebraic equations for each given time point. The program TRAC [142] implemented these techniques. Approximately at the same time, Shichman proposed a second-order implicit-integration scheme that proved a better performance relative to TRAC. This research led to CIRPAC [275] and to other modifications to the method that included variable order and variable time-step implicit integration routines [94].

In the late 1960s, Howard at Berkeley developed a program that solved numerically a set of simultaneous nodal equations. A simple nonlinear device model was used and the equations were linearized at the equilibrium using iterative methods based on Newton-Raphson and excursion limiting techniques [200]. In contrast to the developments at Berkeley, an independent research effort starting from a theoretical base took place at IBM. Hachtel and colleagues proposed a new formulation of network equations based on the sparse-tableau concept [116]. This approach, allowing the use of far more efficient techniques for the solution of large systems of linear equations, led to the development of ASTAP [9].

The experience accumulated from these research teams was eventually incorporated in the CANCER [218] and the SLIC [137] projects. With the formalization of modified nodal analysis and the development of sparsity-aware pivoting and matrix reordering techniques, the CANCER project evolved into the SPICE program [217].

1.3.2 Digital Timing Analysis and Event-Driven Simulation

Following the enormous success of SPICE and the increasing importance of electrical simulation in circuit design, research in the field developed in two main directions: large-scale simulation and optimization. Early techniques, reviewed in [114], gave way to approaches purposely relaxing accuracy to achieve greatly improved simulation speed [43]. These methods, conceived for digital timing analysis, soon showed limitations in accurately simulating the effects of feedback. It was the study of numerical limitations in timing analysis that led to new techniques based on relaxation in both space and time domain. The main advantage of relaxation-based approaches is the ability of exploiting time sparsity, using the event-driven selective trace techniques first developed in digital simulators. Shortly after the development of timing simulation, mixed-mode or hybrid event-driven simulators emerged, resulting in extensive research in the field [63][223][260]. For a review in the field see [264][237][257].

More recently, this work has evolved into the development of techniques to reduce large lumped RCL circuits into a small, more tractable modal approximation of its transfer function. Thus a significantly higher efficiency can be achieved in simulating the network. An good example of this trend is represented by the asymptotic waveform evaluation method (AWE) [243] developed in the late 1980s, which has proven to be a valuable tool mainly in analysis and verification tasks.

1.3.3 Circuit Optimization

Automated design optimization [24] evolved in parallel to circuit simulation. In fact, the idea of using optimization to help design electrical circuits dates back to the early 1950s. DC biasing effects and frequency-domain matching were among the first considerations to be integrated in the optimization process [71][199]. Catalyzed by breakthroughs in simulation techniques and a new formalized representation of circuit optimization as a general nonlinear programming problem, significant effort was devoted to improving optimizer efficiency. A significant step toward achieving the goal is represented by drastic efficiency improvements in the calculation of network sensitivities, necessary for the most useful optimization algorithms. This work led to the development several tools. In the A2OPT project [117] the simulator ASTAP [9] was used in combination with a minimizer based on the rank-one update method [60]. Constraints were considered in the minimization by introducing an additional penalty to the objective function. A second optimization system based on ASTAP called APLSTAP [115] was built as an interactive CAD consultant tool. A linear programming step was used to quantify the best trade-offs between multiple objective and constraint functions to optimally guide the design process.

The above approaches had several disadvantages. These included a lack of flexibility at the formulation and implementation level, the relatively low degree of interactivity, and serious deficiencies at the simulation level. A successful attempt to alleviate these problems was made in DELIGHT.SPICE. The tool resulted from the merger of the optimizer DELIGHT [229] and SPICE, in which an efficient sensitivity analysis package had been incorporated [228]. Other tools followed on the same track, where more attention was given to user-interface and flexibility issues [281]. Despite their success, numerical circuit optimization tools soon became inadequate due to the explosion in complexity of analog and mixed-signal circuits. In addition, researchers realized the enormous influence of physical implementation on performance [279][290][174][143][134], hence the necessity of optimizing at the schematic design and layout synthesis levels simultaneously.

1.4 EARLY WORK IN COMPUTER-AIDED-DESIGN FOR ANALOG ICS

Due to the challenges posed by these problems, several research teams around the world actively began working on the creation of integrated design systems that would attack the problem of analog design in a systematic fashion. Three main schools of thought emerged to attack the problem: silicon compilation, knowledge-based techniques, and algorithmic methodologies.

1.4.1 Silicon Compilation

The first school of thought, introduced in the early 1980s, advocated the use of dedicated *silicon compilers* for the design and physical assembly of relatively complex yet highly specialized applications. The AIDE2 system [4] is one of the early examples of this trend. In AIDE2 the circuit topology, described using the C language, was mapped onto a fixed-floorplan layout based on a library of subcircuits. Libraries or library generators were provided as complements to the compilation system [5]. More recently, AIDE2 evolved to take into account higher-order effects and parasitics during the compilation process. The approach proposed in [132] was aimed at minimizing all parasitic effects by making use of linearized models of performance based on sensitivity. The fixed-

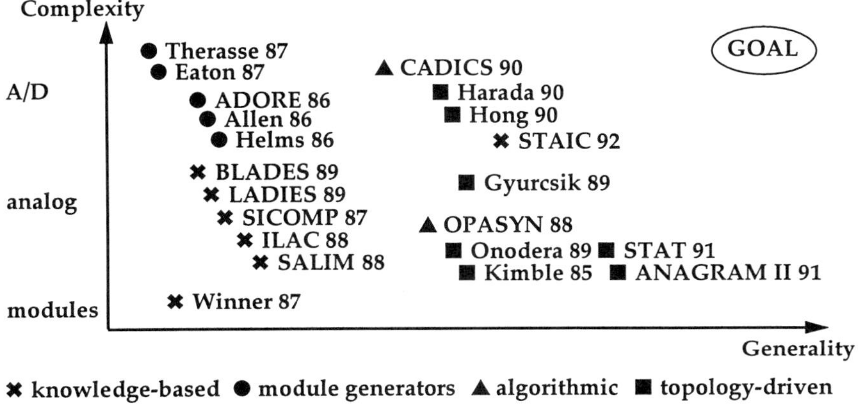

Figure 1.1 Early Work in Computer-Aided-Design for Analog ICs

floorplan layout synthesis style was replaced with a depth-first-search *topological sort* [158] operating on clusters of "sensitive" components, i.e. devices connected to sensitive nets. Numerically computed sensitivities were used to derive a priority schedule for a digital channel router [276].

1.4.2 Knowledge-Based Techniques

PROSAIC [22], the precursor of most knowledge-based systems, was one of the first tools using such an approach. The approach was originally derived from the work on declarative circuit modeling [250] and later [301]. The idea consisted of creating a large database of rules to be used by an inference engine driving a sequence of decisions determining the course of the design. A number of design systems using similar rule-based approaches appeared later in the literature [93][305]. For example, BLADES [76] was based on a conventional expert system consisting of a dedicated knowledge base and an inference engine. A numerical "consultant," generally a simulator, was used for verification purposes. There are several disadvantages associated with rule-based systems. The creation of the knowledge base or of the rule set is generally a relatively complex process requiring the expertise of highly experienced designers. Knowledge bases are very specific to a technology and even a very small class of problems, hence redesign and library synthesis turn-around is often exceedingly time consuming.

1.4.3 Algorithmic Methodologies

Algorithmic design methodologies first appeared in the mid 1980s for the layout synthesis of analog and high-speed digital circuits and soon migrated to schematic design automation and optimization. LTX2 [73][156] is the first example of this direction. In LTX2 the physical assembly problem was partitioned into placement, floorplanning, global and detailed routing, according to a classical scheme derived from the digital world [232][266]. A 2-D placement tool, based on a modification of the Kernighan and Lin algorithm for graph partitioning [151], was used to create divided clusters of analog and digital cells [74]. During placement, separation between sensitive signal nets and large swing analog and digital signals was guaranteed by alternating sensitive and insensitive routing channels in the standard cell floorplan. Detailed routing used shielding to minimize the coupling between sensitive nets residing in the same channel [72]. This technique was successfully applied to relatively simple circuits acting as an interface with digital cores.

1.5 FIRST COMPLETE DESIGN SYSTEMS

1.5.1 IDAC/ILAC

After the initial phase, experimentation gave way to increasingly complex and more flexible systems, designed for larger mixed-signal circuits and a number of technologies. The IDAC [61] system proposed a number of innovations later to be used by other systems. Among the most notable ones were a systematic architecture selection mechanism using simplified equation-based circuit analysis and a set of predefined *synthesis strategies*, a relatively large library of circuit topologies, and a layout synthesizer ILAC [252]. ILAC's main novelty was the classification of each net based on its criticality and minimization of parasitics on sensitive nets and of coupling between noisy nodes during the routing phase. A procedural layout block generator allowed the enforcement of limited geometric constraints, such as symmetry and matching between devices. The detailed routing step, based on a gridless scan-line incremental channel router [232, Chapter 4], was semi-interactive, allowing controlled rip-up options but no spacing. The other layout phases, a slicing-tree floorplanning [236], and a best-first maze algorithm for global routing [232, Chapter 3], reflected a digital-like methodology.

1.5.2 OPASYN

A main limitation of the IDAC/ILAC system was a lack of flexibility of the design process due to the relative simplicity of the models used for circuit characterization and the parasitic approximations used during the layout synthesis. In OPASYN [159] similar analytical models were used, but refinements were made to take into account second-order effects and parasitics that the physical implementation could introduce. The system assumed a synthesis by analysis approach. Optimization was based strictly on analytical models rather than simulation as in [228][159]. In OPASYN layout was generated from a fixed-floorplan arrangement, capturing a set of important considerations in the design of analog circuits. Routing was performed disregarding any analog constraints, using the digital tool, MIGHTY [276]. The approach was strictly non-hierarchical with a number of non-interchangeable circuit topologies. The obvious disadvantage was given by the lack of flexibility within the design and the schematic optimization. A similar synthesis strategy was proposed in OAC [235]. Full performance optimization was performed during design and physical assembly. The system used fixed topology op amps on which it performed nonlinear optimization to roughly size all devices. Detailed design was then carried out to precisely take into account every parasitic component associated with the layout.

1.5.3 OASYS/ACACIA

Improved parasitic analysis techniques guaranteed a better estimation of circuit performance after fabrication. The OASYS design optimization system [123] and the layout synthesis environment ACACIA [56] were built in the late 1980s to utilize these techniques systematically for more diverse and complex circuits. The original concept of OASYS was similar to that used in IDAC, except for the fact that hierarchical decomposition was used during the design as a way of reducing a large, inherently complex optimization problem into a number of simpler ones. Hierarchical decomposition had been proposed before for digital design [266] and, independently, for analog design such as AN_COM [14], however design adjustments were not handled systematically. Hierarchical components were regarded as *template-connected sub-blocks* and top-level specifications were recursively transformed during synthesis until the leafs of the design were reached and the individual sub-block specifications were generated and imposed on the automated layout generator. Some degree of flexibility was allowed in the topology of each block. The equation-based models for each block were used to operate *backtracking* on the hierarchy, for diagnosing design failures, and for proposing reparative strategies. Nonetheless, the rule-based nature of the system limited the exploration of a large set of feasible designs, resulting in a locally but not globally optimized circuit. Another major limitation of the system was the rather weak link between design and physical assembly, where mainly digital-oriented techniques were used in all phases of the layout synthesis.

More recently, OASYS has evolved onto the ASTRX/OBLX system [230], where the rule-based decision process was replaced by a purely numerical optimization approach similar to that of DELIGHT.SPICE except for the use of simulated annealing (SA) [157] as the exclusive optimization engine. The main novelties of ASTRX/OBLX were the relaxation of the requirement that the circuit be feasible, i.e. that Kirchoff's laws be satisfied, at each annealing step and the use of AWE in combination with symbolic analysis to quickly evaluate circuit performance. A limited topology selector based on a branch-and-bound algorithm was later added to the optimization [198].

The ACACIA environment also evolved from the digital-like layout system ANAGRAM into the KOAN/ANAGRAM II place and route system [56][57]. KOAN, a SA based placement tool, could perform device-shaping and abutment on MOS transistors dynamically during the annealing. The enforcement of analog topological constraints such as device symmetry and matching was integrated in the algorithm's cost function. ANAGRAM II, a detailed line-expansion router [232, Chapter 3], supported symmetric differential routing, cross-talk avoidance, and over-the-device routing. Contrary to other approaches [233][80], KOAN/ANAGRAM II did not use compaction as a way of further area reduction and/or performance adjustment or redesign. Recent developments within ACACIA include RAIL [289], a power/ground synthesizer, and WREN [213], a signal global/detailed router. The objective of the tools is the estimation and control of the effects of current injections through the substrate. The substrate is modeled by a simplified network and efficiently analyzed using an AWE simulator. A good review of these methods can be found in [259]. Although a number of algorithms were proposed for the minimization of passive parasitics, performance specifications were never explicitly enforced in the tool and the designer remained a key player in guiding the synthesis by determining the criticality of interconnects.

1.6 EVOLUTION OF APPROACHES

1.6.1 Silicon Compilation

In the mean time, due to the dramatic increase of the complexity of analog circuits of the early 1990s and the emergence of new mixed-signal circuits, silicon compilation was still regarded as an effective and powerful tool for schematic design and physical assembly. During this time the original tools migrated towards new domains of application [320][306][318][75][282][8][221]. New systems based on a standard cell approach, e.g. [32], had been refined to support large and possibly mixed-signal designs [32]. At the same time, techniques for the routing of analog components in the presence of digital signals [156][187][113][175], in combination with a traditional semi-fixed floorplan paradigm allowed the creation of compilers where some low-level parasitic issues were addressed. In CONCORDE [128], a compiler for successive approximation analog-to digital (A/D) converters using a set of high-performance pre-designed analog circuits was created. In MxSICO [13], a 2nd order Σ-Δ modulator compiler, modified vertical/horizontal constraint graphs were used for cross-over balancing in sensitive nets during the routing of channels. SCF [11] proposed a more modular approach with integrated module generation and physical assembly. The suggested approach clearly goes towards a more general and flexible design system. CADICS [146], a compiler for cyclic A/D converters, introduced the need to support the synthesis process with a behavioral model and a set of performance-driven layout tools for the generation of circuit components, as well as floorplanning and detailed routing. Thanks to behavioral modeling and simulation, the digital-to-analog (D/A) converter performance could be quickly estimated at each stage of the optimization, thus ensuring a much broader and more systematic exploration of the design space. One level of hierarchy was employed and a number of critical non-idealities were considered during the top-down synthesis. Careful parasitic extraction during the bottom-up verification phase provided accurate and reliable verification.

The CATALYST design system for switched-capacitor (SC) converters [310] was essentially an extension to dedicated silicon compilation with the incorporation of architecture selection mechanisms based on figure-of-merit. Hierarchical system partitioning and macromodels were used for figure-of-merit calculations as well as bottom-up performance evaluations during the verification phase.

Today, silicon compilation for analog and mixed-signal applications occupies an important niche in the vast panorama of design systems. It has been shown to be well suited for specific applications, and in specific cases it could be even preferable to more general approaches. A number of surveys have appeared on the subject of compilation and module generation for specific circuits and, in particular, data converters [3][15].

1.6.2 Knowledge-Based Systems

Due to the increasing success of algorithmic-based tools and the superior performance of silicon compilers in dedicated applications, knowledge-based systems gradually became the environment for a set of algorithmic tools and compilers. SALIM [150][242], for example, was a rule-based design system governing a set of layout algorithms, some of which were derived from the digital domain. A PROLOG-like hardware description language (HDL) was used to represent the design problem in a procedural fashion and through inference rules. The language was the basic glue between specifications and layout synthesis algorithms. The generation process was bottom-up, starting from the transistor schematic and continuing through the grouping of analog functions into library blocks until the complete circuit was generated.

The design environment STAIC [127] mapped structural and performance specifications onto a layout description language, SPICE netlists, and a data sheet. A number of intermediate and complementary descriptions were used to guide the user and the optimization tools throughout the design path. The final code, compiled by ICEWATER [244], was executed to generate the complete layout. The design methodology proposed in STAIC made extensive use of hierarchy, analytical model generation, and successive refinements at each stage of the synthesis. Both knowledge-based and numerical methods were used for topology selection and semi-automated design of simple circuits. The layout generation was guided by coded rules and by pre-defined floorplans, as in OPASYN.

The knowledge-based design system proposed in [274] and [84] was also based on an expert system that operated directly on the circuit primitives. The primitives were extracted from an initial schematic by means of a rule-based scheme. An iterative equation-based routine improved the circuit performance by performing a series of substitutions in the circuit topology guided by the expert system and/or by human interaction. The layout synthesis system SLAM [44][45] used knowledge of the primitives, in combination with qualitative sensitivity analysis, to create a priority schedule for floorplanning and routing. The slicing-structure-based floorplanning algorithm made use of sensitivity information to define highly sensitive zones near which devices should be placed. The channel router used again a priority schedule for an ordered generation of nets, starting from critical ones.

1.6.3 Hybrid and Human-Driven Systems

To cope with increasingly sophisticated circuits, alternative hybrid systems involving a rule-based approach to design and silicon compilation for physical assembly have appeared. In C5 [162], for example, all phases of the synthesis process were functions in a C-like HDL, while ALSYN [16] enforced additional user-determined analog-specific rules by incorporating them directly into the object-oriented circuit database. In SEAS [226] a *seed circuit* was used for initializing a simulated evolution engine, which generated a set of feasible variants or *mutations* to the seed. The algorithm terminated when the *score* associated with the current circuit could not be further increased.

In [58] the knowledge base was present in the form of a circuit example used as a starting point for an improvement-based synthesis. In these systems floorplan and placement were generally performed using modified versions of the min-cut algorithm for slicing structures [170][236]. The final layout was obtained through compaction-free maze routing [232, Chapter 3] or routing-free symbolic compaction.

In the late 1980s extensive experimentation in semi-automated analog layout systems led to increased human presence in the design loop. In LADIES [215][277], for example, a knowledge-based combined with an algorithmic approach to the analog synthesis problem was proposed. The system used a number of techniques directly imported from the design automation of high-performance digital circuits [278]. A set of simple rules were used for the analysis of the schematic and the generation of constraints on the layout geometries. Matching, design rule checking (DRC), and critical coupling were generated in this way. Layout was generated by maintaining a physical topology equivalent to that of the schematic itself. Algorithmic optimization tools for placement [161], global routing, and detailed routing were used to generate the initial layout, which was subsequently improved using another set of rules designed to enforce the original constraints while minimizing area and wire length. Later implementations of a similar methodology, such as ALE [139] improved the refinement phase and gradually increased the importance of algorithmic operations in the system, thus obtaining more compact layouts and significantly higher flexibility in the creation and enforcement of analog-specific constraints.

The CHIPAIDE system [304][189], on the contrary, proposed a top-down methodology based on hierarchical decomposition and qualitative reasoning at the schematic level. Physical assembly was performed mainly using *ad hoc* generators. The synthesis environment ISAID [188] propagated specifications throughout the design hierarchy using rough parasitic estimates along the way. At early stages of the design, macromodeling [160] was used to allow specification-driven architectural selection, based on a ranking system similar to [61]. At later design stages, models were used mainly to speedup the synthesis process. Synthesis was followed by an improvement phase based on the principles of qualitative reasoning applied to MOS design [316].

The RACHANA package was responsible for the layout generation process [103]. Using a rule-based algorithm, instances of primitives were automatically recognized from the schematic and realized using a parametrized module generator. A conventional floorplan algorithm was followed by an iterative place-and-route procedure. Module placement followed the topological order of the schematics as in [277]. Routing was performed by an area router which minimized inner-resistance, number of bends, and capacitive parasitics.

Despite satisfactory results obtained in recent years, knowledge-based systems still lack of the necessary flexibility for today's complex designs. In addition, it is still not clear how quickly these systems can migrate to new technologies and higher frequencies, where interactions within the chip are more complex and critical.

1.7 CONSTRAINT-BASED APPROACHES

1.7.1 Foundations

The constraint-based approach to design, due to Choudhury and Sangiovanni-Vincentelli in 1990, originated from the research on parasitic-aware channel routing [49]. A typical constraint-driven approach to layout consists of two phases. First, performance specifications are mapped onto bounds on all physical parasitics relevant to the implementation. Then, each bound is enforced during physical assembly, hence guaranteeing the satisfaction of the original specifications. Bound generation is a complex process consisting of a performance modeling and an optimization phase. The dependence of performance from parasitics is generally evaluated using sensitivity analysis, while the actual bound generation is performed by constrained optimization [51][50].

1.7.2 First Constraint-Driven Design Tools

The approach in its original formulation was used to determine the weights of the edges of a constraint-graph [232, Chapter 4] representing a channel where critical nets needed to be implemented. The original approach soon migrated to maze routing [191], placement [39], and compaction [80] tools, all integrated in the OCT-VEM environment [125][126][190]. A similar sensitivity-based constraint generation scheme was proposed [87][88] and applied to the placement problem [86].

1.7.3 Later Implementations

Using similar constraint-based approaches, others proposed to solve specific problems in physical and schematic design. In STAT [203][204], for example, a semi-automated layout synthesis approach with enforcement of geometric analog-specific constraints was presented. Symmetry and matching constraints were annotated directly on the schematic as "related_to" properties. A graph, derived from the schematic based on these relations, was the starting point for the placement algorithm, which was based on a conventional *topological sort* [158]. A maze router [232, Chapter 3] was modified to control wiring resistance and to prevent electromigration. Technology-independent parametrized module generation completed the layout system.

In LIBRA [120][121][231] constrained optimization and sensitivity analysis were combined to obtain compact layouts while enforcing a small set of performance specifications. Models for worst-case performance degradation due to technology deviations and resistive parasitics were derived. Specification violations were evaluated and their elimination was attempted at each stage of the layout by building appropriate cost functions.

1.8 MIGRATION OF CONSTRAINT-DRIVEN PARADIGM TO SYSTEM DESIGN

From the first promising results of constraint-based approaches for physical assembly, we have extended the paradigm to methodologies for analog and mixed-signal system design [37]. We describe this new design methodology and the necessary set of tools that support it in this book. The methodology has two basic goals: (1) making the design cycle robust by use of hierarchical partitioning, behavioral modeling, and specification propagation; (2) drastically reducing the number of design iterations by use of accurate performance evaluation and early error diagnosis. The key points of this methodology are:

- top-down hierarchical process starting from the behavioral level based on early verification and constraint propagation;

- bottom-up accurate extraction and verification;

- automatic and interactive synthesis of components with specification constraint-driven layout design tools;

- maximum support for automatic synthesis tools to accommodate users of different levels of expertise but not the enforcement of these tools upon the user; and this is not an automatic synthesis process;

- consideration for testability at all stages of the design.

This work is part of an on going research effort at the University of California at Berkeley in the Electrical Engineering and Computer Sciences Department. Many faculty and students, past and present, are working on this design methodology and its supporting tools. Our principle goals are: (1) developing the design methodology, (2) developing and applying new tools, and (3) "proving" the methodology by undertaking "industrial strength" design examples. The work presented here is neither a beginning nor an end in the development of a complete top-down, constraint-driven design methodology, but rather one step in its development.

1.9 BOOK ORGANIZATION

This work is divided into three parts. Chapter 2 presents the design methodology along with foundation material. Chapters 3-8 describe supporting concepts for the methodology, from behavioral simulation and modeling to circuit module generators. Finally, Chapters 9-11 illustrate the methodology in detail by presenting the entire design cycle through three large-scale examples. These include the design of a current source D/A converter, a Σ-Δ A/D converter, and a video driver system. Chapter 12 presents conclusions and current research topics.

2

DESIGN METHODOLOGY

2.1 INTRODUCTION

The first component necessary for a top-down design process is a well-defined behavioral description of the analog function. The behavioral characterization of an analog circuit is quite different from the characterization of a digital circuit; analog characterization is composed of not only the function that the circuit is to perform, but also the second-order non-idealities intrinsic to analog operation. In fact, errors in the design often stem from the non-ideal behavior of the analog section, not from the selection of the "wrong" functionality. To shorten the design cycle, it is essential that design problems be discovered as early as possible. For this reason, behavioral simulation is an essential component of any methodology. This simulation can help in selecting the correct architecture to implement the analog function with bounds (constraints) on the amount of non-idealities that are allowable given a set of specifications at the system level.

Constraints on performance specifications of the selected architecture are propagated down to the next level of the hierarchy onto the components that can be designed following the same paradigm until all the leaves of the design space are reached. These leaves can either be transistors, other atomic components, or library objects. Since models have to be estimated at high levels in the hierarchy, a bottom-up verification is also essential to fully characterize components, interconnects, and parasitics.

The physical assembly of basic blocks at all levels of the hierarchy is time-consuming and rarely very creative. This step can be effectively accomplished with automatic synthesis tools, recognizing that the layout parasitics do affect the behavior of analog circuits and as such have to be controlled carefully. The amount of parasitics allowed on the interconnects is often estimated by the designer, who usually is not able to guarantee the accuracy of the estimation when presented with the final circuit. To guarantee proper functionality, designers will often overconstrain the allowed parasitics. Proposed here is an approach where layout tools are directly driven by constraints on performance specifications of the design components.

The testing of analog circuits requires a great deal of time as well as expensive equipment. This problem increases with the complexity of the circuits. It can be solved in part by taking into account the testing problem during all stages of the design, unlike the common practice of considering testing only after the design is finished.

Finally, it is not believed that full automation is achievable for all analog circuits. The amount of creativity and complexity needed to master the design of analog circuits is high. It is believed, however, that the creative task of the designer can and should be fully supported by a set of automatic and interactive tools that allow him/her to explore the design space with ease and full understanding of the trade-offs involved. Analytical tools play an important role in our methodology. It is also maintained, though, that some components of an analog design could indeed come from module generators and some from libraries, both of which embody the experiences of other designers. Thus, this methodology does accommodate tools that favor design-reusability in its general framework.

2.2 ANALOG DESIGN

Many methodologies exist for analog integrated circuit design. In any design, a methodology is either explicitly stated or implied. On the surface most design paradigms are similar. All follow the general design flow illustrated in Figure 2.1.

The designer begins with a set of specifications. These have either been provided by a customer, or these are the requirements for proper operation in a larger system. Specifications generally include not only the functional and performance criteria for the product, but also data on the target process and an overall engineering objective. For example, minimizing production cost is often the objective. This could translate to minimizing area, which maximizes yield, which, in turn, minimizes cost.

Design synthesis takes two inputs: (1) the specifications and (2) an architectural library. The output is a schematic which contains not only a netlist of the basic circuits elements—transistors, resistors, capacitors, etc.—but also their parametric values. The

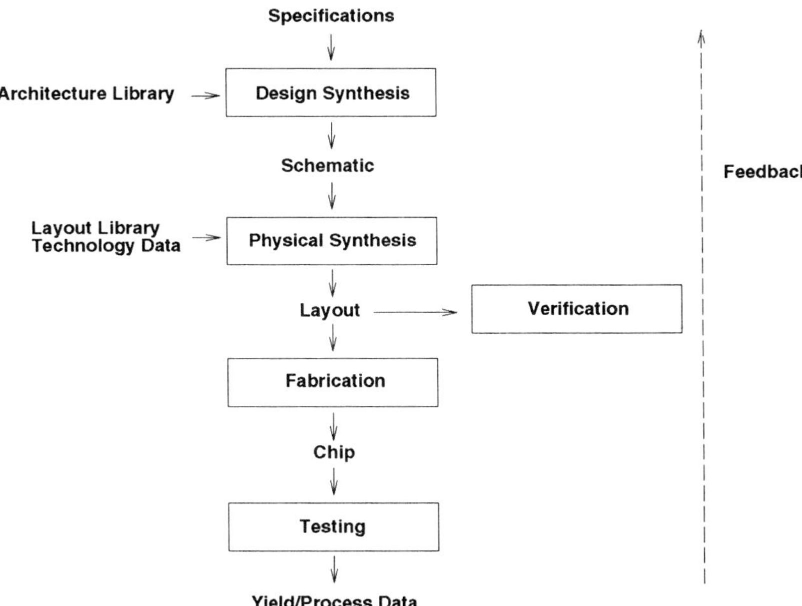

Figure 2.1 General Design Flow

architectures in the library are either templates for the schematic, or they are starting points from which new architectures can be developed.

Until now, analog "hand" design has been considered an art. Rigid methods based on automatic synthesis have yet to be used in practice. Most of the tools commonly used are analysis tools for the evaluation of designs. In addition, "high-level" analysis is done on an *ad hoc* basis since a formal method for abstraction is missing. High-level models are not "precise" enough for analog design where second-order effects are important. Hence circuit simulators such as SPICE and prototyping on bread boards [85] is almost exclusively used for the determination of these effects. Unverifiable top-down decomposition has resulted in an unsystematic bottom-up design style. Designers overdesign low-level components to compensate for unpredicted non-idealities. This overdesign is time consuming. Bottom-up designs also imply that sub-systems cannot be verified until all of their components have been designed. This results in further time consuming iterations when errors are detected. When verifying the system, circuit simulation is often too CPU expensive or infeasible. This results in systems which are not fully verified until testing.

Physical synthesis is the next phase. It takes as inputs: (1) the schematics, (2) a layout library, and (3) data on the technology. The output for this step is layout. Varieties of methods exist for layout generation. Layouts can be copied from a layout library. Layouts can be modified from an existing library entry. Layouts can be generated from scratch either manually or automatically, although designers currently resort to manual techniques almost exclusively. Automatic synthesis techniques have found little acceptance.

The next step is to verify that the layout meets specifications. Because of the sizes of the circuits involved, often verification cannot be accomplished in a single step nor by any single tool. A variety of methods have been developed. One method is simulation. In circuit simulation sets of test vectors are supplied to the system and simulated to verify functionality. Behavioral simulation verifies performance specifications directly, but detection of differences between the schematic and the layout is not guaranteed. An approach to solve this problem is to compare directly the schematic and the layout, element by element, connection by connection. This verification method, however, cannot find performance degradations due to layout parasitics as simulations can. A combination of the two methods is usually used.

Once the layout has passed verification, it is sent for fabrication. This process typically requires two to six weeks. When the chips are received, the parts are tested against the specifications for function and for performance. Typically, yield and process information are also gathered from the parts to determine the actual cost of the product and to collect data to improve future designs. Though rigorous methods for testing have been developed [212][210], they are also not used in practice.

Feedback is implied at any step in the design flow. At any time an output fails to meet the specifications, design steps must be re-iterated. Feedback loops can be small or large depending on how much redesign is required.

Many extensions exist to the general design flow. One extension is the use of hierarchy. This is illustrated for the design and physical synthesis phases in Figure 2.2. Systems which are too large to design as one individual unit are divided into sub-systems. These smaller sub-systems can then be designed independently or if they are still too large, can be further subdivided. The interactions among the sub-systems are approximated to allow a certain independence of design.

In the design synthesis phase, not only is the architecture subdivided, but also the system specifications are decomposed into specific specifications for each sub-block. This process of decomposition is often defined as the "top-down" design phase. Its counterpart, the "bottom-up" design phase refers to the sub-block design process in which schematics are generated. In analog design, the top-down phase is often very difficult. Many designs are considered "bottom-up designs," because the emphasis has been placed on this step. Circuit verification can also be accomplished hierarchically. Sub-blocks are simulated and verified. They are replaced with abstracted models which characterize their function and/or performance specifications. These models are combined and simulated to verify the overall system function and/or performance. Layouts are generated hierarchically as well. First, layout requirements for each sub-block are generated in a floorplanning step, e.g. aspect ratio, bus and power line placement, pad placement. After the sub-blocks have been generated, they are combined level by level until the layout for the system is complete. Once again, feedback is implied for all of the design steps.

Though most methodologies do follow the design flow in Figure 2.1 and consider the use of hierarchy, they differ in the details.

2.3 RESEARCH IN ANALOG DESIGN METHODOLOGIES

A great deal of research has focussed on solving the problems faced in analog design (Section 1.3). Most research has emphasized specific tools and/or sets of tools. In all cases an underlying methodology is implied, but seldom discussed. Furthermore, rarely is the entire design flow considered. Many have made claims for automatic

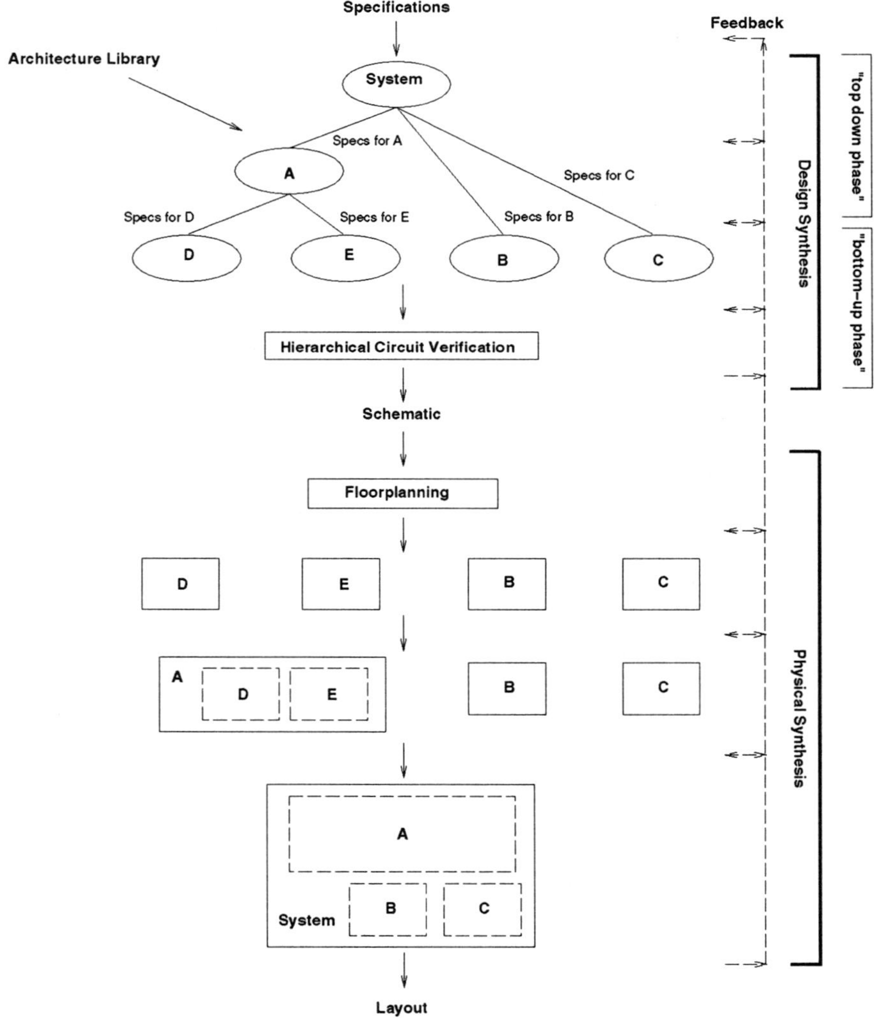

Figure 2.2 Design Using Hierarchy

synthesis of large systems, but none have fabricated and tested these designs. Our emphasis is to explicitly develop the methodology, then develop the tools to support it. And furthermore, we consider the entire design flow, "proving" the methodology using industrial strength design examples.

Recent proposals to address the design problem have also followed in this vein by explicitly stating a design methodology. ARIADNE [296][298] was developed for the generation of analog and, more recently, mixed-signal circuits [70] from specifications to layout. The ARIADNE approach was based on symbolic analysis proposed by [100][99] and independently by [271] for the modeling of circuit components, the generation of constraints on the design space and the general topology selection procedure [295]. Symbolic models, in combination with multiple objective optimization [101][297], were used to size circuits, as in [159]. Hence, new topologies could be quickly characterized and incorporated into the database, allowing considerable design flexibility. As a byproduct of the optimization, constraints on layout geometries and parasitics could be obtained. Physical assembly and testing tools [102], completed the system. They have also called for the inclusion of high-level description languages [70].

However, lacking from this system and others which addressed system design issues (Section 1.3), is a precise method to exploit hierarchy to cope with system complexity. Behavioral simulators are not provided, and constraint propagation is not clearly defined. For this reason, many of these systems have been fundamentally limited to the design of low-level circuits. This is clearly shown by the design examples: the focus is almost always operational transconductance amplifiers (OTAs) or comparators. Some of these works do present larger systems as examples; however, the high-level simulators which are employed lack the ability to accurately capture critical performance specifications, and, therefore, constraint propagation is not clearly defined. One of the major contributions of this work is to clearly define how hierarchy is used, and to use it in all of the design examples.

2.4 ANALOG DESIGN CONCEPTS

We use an example to illustrate the issues that must be addressed in analog design. Current source D/A converters have been selected because they will be one of the circuits used to illustrate the design methodology (see Chapter 9).

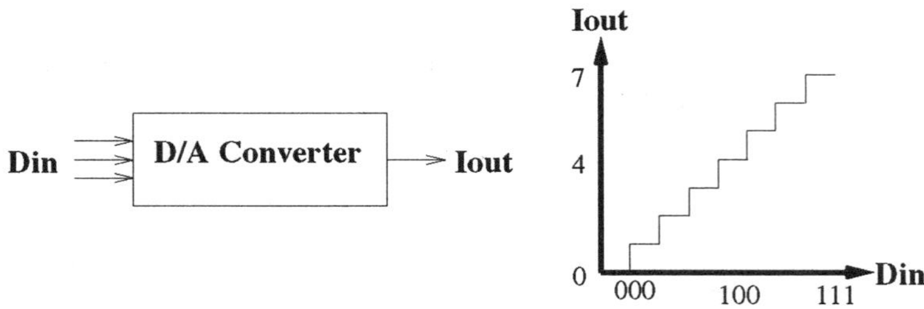

Figure 2.3   D/A Input/Output Relationship

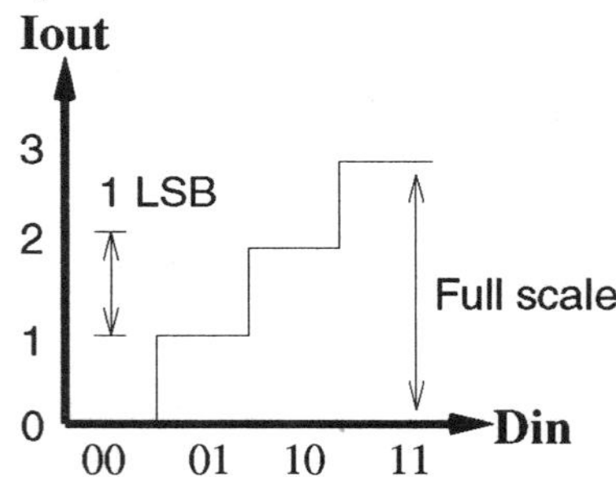

Figure 2.4   Definition of LSB and Full Scale

2.4.1   I/O Relationship

Figure 2.3 shows the *functional* (input/output) relationship of a D/A converter. It converts an n-bit digital word into a single analog signal. A three-bit ($n = 3$) converter is shown in the figure. There are three input bits and one current output.[1] In the graph, the x-axis contains the 8 (2^n) possible input codes while the y-axis shows the corresponding current output.

Two important metrics in characterizing D/A converters are illustrated in Figure 2.4. One least significant bit (LSB) is the ideal difference in current between any two adjacent input codes. The *full scale* range is the ideal difference between highest code ($2^n - 1$) and the lowest code (0). In the figure, 1 LSB equals 1 unit of current, and the full scale range is 3 units of current. The unit amount is usually set by a reference. Typical reference values range between 1 μA and 1 mA.

---

[1] Voltage outputs can also be used.

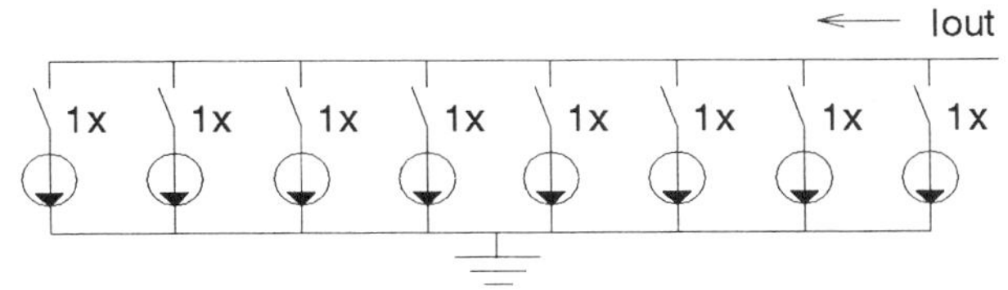

Figure 2.5 Three-Bit Current Source D/A

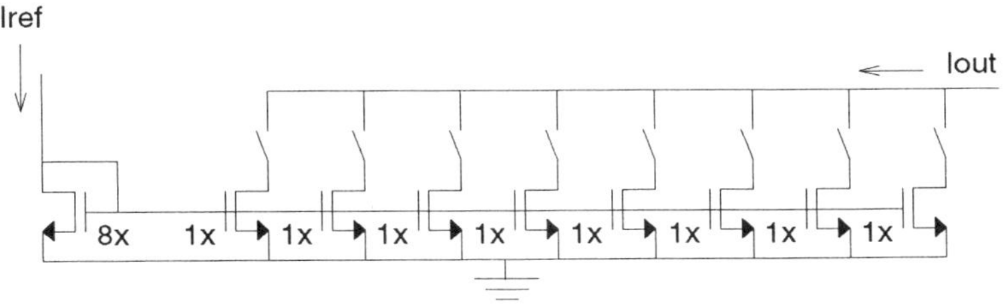

Figure 2.6 Three-Bit Current Source D/A Implementation

2.4.2 Linear Array Architecture

One architecture for a D/A converter is a *linear* or *unit* architecture. This is shown in Figure 2.5. Identical current sources are placed in an array with their outputs connected via switches to a single output. Since current sources have high impedance outputs, the currents sum at the output node to produce the desired current. Therefore, the output is proportional to the number of switches turned on. To control the switches, the three-bit input word is thermometer decoded. For example, the input code 110 translates to 11111100 turning on the first six switches.[2]

One possible implementation is shown in Figure 2.6. Each current source (unit element) is implemented by a transistor. Pulling a reference current, a diode connected transistor eight times the size of the unit elements sets up the gate voltages. Each current source produces 1/8 of the reference current. An LSB is $1/8\ I_{ref}$. The full scale range is $7/8\ I_{ref}$.

---

[2] 000 converts to 00000000. The 2^nth unused current source is often kept for matching and symmetry reasons.

2.4.3 Non-Ideal Performance

If the transistors in Figure 2.6 were truly identical and had infinite output resistances, then this technique could be applied to build arbitrarily accurate D/A converters. Unfortunately, even when drawn identically, i.e. same gate width and length, same diffusion areas, etc., and placed within close proximity, transistors do not behave identically. There are many reasons for this; e.g. gate oxide thickness variations, lithographic variations, substrate doping variations, surface gradients, and temperature gradients [240]. This limits the number of bits that can be achieved for this type of D/A converter. Fortunately, in design, variations between transistors can be controlled using device sizing and placement techniques. Ultimately, however, typical maximum resolutions using this method lie between 10 and 12 bits for current technologies.

There are two main lumped causes for transistor mismatch. The MOS equation for a long channel transistor in saturation is

$$I_d = \frac{1}{2}k(V_{gs} - V_T)^2 \quad \text{where} \quad k = \mu C_{ox}\frac{W}{L} \tag{2.1}$$

Taking a derivative and applying the chain rule, we have

$$dI_d = \frac{1}{2}(V_{gs} - V_T)^2 dk - k(V_{gs} - V_T)dV_T \tag{2.2}$$

Then dividing, Equation 2.2 by Equation 2.1, we obtain

$$\frac{dI_d}{I_d} = \frac{dk}{k} - 2\frac{dV_T}{V_{gs} - V_T} \tag{2.3}$$

dI_d/I_d represents the amount of mismatch between two transistors. Variations in k and V_T cause the mismatch. One method for reducing mismatch is to increase $(V_{gs} - V_T)$. However, this technique is limited by the supply voltage and does not address the k term.

An alternative technique is to control the variations in k and V_T directly. Assuming k and V_T are statistically independent [240] [208], we can rewrite Equation 2.3 as

$$\left(\frac{dI_d}{I_d}\right)^2 = \left(\frac{dk}{k}\right)^2 + 4\left(\frac{dV_T}{V_{gs} - V_T}\right)^2 \tag{2.4}$$

Functional forms for the k and V_T variances have been proposed [240][208]:

$$\sigma_k^2 = \frac{A_k^2}{WL} + S_k^2 D^2 \tag{2.5}$$

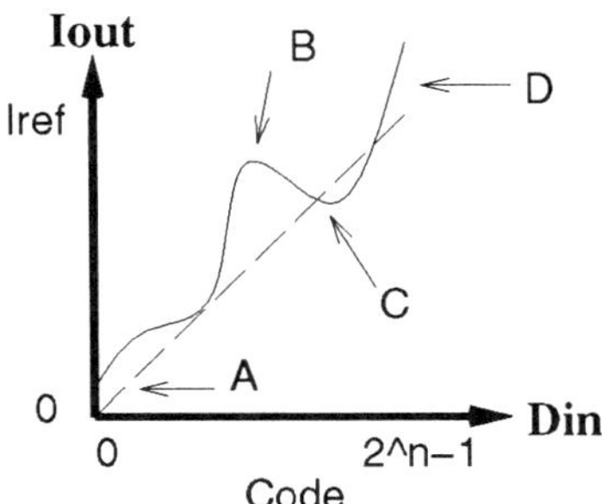

Figure 2.7   I/O Characteristics of a Typical D/A

$$\sigma_{V_T}^2 = \frac{A_{V_T}^2}{WL} + S_{V_T}^2 D^2 \tag{2.6}$$

A_k, S_K, A_{V_T}, S_{V_T} are process dependent constants. W and L are the width and the length of the transistor. D is the distance between the two transistors. Thus, increasing the transistor size and placing them close together can reduce mismatch. This is an effective way to increase matching.

Refinements to Equations 2.5 and 2.6 based on our experimental results (Chapter 9) have been made [81]. However, the trend that mismatch is reduced by using larger transistors and by placing them close together still remains true.

These mismatches result in non-ideal *performance*. "Non-working" parts are primarily due to these second-order effects. An exaggerated possible transfer curve is shown in Figure 2.7 by the solid line. The dotted line represents the ideal behavior. Problematic points are indicated on the curve. Point A shows that there is no "off" state. Point B illustrates an extreme deviation from the ideal. In a signal processing system, this could result in large unwanted harmonics. Point C shows an area of non-monotonicity. In a feedback system, this could result in a disastrous reversal of the feedback polarity. Point D shows an extremely high output. This could overload the next stage. Methods for characterizing these non-idealities have been developed. Figures 2.8 and 2.9 describe this characterization.

Figure 2.8 illustrates the linear components of error. Offset error is a constant difference between the ideal and the actual output. Gain error is a measure of the constant difference between the slopes. In general, linear errors are not critical; preceding or following stages can often compensate for these errors.

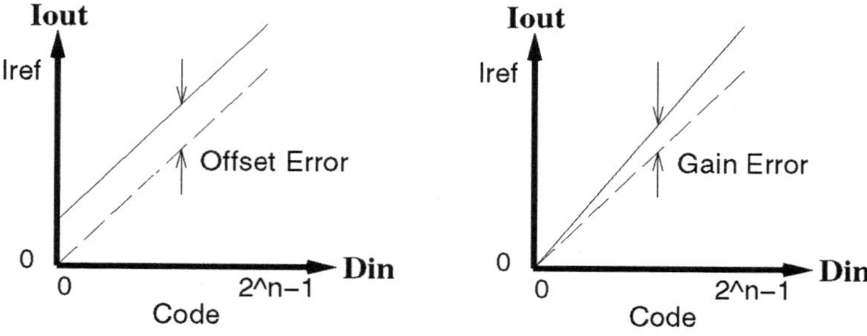

Figure 2.8　Definition of Offset and Gain

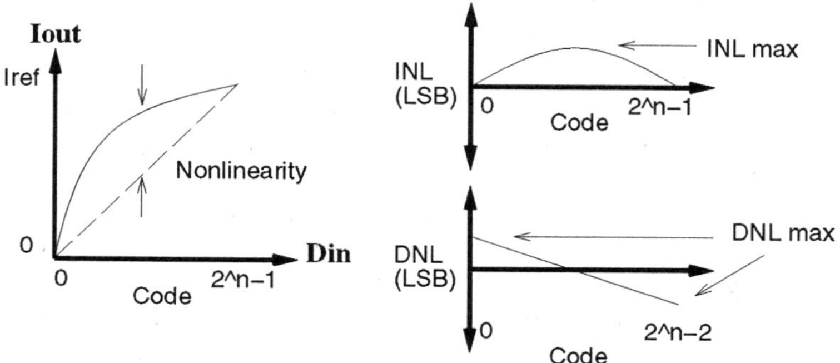

Figure 2.9　Definition of INL and DNL

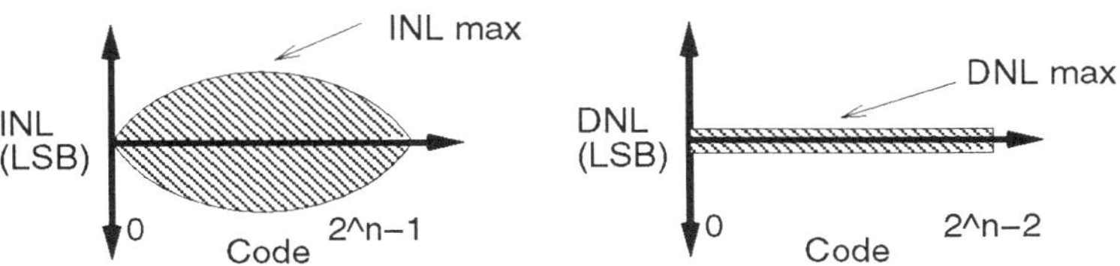

Figure 2.10 One σ Bounds for INL and DNL

More critical, however, are nonlinear errors. Two metrics for their characterization are integral nonlinearity (INL) and differential nonlinearity (DNL), as shown in Figure 2.9. INL is the difference between the ideal and the observed output after gain and offset errors have been corrected. DNL is the difference between the actual step and the ideal step for two adjacent codes. See Section 3.3 for details.

2.4.4 Analysis

A simple model for the circuit is to assume that the current sources mismatch randomly with error, σ_e, and assume that errors due to finite output resistance are negligible. DNL can be represented for all input codes as

$$\sigma_{DNL} = \sigma_e \tag{2.7}$$

since all current sources are unity weighted. The DNL is graphed in Figure 2.10. The computation for INL is more complicated. It is graphed for all codes in the figure on the left. The maximum value is given by [19]:

$$\sigma_{INL_{max}} = 2^{\frac{n}{2}-1}\sigma_e \tag{2.8}$$

Typical requirements for INL and DNL range between 0.5 LSB and 2.0 LSB. As n increases, INL requirements become increasingly difficult to achieve while there is no change in the difficulty of the DNL requirement.

2.4.5 Architectural Alternatives

An alternative architecture for current source D/A converters is a *binary* weighted architecture. An example three-bit converter is shown in Figure 2.11. In CMOS, the

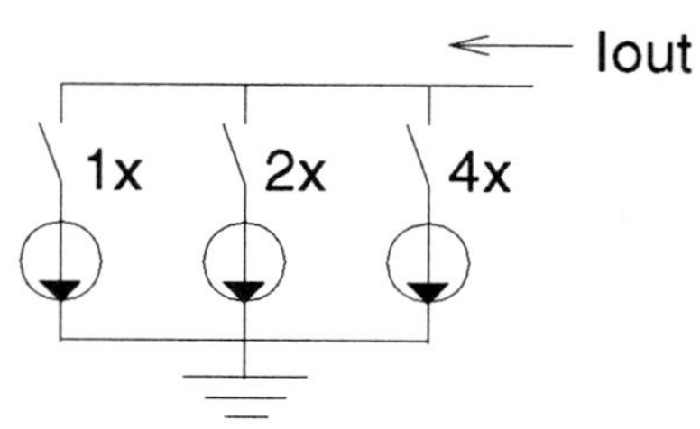

Figure 2.11   Three-Bit Current Source D/A—Binary Weighted

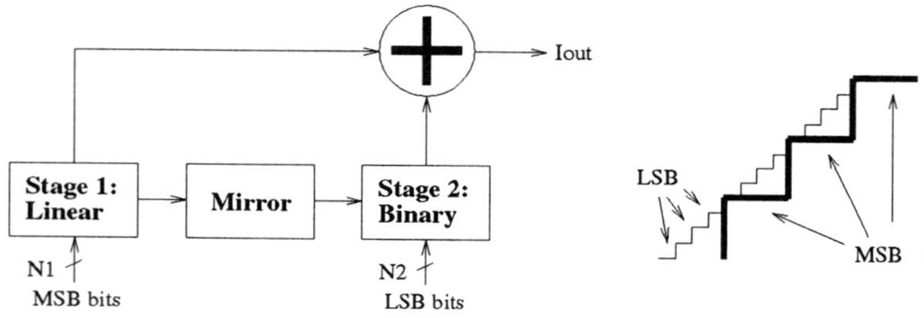

Figure 2.12   Two Stage D/A Architecture

principal advantage of this architecture over the linear architecture is that there are fewer switches. Since fewer switch control signals are required, there is a tremendous reduction in routing complexity, and this translates into a huge savings in layout area. There are no savings in active transistor area since an nx sized transistor is implemented by n $1x$ transistors in parallel. Another advantage is that a thermometer decoder is not required. The input bits directly control the switches. For example, code 011, activates the $1x$ and $2x$ current source to produce a $3x$ output.

The INL for the binary and linear array architectures is the same [19]. The DNL, however, is much worse. The maximum DNL is

$$\sigma_{DNL_{max}} = \sqrt{2^n - 1}\, \sigma_e \qquad (2.9)$$

A reduction in layout area is traded for increased DNL.

Often this DNL penalty is too large. A compromise is to combine these two architectures. A block diagram and I/O graph for a two stage architecture are shown in Figure 2.12. The first n_1 most significant bits (MSB) are decoded in a linear first stage. The n_2 least significant bits (LSB) are decoded in a binary second stage. The first stage generates a granular staircase as shown in bold in the graph in the figure.

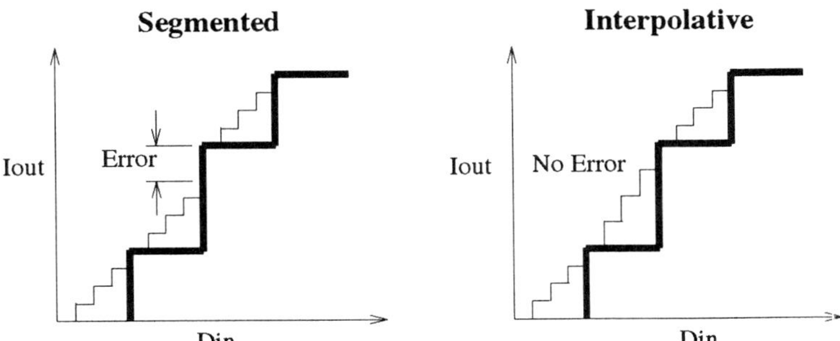

Figure 2.13   Segmented vs. Interpolative Two Stage Architecture

Using one of the unit current sources from the first stage as reference, the second stage divides this current to produce the fine step sizes shown. This produces $n = n_1 + n_2$ bits of resolution.

Two stage D/A converters can be implemented in either of two basic methods. The simpler of the two is a *segmented* approach. One *dedicated* unit element of the linear array is used as reference for the binary array. The disadvantage of this approach, as shown in Figure 2.13 on the left, is that if there is a large difference between the reference unit element and the other unit elements, then this results in large DNL errors when the second stage switches from all bits on to off. The second approach, an *interpolative* architecture, solves this problem. The reference element for the second stage is always the next unit element to be turned on in the first stage. In this way, the second stage transition point error is eliminated. The cost of the latter approach is more switches which results in larger layouts.

The maximum INL for both two stage architectures remains the same as before (Equation 2.8) where $n = n_1 + n_2$. The maximum DNL for the segmented D/A is

$$\sigma_{DNL_{max}} = \sqrt{2^{n_2+1} - 1}\, \sigma_e \qquad (2.10)$$

The maximum DNL for the interpolative D/A is

$$\sigma_{DNL_{max}} = \sqrt{2^{n_2} - 1}\, \sigma_e \qquad (2.11)$$

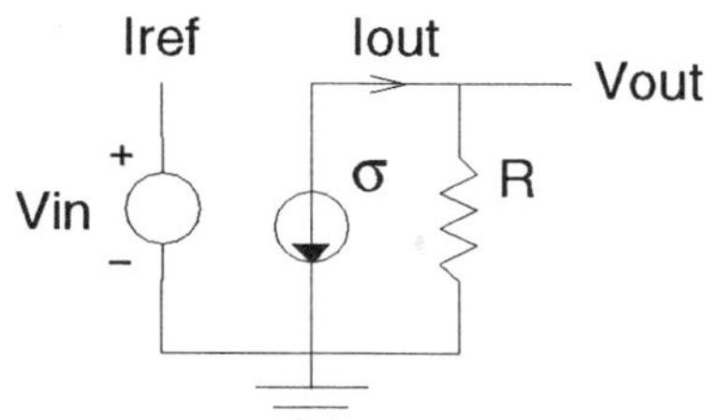

Figure 2.14   Behavioral Model for a Unit Current Source Element

2.4.6   Simulation Methods

Besides transistor mismatch, other effects such as voltage drops and finite current source output resistances cause nonlinearities. To obtain more accurate INL and DNL estimates, simulation techniques can be applied.

The traditional approach is to use SPICE, a circuit simulator. Unfortunately, SPICE can only give the *functional* ($\sigma_e = 0$) characteristics. Though this is useful for verifying circuit functionality, it provides no information on INL and DNL. This simulation is also CPU intensive. 2^n operating point (".op") analyses are required. Experiments have shown that for a 10-bit converter, this requires approximately 1 CPU day. Applying Monte Carlo techniques by varying transistor models in simulation can be used to find INL and DNL, but is virtually impossible because of this lengthy CPU time.

A newer technique [177] based on *behavioral simulation* directly calculates the circuit *performance specifications*, INL and DNL. The simulator, given a circuit description (Figure 2.14) and a list of random variables, can calculate the INL and DNL on the order of CPU minutes with an accuracy difference compared to using SPICE Monte Carlo of less than 0.05 LSB.

2.5   TOP-DOWN, CONSTRAINT-DRIVEN DESIGN METHODOLOGY

Figure 2.15 shows the standard design hierarchy. At the top of the design hierarchy is a top node which could represent an entire chip or just part of a chip. From this top node there is a set of direct descendent lower nodes, representing a first-level decomposition of the upper node. From these lower nodes the graph continues to be expanded. The

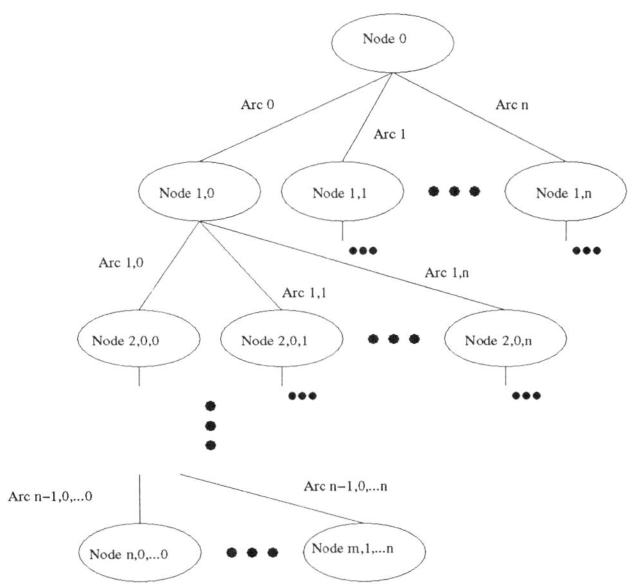

Figure 2.15   Design Hierarchy

decomposition may cease at any given node at any given level. The graph is completed with the decomposition of all of the nodes.

An example decomposition is shown in Figure 2.16. Here, a mixed-signal chip is our top node. This diagram shows a way in which this top node can be decomposed. It also shows how decomposition may cease at any level. For example, a voltage reference circuit is found in our cell library. Thus, the decomposition of this node is not necessary. A solid line under the terminating node indicates this. In some cases the decomposition does not cease until the transistor or passive element level, as indicated by the capacitors and the MOSFETs in this figure.

The methodology, illustrated in Figure 2.17, assists the hierarchical generation of the design by providing a rigorous procedure based on interactive and automatic tools. The large bubble encompassing most of the diagram represents the mapping. Given a set of circuit specifications (circuit characteristics, the design rules, the technology, and user options), a mapping is made to schematics or to layout.

Given a library of *n* architectures, the first operation that must be performed is architectural selection. Simulators and optimizers are used to aid the decision-making

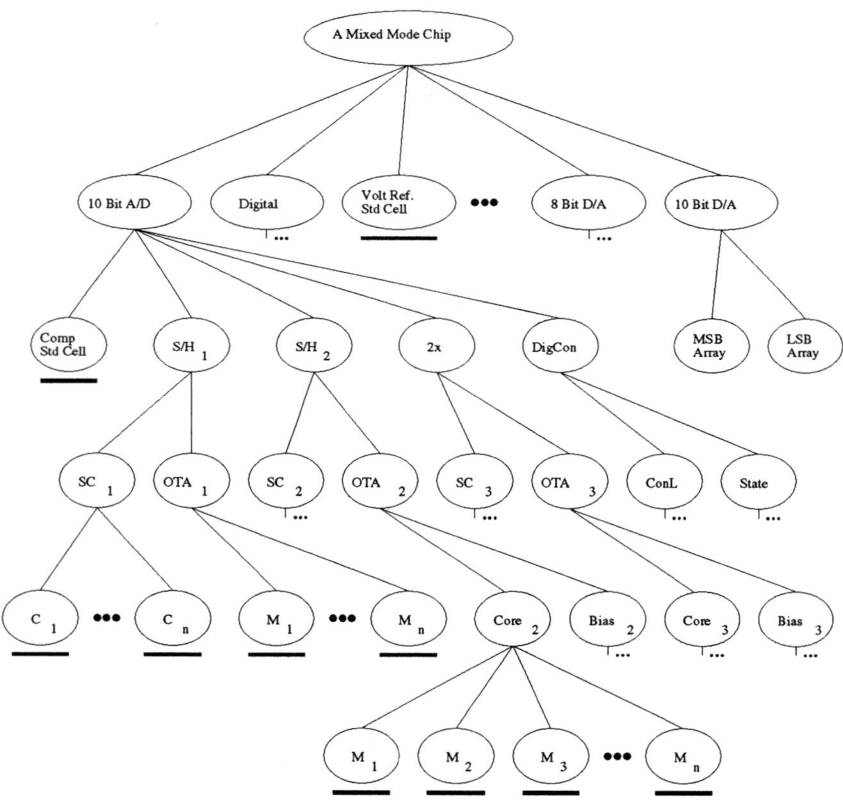

Figure 2.16   Example Design Hierarchy for a Chip

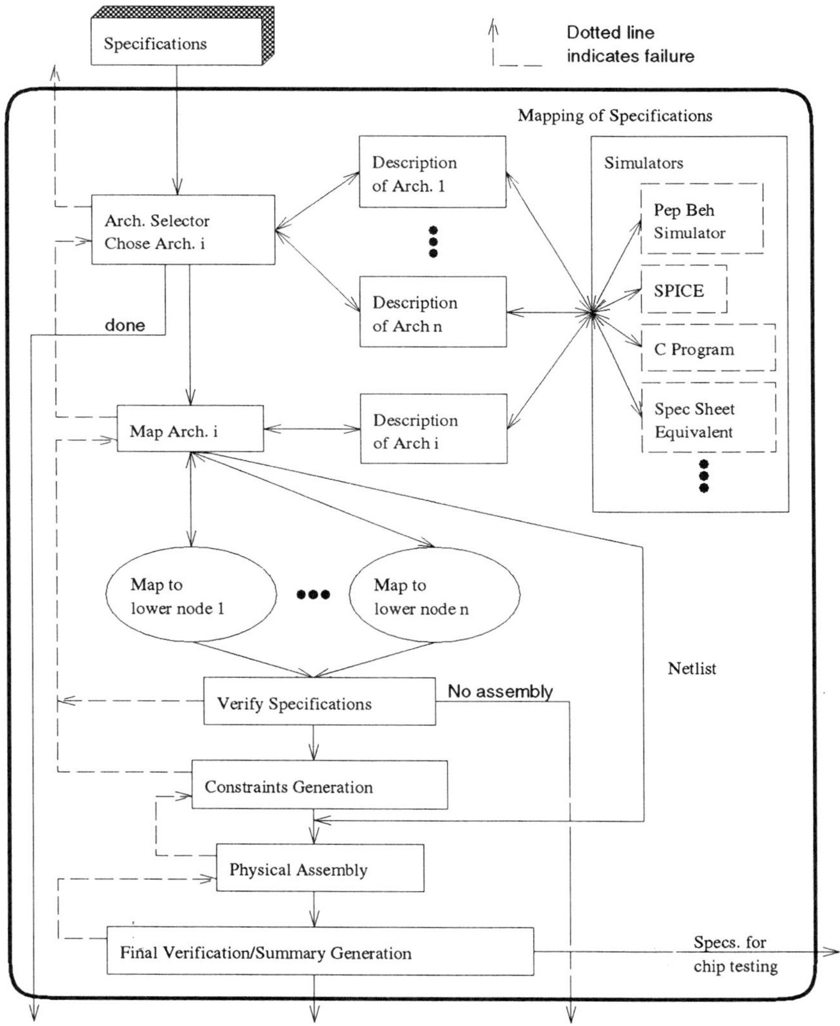

Figure 2.17 Design Methodology

process. For very high-level blocks, a behavioral simulator may be employed. For low-level circuits, a circuit simulator such as SPICE may be run. For an architecture where no simulators exist (e.g. a pre-made cell) the "simulator" could just be a list of performance specifications. If a suitable architecture cannot be found, then this selector must return to the upper node the fact that the selection has failed. This would mean that the specifications cannot be met; they must be changed in order to continue. If a standard cell (pre-designed cell) was chosen then a successful return to the upper node is made.

Given specifications for a particular architecture the next step is to map the chosen architecture to the detailed specifications of the component (lower) blocks. This task can be very difficult. Architectures can be mapped automatically using nonlinear optimizers; however, when this procedure fails the user must do the partitioning. If a solution to the mapping problem cannot be found, another architecture must be chosen.

The lower blocks are expanded in the same manner as this diagram depicts, thus recursively expanding the design hierarchy. From these lower blocks a set of the component specifications is returned. If the returned specifications fail to meet the criteria set by the mapping function, then the flow control is returned to the mapping function.

On a successful return from all of the components, the component specifications are compiled to form a list of specifications corresponding to the original list. If this compiled list fails to meet the expected specifications, then a new mapping is attempted. If it is successful, there are two options, either proceed with the layout or stop here returning only the schematics.

In creating the layout, the first step is the generation of the layout constraints for the assembly. This can be done either manually or automatically. As before, if this step fails, the flow control is returned to the mapping function. If it is successful, then a constraint-driven physical assembly is performed. If this fails to meet the requirements then an alternate set of assembly constraints is derived. If the physical assembly is successful, then the final verification step/summary generation step is required. This step includes the extraction of the circuit and simulation with the same simulator used in architectural selection. The "extraction" process spans the entire range from a simple net-list extraction to a complete extraction with parasitics. Finally, a summary of the expected performance specifications is generated. If all of the specifications are met, the flow control is returned to the upper node with the generated specifications.

This final block also formulates the test set and methodology, based on the extracted specifications, technology considerations, and the final layout. If the desired test coverage cannot be obtained then possible hardware modifications or alternative architecture suggestions are fed back to the mapping function or the architectural selection function, as appropriate. Since the majority of analog circuit failures are due to parametric rather than catastrophic device malfunctions, determining the test set is an essential part of the design methodology which cannot be split into a separate post-processing step. Testing is based on the same functional model as the high-level behavioral simulation. This model allows the circuit to be parametrically characterized by a minimum number of well-chosen tests. The test set is derived and ordered to minimize test expense (time and equipment) while meeting the test coverage constraints. The results of the high-level behavioral simulation are used to determine the relative value and cost of testing each behavioral component of the circuit. The test coverage constraints are formulated to include both catastrophic and parametric failures. Test specifications and constraints are thus incorporated into the design of the circuit in a manner very similar to the manner in which performance constraints are accommodated. An appropriately optimized test set is part of the final chip specifications.

2.6 CONCLUSION

Our basic design methodology has been presented. Having formulated the design flow, the true test, then, is in actually using it—to show that it can be *systematically* applied, and to show that it is *effective*. Our work on design methodologies has focussed on this aspect. Our design examples are presented in Chapters 9, 10, and 11.

3

SIMULATION AND BEHAVIORAL MODELING

Simulation with *accurate* and *realistic* models has been an invaluable tool for the verification of the electrical performance of integrated circuits. Simulators hold a particularly important place in the world of analog and mixed-signal design tools. Primarily, they verify that the circuit as designed will perform as expected. They are also used in design space exploration. It is most important that a simulator be trustworthy, but it is also important that the simulator be fast and robust. When designing very high-performance circuits, accurate prediction of the time and frequency-domain behavior of the circuit is needed. Such prediction can only be achieved by the use of circuit simulators that analyze detailed waveforms. These simulators (e.g. SPICE [217][248]) can provide accurate results, but sometimes at the expense of very long (possibly unaffordable) computing time. In particular, when complex analog and mixed-signal systems are being designed, accurate circuit simulation of the entire circuit is out of the question. The complexity in terms of the number of components, and of the types of analyses makes the use of SPICE too time consuming. Moreover, it may not be possible to use a circuit simulator to make design "decisions" at the "higher" levels in a top-down design process, because the detailed circuit implementations of the sub-blocks may not be available at that particular level of the design hierarchy.

Our approach to designing complex analog and mixed-signal systems relies on realistic circuit modeling at each step and the ability to propagate constraints step by step. We do not need to verify top level system performance from low-level implementation details with a SPICE simulation of the entire system, but we do need to guarantee correct translation of one level of constraints to a set of lower level constraints. The tools for constraint translation are the behavioral models at each level, the behavioral simulator, and optimization, which will produce constraints for the lower levels, using the behavioral models and the simulator.

3.1 BACKGROUND

3.1.1 Circuit Simulation

SPICE [217][248], which was developed at U.C. Berkeley, is recognized worldwide as the standard in circuit simulation. SPICE-based or SPICE-like simulators are used extensively throughout the analog design community for design space exploration, and for the final prefabrication verification of the detailed electrical behavior of the sub-blocks and the system.

In SPICE-like simulators, circuits can contain resistors, capacitors, inductors, independent/dependent voltage/current sources, transmission lines, switches, distributed RC lines and semiconductor devices (diodes, BJTs, JFETs, MOSFETs, MESFETs) [141]. There are built-in models for the semiconductor devices which are composed of basic circuit primitives. Circuit simulators provide the following basic types of analyses:

- Nonlinear DC analysis to obtain the quiescent operating point of nonlinear circuits.

- Small-signal AC analysis to obtain the frequency domain response of a nonlinear dynamic circuit *linearized* at a *fixed* operating point.

- Transient analysis to obtain the time-domain behavior of a nonlinear dynamic circuit starting from a DC operating point obtained by nonlinear DC analysis or set by the user.

In addition to the basic types of analyses described above, circuit simulators provide specialized types of analyses which are especially useful and essential in high-performance analog design:

- Noise analysis with linear time-invariant transformations to obtain the frequency domain noise performance characterization of a nonlinear dynamic circuit *linearized* at a *fixed* operating point.

- Pole-zero analysis to obtain the poles and/or zeros in the small-signal AC transfer function of a nonlinear dynamic circuit *linearized* at a *fixed* operating point.

- Fourier analysis on the waveforms generated by the transient analysis.

- Steady-state analysis [166] to obtain the sinusoidal, periodic or quasi-periodic steady-state response of a nonlinear dynamic circuit (possibly containing distributed elements).

- Distortion analysis based on Volterra series [254].

In the design of today's complex analog and mixed-signal systems, use of SPICE simulation for design space exploration or performance verification of the whole system, or even some sub-blocks, e.g. a phase-locked loop frequency synthesizer in a radio-frequency (RF) receiver, is simply not possible. Although the algorithms behind the various types of analyses in SPICE were developed and refined over the past 25 years, they are still very inefficient for use with a complex (in terms of the number of components and the types of analyses required) design. The solution of sparse linear system of equations (i.e. sparse Gaussian elimination [164]) lies at the core of the basic SPICE analysis algorithms. It has been shown that the memory and computing time required by SPICE grow super-linearly with the circuit size. In some cases, even if the circuit size is small, it might still be impossible to do a detailed SPICE simulation. For instance, in phase-locked loop (PLL) circuits, the period of the voltage-controlled oscillator (VCO) waveform is much smaller than the loop time constant. As a result, simulation has to go through many cycles of the VCO to get an idea about the behavior of the circuit, which makes SPICE simulation impractical. During the design space exploration phase of a new design, the implementation details of the sub-blocks of the system are not available. As a result, it is not possible to use detailed circuit simulation at this phase of the design.

To solve these problems that have arisen in the use of circuit simulation in analog design, several approaches have been taken. We now discuss these approaches with their virtues and deficiencies.

Relaxation-Based Electrical Simulation

In order to reduce the simulation time for large circuits, new circuit simulation techniques were developed, which go beyond the approach (sparse Gaussian elimination, Newton iteration, stiff implicit time integration) taken by "second-generation" circuit simulators (e.g. SPICE2), and are referred to as the "third-generation" simulation techniques [118]. These techniques concentrated on the time-domain transient analysis of nonlinear dynamic circuits, and they employed "tearing" decomposition or "temporal" decomposition (relaxation) methods at one of the three main levels of circuit simulation, namely the differential equation level, nonlinear algebraic equation level and the linear equation level. For instance, in *iterated timing analysis* [222], iterative relaxation methods (Gauss-Jacobi or Gauss-Seidel) are used to solve the nonlinear algebraic equations obtained by applying a stiffly stable integration formula to the differential equations that describe the circuit behavior. The iterative relaxation techniques can also be applied at the differential equation level [315], resulting in a family of techniques called *waveform relaxation*. Iterated timing analysis and waveform relaxation were successful in significantly speeding up the timing simulation for digital MOS integrated circuits, but this advantage disappears with tightly coupled analog circuits with feedback. Hence, relaxation-based circuit simulation is not suitable for use in high performance analog design.

Macromodeling

Words such as "macro," "high-level," "behavioral," "functional," etc. are used interchangeably to explain or discuss several related concepts. When we use one of these words, we will try to explain what concept we are referring to as clearly as possible. Traditionally, *macromodeling* for analog circuits is used to mean creating a model of a circuit block from the implementation details (transistor-level circuit configuration) using pre-defined circuit primitives available in a circuit simulator such as linear/nonlinear controlled sources, resistors, capacitors, switches, etc. Actually, all the semiconductor device models available in SPICE are implicitly composed of the same circuit primitives, which makes them conceptually equivalent to a macromodel. The transistor-level representation of a circuit block can be thought to be a very detailed "macromodel." In a macromodel, one tries to capture *much more* "functionality" than there is in a single semiconductor device, with *far less* implementation details. This suggests a definite trade-off between accuracy and the complexity (number of components) of the macromodel. Macromodels in circuit simulators are used to

- reduce the simulation time (because of reduced number of nodes and complexity) and

- simulate circuits with sub-blocks without implementation details.

Macromodel creation (for a specific circuit block described at the transistor level) for nonlinear circuits is done by "iterating" over the following two steps:

- First, a parameterized model (an interconnection of circuit primitives) which captures the "functionality" of the circuit block being modeled is created. There is no systematic way of creating these models and this step heavily depends on the experience of the user. Creating an accurate and compact (with much fewer components than the actual circuit) macromodel can be very hard.

- Then, the parameters of the created model is optimized (manually or automatically) to meet some specifications.

[33] and [145] propose techniques to automate this process partially. Casinovi in [33] developed a general-purpose rigorous algorithm for macromodel optimization. In

[145], the designer provides a transistor-level description of the circuit block under consideration, a time-domain macromodel with a number of adjustable parameters, and a set of target specifications. The input excitations used to verify the transistor circuit and a reasonable range for each parameter are also provided by the user. Parameters of the macromodel are optimized to minimize the "difference" between the waveforms obtained from the macromodel and the transistor-level circuit. These approaches facilitate macromodel creation for nonlinear circuits to a certain degree, but the accuracy of the macromodel-based circuit simulation approach still heavily depends on the macromodel architecture and specifications provided by the user.

When the circuit block to be macromodeled is linear time-invariant (by construction, or the small-signal equivalent circuit for a nonlinear circuit that has been linearized at a fixed operating point) then the task is relatively easier. Macromodels for linear time-invariant circuits can be generated through the use of techniques such as symbolic analysis [99], asymptotic waveform evaluation [249] and block Lanczos algorithm [83][78]. However, we do not believe that automatic generation of macromodels for nonlinear circuits in general is feasible.

Macromodels composed of interconnections of basic circuit primitives available in circuit simulators are convenient in the sense that a circuit simulator such as SPICE can be used without any changes.

3.1.2   Analog "Multi-Level" Simulation

The flexibility of the macromodeling approach can be increased by adding some additional features in the simulator. For instance, in iMACSIM [283], circuit blocks described by s-domain or z-domain transfer functions and macromodels written in "C" are allowed. Simulators having similar features are referred to as "Analog 'Multi-Level' Simulators" [261]. There are quite a few commercial simulators which have similar capabilities: PSPICE (MicroSim), HSPICE (MetaSoftware), SABER (Analogy), SPECTRE (Cadence), etc. [261].

In analog "multi-level" simulators, "behavioral" models are described in terms of s-domain transfer functions, system of differential equations, z-domain transfer functions, difference equations, "C" code, and "analog hardware description language (AHDL)" [261]. There are several technical issues in the implementation of an analog "multi-level" simulator, such as synchronization between the simulation of continuous-time and discrete-time portions of the circuit, convergence in the presence of discontinuities, computation of first-order derivatives for user-defined nonlinear models. Synchronization between the simulation of continuous-time and discrete-time portions of the circuit is maintained following an event-driven paradigm [261]. First-order derivatives for nonlinear models are obtained using techniques such as automatic symbolic differentiation [186] and simplicial subdivision [196], or Broyden's method is used to calculate modified nodal analysis (MNA) stamps for nonlinear models [34].

As it was pointed out at the beginning of this section, these new features provided in analog "multi-level" simulators are convenient extensions to the traditional macromodeling approach discussed in the previous section. But, we believe that, they do not provide any conceptual novelty in analog "high-level" modeling over the traditional macromodeling approach, and suffer from the same problems. For instance, Liu in [176, p. 10] shows that a model of a comparator using an AHDL (with differential, algebraic equations and table look-up functions) is equivalent to a traditional macromodel (composed of basic circuit primitives) of the comparator. In the remainder of the text, the traditional macromodeling with basic circuit primitives and the additional features provided in analog "multi-level" simulators will be referred to as the *macromodeling* approaches. The phrase "behavioral modeling" will be reserved for a new "high-level" modeling concept that is significantly different from macromodeling.

Some attributes of macromodeling are listed below:

- Macromodels reduce simulation time because of reduced circuit size, but a macromodel accurate for a variety of input conditions has from a third to a half of the number of elements of the original circuit [176]. As a result, simulation time reduction is not satisfactory.

- Creation of macromodels for general nonlinear circuits depends heavily on the experience of the designer, even though there are some techniques [33][145] which facilitate macromodel verification and optimization.

- Macromodels are useful when the detailed implementation of a circuit block (such as op amps, phase-frequency detectors, etc.) is not available. A library of well-built, thoroughly tested macromodels represented in some standard AHDL could be useful.

- Macromodels are deterministic and mostly concentrate on time-domain behavior. It is very difficult to simulate frequency domain performance, *noise* performance and effects due to *statistical parameters* arising from process variations [176].

We believe that macromodels would be most useful in simulating mostly digital designs with some non-critical analog sub-blocks. We do not believe that they are satisfactory for use in high performance analog and mixed-signal design, where *analog second-order effects* such as noise, distortion, and statistical variations become critically important.

3.1.3 System-Level Functional Simulation

The prohibitive CPU time cost of circuit simulation and macromodeling approaches forced analog designers to use simpler simulation techniques with idealistic models to do system-level functional simulations to get a first-order idea about the performance of the system. Some of these simpler simulation techniques took the form of customized simulators for specific types of circuits. The conspicuous ones are:

- SWITCAP [77] for time and frequency domain analysis of linear switched-capacitor circuits modeled with ideal switches, linear voltage-controlled voltage sources, linear capacitors and independent voltage sources. The simulation algorithm is discrete-time and solves charge conversation equations with the assumption that charge redistribution happens instantly after clock transitions.

- MIDAS [317] for discrete-time functional simulation of mixed digital and analog sampled-data systems.

For other specific types of circuit, for which special-purpose simulators do not exist, circuit designers write customized simulators (using "C" code) to do system-level simulations using idealistic models (possibly including some non-idealities) for the system components. Use of numerical math software, such as MATLAB from Math-Works, for various high-level simulation tasks is also common. SIMULINK (comes with MATLAB) is an interactive system for simulation of general nonlinear dynamic systems, and provides a similar "high-level" simulation environment like the ones in analog "multi-level" simulators. User creates the model of a system building an interconnection of "blocks" from a library using a graphical language. Various types of linear, nonlinear, continuous-time, discrete-time modeling blocks, signal sources and post processing functions are available.

3.1.4 Analog Hardware Description Languages

Currently, there is a great deal of activity to define and develop a standard AHDL [261][262]. The MIMIC Hardware Description Language (MHDL) [207] is intended to address the domain of analog/microwave hardware [262]. Parallel efforts are directed towards extending VHDL and Verilog to support analog circuits, which are called VHDL-A [138] and Verilog-A, respectively. MAST from Analogy [196] has been around for a while, and it is used (together with the simulator SABER from Analogy [196]) widely even outside the electrical system design community. Other commercial simulators such as PSPICE (MicroSim), SPECTRE-HDL (Cadence) also come with AHDLs.

Development of a standard AHDL, in which models from all levels of design hierarchy (layout, transistor-level, macromodels, "behavioral," etc.) can be represented, is surely necessary. Such a language can serve as the standard exchange medium between designers. Libraries of models represented in the standard AHDL can be developed and shared between designers.

Most of the currently available AHDLs are associated with an "Analog 'Multi-Level' Simulator." They provide convenient constructs for representation of models from various level of the design hierarchy, including *macromodels* discussed before. We believe that an AHDL for high-performance analog and mixed-signal design must provide full support to represent the *analog second-order effects* such as noise, distortion and statistical variations. There should be constructs for representation of models to be used in *statistical/noise simulations* using Monte Carlo and especially *non*-Monte Carlo techniques.

3.1.5   Specialized Simulation Techniques for Analog Circuits

Nonlinear DC analysis for the quiescent operating point and *transient analysis* for the time-domain behavior of nonlinear dynamic circuits are the most used forms of analyses available in circuit simulators. One reason for this is the fact that time-domain is the main simulation domain in the design of digital integrated circuits. On the other hand, linear frequency-domain performance characterization (for nonlinear circuits linearized at a fixed operating point) in the form of small-signal AC analysis, pole-zero analysis, noise analysis with linear time-invariant transformations [253] are essential and widely used in linear analog integrated circuit design. Fourier analysis of time-domain waveforms obtained with transient analysis is used to calculate various distortion criteria.

Analog and mixed-signal designs are becoming highly integrated and smaller in form factor, and constraints on power, size, performance are becoming tighter. As a result, analog second-order effects are becoming even more critical. More advanced and accurate specialized analysis, modeling, and simulation techniques are needed to predict the performance of designs in the presence of second-order non-idealities.

In the specialized analysis domain for analog circuits, more problems arise from to the difficulty and high complexity of the required analysis and modeling tasks than from the large size of the circuit.

More accurate and advanced specialized analysis, modeling and simulation techniques are being developed to address critical problems which arise in analog design:

- [166] presents steady-state analysis methods to obtain the sinusoidal, periodic or quasi-periodic steady-state response of a nonlinear dynamic circuit (possibly containing distributed elements) efficiently. Authors present time domain (finite-difference methods, shooting methods), frequency domain (harmonic balance) and mixed-frequency-time methods and compare them. [205] presents an efficient multi-tone distortion analysis technique based on the harmonic balance method using preconditioned iterative linear system solution. The steady-state analysis technique in [302] is based on the matrix-free Krylov-subspace method.

- Kundert in [165] presents an accurate Fourier analysis method for circuit simulators based on the direct computation of the Fourier integral rather than on the discrete Fourier transform.

- A problem that arises with higher integration is the undesired interactions through substrate and package coupling [108]. A substrate and package modeling and simulation methodology for accurate prediction of these coupling effects prior to fabrication is definitely needed [108]. Recent work on this problem can be found in [96][308][294].

- On-chip inductors (bond-wire or spiral) are becoming essential components for low-power RF design. This created the need for accurate models and design tools for integrated inductors [108].

Still, these existing techniques need to be refined and new analysis, modeling and simulation techniques need to be developed to address other special problems that arise in analog design. For instance, one critical non-ideality in analog circuits is electrical noise. It is caused by the small current and voltage fluctuations that are generated within the integrated-circuit devices themselves. Accurate prediction of the noise performance of the analog blocks in a mixed-signal system design is crucial to ensure that the overall system will work correctly. It is desirable to have a noise analysis technique which can predict the noise performance of *nonlinear dynamic circuits* with arbitrary excitations. Such a noise simulation method will be presented in Chapter 6.

3.1.6 Summary

The complexity of electronic systems being designed today is increasing in many dimensions: on one hand the number of components is growing constantly, on the other several radically different functions must be integrated. Circuit simulation with accurate and realistic models has been an invaluable tool for the verification of the electrical performance of integrated circuits. When complex analog and mixed-signal systems are being designed, accurate circuit simulation of the entire circuit is out of the question. The complexity in terms of the number of components and of the types of analyses makes the use of a circuit simulator impractical.

Relaxation-based electrical simulation and macromodeling were proposed to address the problems that arise with traditional circuit simulation. Relaxation-based electrical simulation is not suitable for use in analog design. The level of efficiency acquired and the types of analyses available with macromodeling approaches are not satisfactory. A new concept in "high-level" modeling and simulation of analog circuits is needed to address the specific problems that arise in analog design.

There is a great deal of activity to develop a standard analog hardware description language. Most of the analog "multi-level" simulators are associated with an AHDL. An AHDL for high-performance analog and mixed-signal design must provide full support to represent the *analog second-order effects* such as noise, distortion and statistical variations. There should be constructs for representation of models to be used in *statistical/noise simulations* using Monte Carlo and especially *non*-Monte Carlo techniques.

Specialized modeling, analysis and simulation techniques such as steady-state analysis, Fourier analysis, substrate and package modeling and simulation are essential in analog design. With higher integration levels and tighter performance constraints, existing techniques need to be refined, and new analysis, modeling and simulation techniques need to be developed to address other special problems that arise in analog design, such as noise analysis for nonlinear circuits.

3.2 STRATEGY

The objective of *behavioral modeling*, in general, is to represent circuit functions with *abstract mathematical models* that are independent of circuit architectures or schematics. In top-down design, designers can verify the system design before investing time in detailed circuit implementation, which enables them to explore the system design space rapidly. In bottom-up design verifications, designers can verify complex system behavior efficiently, because evaluations of behavioral models are computationally cheap, resulting in fast system simulations. Based on the success of behavioral modeling for digital circuits, there is great interest in extending the idea of behavioral modeling to analog circuits.

For digital circuits, behavioral modeling and simulation can be performed using hardware description languages such as VHDL or Verilog. These languages do not provide the features needed for an adequate simulation and modeling of the analog blocks in a mixed-signal system design. While there is a great deal of activity to define and develop a standard analog hardware behavioral description language, there is still *little* work being done to develop analog behavioral models and specialized simulation techniques to exploit the particular domains of interest in mixed-signal system analysis. Behavioral models and simulation techniques are receiving more and more attention because of the trend in ASICs and VLSICs to integrate total systems, including both analog and digital blocks, onto a single chip. Another reason is that even "simple" analog blocks are too complicated for a complete transistor-level simulation within reasonable amounts of CPU time.

The top-down design process implies a well-defined behavioral description of the analog function. The behavioral characterization of analog circuits is quite different from the digital one; the analog characterization is composed of *not only* the function that the circuit is to perform, but also the *second-order non-idealities* intrinsic to analog operation. In fact, errors in the design often stem from the non-ideal behavior of the analog section, not from the selection of the "wrong" functionality. To shorten the design cycle, it is essential that design problems be discovered as early as possible. For this reason, behavioral simulation is an essential component of any methodology for the design of analog and mixed-signal systems. This simulation can help in selecting the correct architecture to implement the analog function with bounds (constraints) on the amount of non-idealities that is allowable given a set of specifications at the system level.

Our approach to designing complex analog and mixed-signal systems relies on realistic circuit modeling at each step and the ability to propagate constraints step by step. We do not need to verify top level system performance from low-level implementation details with a SPICE simulation of the entire system, but we do need to guarantee correct translation of one level of constraints to a set of lower level constraints. The tools for constraint translation are the behavioral models at each level, the behavioral simulator, and optimization, which will produce constraints for the lower levels, using the behavioral models and the simulator. The goal of our modeling and simulation research is to

- develop behavioral models and simulation techniques for the design of complex analog sub-systems such as A/D and D/A converters, phase/delay-locked systems, RF transceivers,

- develop specialized modeling, analysis and simulation techniques for specific problems that arise in analog design, such as a noise analysis method for nonlinear dynamic circuits with arbitrary excitations,

- and integrate these techniques within the framework of our top-down constraint-driven design methodology for analog and mixed-signal systems.

We do not plan to work on the development of an analog hardware behavioral description language. Instead, we will work closely with industry to use whatever language will be considered standard to describe our models. We will provide feedback to define some aspects of the language that may make models easier to describe and manipulate. Our emphasis is on the *mathematical aspects* of analog behavioral modeling and simulation. In analog behavioral modeling and simulation research, we believe that the *real added value* is in creating realistic models of circuit components which capture analog second-order effects using the most suitable mathematical techniques. Frameworks, description languages and simulation environments without good models and specific simulation techniques are ineffectual.

For the behavioral simulation and modeling of analog blocks in a mixed-signal system design, the following features are essential:

- The simulator and the behavioral models have to be general. The behavioral model of a given analog block must describe the behavior of that block considered as a black box, describing its input-output behavior in terms of a set of model parameters to be supplied by the designer. It also has to hide all the internal architectural details as much as possible, resulting in generic models. The simulation engine must be independent of any particular model, so that it is possible to simulate different architectures in the same environment instead of having a different dedicated simulator for each specific architecture.

- The behavioral models for the analog blocks must include *not only* the first-order behavior of the circuit, but also the *analog second-order effects*, such as *noise* and *distortion*, in order to get a realistic idea of the performance of the overall system. Also, the *statistical variation* of the circuit performance has to be taken into account.

- The behavioral simulation is to be done in time or frequency domain or a mixture of both depending on the nature of the problem. There should be methods to switch between these domains, and to deal with noise, distortion and statistical parameters.

In order to realize a design environment having the features discussed above, our strategy is to

- Find the best *abstract mathematical representations* for specific types of analog circuits, and develop realistic models by using appropriate

    - algebraic expressions,
    - ordinary differential equations,
    - state-space representation,
    - difference equations,
    - s-domain or z-domain transfer functions,
    - probability distributions,
    - random processes,
    - stochastic differential equations,
    - look-up tables,
    - and other suitable mathematical constructs.

- Develop behavioral simulation techniques based on the behavioral models developed.

- Develop techniques to extract behavioral model parameters from lower level models of the circuit (bottom-up extraction). This allows us to investigate the performance of different architectures, in order to be sure that the generic models cover all (or as much as possible) architectures and as much as possible second-order effects in a realistic way. Bottom-up extraction of behavioral model parameters from a lower level description is essential in verifying the performance of a completed design.

In the remainder of this chapter, we present examples of our work on behavioral simulation and modeling for analog circuits. In particular, we will present behavioral models and simulation techniques for Nyquist-rate data converters, noise in mixed-mode sampled data systems, and phase-locked systems.

3.3 NYQUIST-RATE DATA CONVERTERS

3.3.1 Background

Data converters convert the continuous-time, continuous-amplitude *analog* signals in the outside world to discrete-time, discrete-amplitude *digital* signals in electronic systems, and vice-versa. The behavior of a converter is affected by two basic statistical effects: (1) *noise*, (2) *process variations*. Noise can be of different sources, e.g, thermal noise, flicker noise, shot noise, and noise coupled from digital circuitry. Noise can cause the same chip to behave differently even if the same inputs are applied. Process variations cause different chips of the same circuit to have different behavior. To characterize these unintended, *second-order effects*, traditional user specifications include static specifications such as integral nonlinearity (INL), differential nonlinearity (DNL), gain error, offset error, and probability of error due to noise, and dynamic specifications such as signal-to-noise ratio as a function of input frequency or input amplitude. The static specifications are for moderate speed converters that can be treated as *memoryless* (e.g. output does not depend on past inputs), while the dynamic specifications are important for high speed converters, where output can depend on past inputs. In this research, we focus on modeling a memoryless converter for static performance specifications.

Previously, Ruan [255] proposed different behavioral models for different types of A/D converter architectures. One drawback is that the architectural dependence necessitates derivation of a new model for any changes in the converter architecture. Another drawback is that the model is deterministic. For example, a deterministic model can be expressed mathematically as

$$code = A(V_{in}, \hat{v}) \tag{3.1}$$

where *code* is the output code, A is a function, V_{in} is the input, and $\hat{v}$ is a set of m parameters. There are two problems with this behavioral modeling approach; namely,

- Worst case analysis must be used to find worst case converter performance. For example, 2^m simulations are needed to find the worst case. For a typical converter, m can be very large, making worst case analysis infeasible.

- Noise effects are not modeled, so signal-to-noise ratio cannot be calculated.

3.3.2 New Data Converter Behavioral Model

In contrast to the traditional deterministic model, we derive a stochastic converter model that include noise and process variation effects as follows. A memoryless Nyquist A/D operating at a certain environment such as fixed sampling frequency and temperature can be described by its transfer curve, which plots the output code on the range against the input continuous value on the abscissa. Assuming an A/D has N bits of resolution, is designed to be monotonic and has no missing codes, the transfer curve can be characterized by $2^N - 1$ real numbers in a vector, t, representing the values of the transition points t_i, $i = 1 \ldots 2^N - 1$. Due to statistical process variations, the transfer curve t has a statistical distribution. Due to noise, an A/D produces a wrong output with a non-zero probability. For example, if the input is near a transition point, the probability of getting a wrong output code, the code at the other side of the transition point, is non-zero. As a result, the noise effects depend on the input, and they become more significant as the input gets closer to a transition point. Mathematically, this is represented by a probability distribution function of the output codes for any input value. Our behavioral model captures this noise effect by a joint probability density function of the input V_{in} and the output code for a given realization of the transfer curve $\hat{t}$,

$$f(code, V_{in}, \hat{t}) \tag{3.2}$$

From the above equation, which describes the behavior of any memoryless A/D due to noise and process variations completely, we can extract all information about the converter.

In our model, noise and process variation effects are separated. The distribution t captures process variation effects, while the joint density function $f(\cdot)$ captures noise effects. The function $f(\cdot)$ is computed using either direct techniques [181] or Monte Carlo techniques. The process variation effects can be derived by making assumption:

Assumption 3.3.1 *Process variations are the cumulative effects of many integrated circuit fabrication steps. Thus, we assume that the distribution of the process variations, v, is multivariate normal.*

Next, we propose a Gaussian model, as well as a non-Gaussian model for the distribution t. The Gaussian model is derived from making the following assumption:

Assumption 3.3.2 *In general, an A/D is designed to be insensitive to process variations which are relatively small. Thus, the transition points, t, change relatively little with respect to process variations. Hence, we assume that t changes linearly with* *respect to the process variations:*

$$t = \frac{\partial t}{\partial v} v + \mu_t \tag{3.3}$$

where v is a random vector of parameters corresponding to process variations, such as offsets and mismatches in components, $\frac{\partial t}{\partial v}$ is a sensitivity matrix, and μ_t is a vector of the nominal transition points.

From assumption 3.3.1 and (3.3), the covariance matrix of t is given by

$$\Sigma_t = \frac{\partial t}{\partial v} \Sigma_v \frac{\partial t}{\partial v}^T \tag{3.4}$$

where Σ_v is the covariance matrix of v. Thus, the distribution of t is

$$t \sim normal(\mu_t, \Sigma_t) \tag{3.5}$$

In general, the rank r_t of Σ_t is much less than full rank since not all the transition points are independent. In other words, any realization of t must fall into a space spanned by r_t independent basis vectors. A set of such independent basis vectors can be found by the singular value decomposition of Σ_t. Mathematically,

$$\Sigma_t = U_t \Sigma_{ct} U_t^T \tag{3.6}$$

where U_t is an $2^N - 1$ x r_t transformation matrix, Σ_{ct} is an r_t x r_t diagonal matrix, and r_t is the rank of Σ_{ct}. The columns of U_t are called the *error signatures* [288] and form a set of orthonormal vectors spanning the *error space*. For storage and computational efficiency t is expressed as the following equation instead of (3.3),

$$t = U_t c_t + \mu_t \tag{3.7}$$

where $c_t \sim normal(0, \Sigma_{ct})$ is a zero mean multivariate normal distribution with r statistically *independent* variates. Typically, $r \ll m$, so (3.7) uses less computational resources than (3.5). The parameters U_t and μ_t are unique for each A/D architecture, while the distribution c represents a mixture of the relevant process variations.

Summarizing, our entire Gaussian A/D model consists of three equations,

$$f(code, V_{in}, \hat{t}), \hat{t} = realization(t) \tag{3.8}$$

$$t \sim normal(\mu_t, \Sigma_{ct}) = U_t c_t + \mu_t \tag{3.9}$$

$$c_t \sim normal(0, \Sigma_{ct}) \tag{3.10}$$

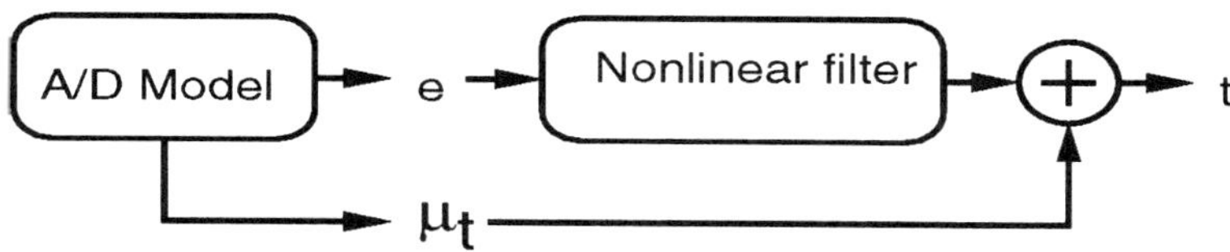

Figure 3.1 Nonlinear model for data converters

For cases where assumption (3.3.2) does not hold, we propose a non-Gaussian model. In such cases, the Gaussian model can be extended to a non-Gaussian model by appending a nonlinear *filter* to create the non-Gaussian distributions [27][26]. Figure 3.1 shows a block diagram of a nonlinear model where the zero mean Gaussian error, $e = U_t c_t$, is being filtered to model a non-Gaussian error. For example, to model an error with third order non-Gaussian statistics [303], a quadratic filter b is used in the following equation for t as a replacement for (3.9),

$$t = \mu_t + U_t c_t + \int_1^{2^N-1} \int_1^{2^N-1} b(t-u, t-v)e(u)e(v)dudv \qquad (3.11)$$

where $e = U_t c_t$, b is the two-dimensional impulse response of the quadratic filter, and $2^N - 1$ is the number of transition points. Although the non-Gaussian model is more general, its use is limited because from our experience of modeling many converters most converters are adequately represented by a Gaussian model. Secondly, a Gaussian model is computationally more efficient than a non-Gaussian model.

After presenting the complete A/D model, we develop a similar D/A model using the same strategy. The resulting Gaussian D/A model is,

$$t = U_t c_t + \mu_t + z \qquad (3.12)$$

$$c_t \sim normal(0, \Sigma_{ct}) \qquad (3.13)$$

$$z \sim normal(0, \Sigma_z) \qquad (3.14)$$

where t is an n-dimensional vector and z is an n-dimensional vector representing additive noise in D/A outputs. To extend the Gaussian D/A model to a non-Gaussian model, we replace (3.12) by the following,

$$t = \mu_t + U_t c_t + z + \int_1^{2^N} \int_1^{2^N} b(t-u, t-v)e(u)e(v)dudv \qquad (3.15)$$

where 2^N is the number of D/A input codes.

3.3.3 Calculation of System Performance

Using the model, we propose a novel strategy to calculate system performance. The performance of a converter is computed in two steps. First, the converter model parameters are extracted from the circuit. Then, the converter performance is computed using only the model parameters since the model captures the converter behavior. The advantage of this strategy is that the model parameters are only extracted once, but used many times to calculate system performance. The following definitions illustrate how system performance parameters can be computed from the proposed D/A model parameters.

Definition 3.3.1 *The ideal output vector, L, of a D/A converter is an n-dimensional vector, where the ith component of L, $L(i)$, represents the desired output for input code i, $n = 2^N$, and N is the number of bits.*

Definition 3.3.2 *The output vector, t, of a D/A converter is an n-dimensional vector, where the ith component of t, $t(i)$, is the output for input code i.*

The output vector, t, is identical to that in (3.12) with noise $z = 0$. Theoretically, t is sufficient to describe the converter behavior due to process variations. Nevertheless, designers traditionally used a different set of specifications which can be derived easily from t. All of them are described as follows,

Definition 3.3.3 *Offset error is $t(1) - L(1)$.*

Definition 3.3.4 *Full scale gain error is $t(n) - L(n) + L(1) - t(1)$.*

The integral nonlinearity is the deviation of the output from the ideal value *after* compensation for the gain and offset errors.

Definition 3.3.5 *The integral nonlinearity vector, s, of a D/A converter is an n-dimensional vector*

$$s(i) = at(i) + b - L(i) \tag{3.16}$$

where a, b are constants such that $s(1) = s(n) = 0$.

Intuitively, Definition 3.3.5 means that we transform t using constants a and b such that the two end-points are ideal. In this case, we have adopted the more popular end-point method for linear error compensation [68], instead of the less popular least square method. The motivation behind the transformation is to compensate for the offset and gain errors before calculation of nonlinearity. In many applications, offset and gain errors are irrelevant as they do not contribute to distortion.

Lemma 3.3.1 *Integral nonlinearity is given by*

$$s(i) = \left[\frac{L(n) - L(1)}{t(n) - t(1)}\right](t(i) - t(1)) + L(1) - L(i) \tag{3.17}$$

The sensitivity matrix of the integral nonlinearity to output vector is given by

$$\frac{\partial s}{\partial t}(i,j) = \begin{cases} \frac{(L(n)-L(1))(t(i)-t(n))}{(t(n)-t(1))^2} & i \neq 1, n \quad j = 1 \\ 0 & i \neq 1, n \quad j \neq 1, i, n \\ \frac{L(n)-L(1)}{t(n)-t(1)} & i \neq 1, n \quad j = i \\ 0 & i = 1, n \quad j = 1 \ldots n \\ \frac{(L(n)-L(1))(t(1)-t(i))}{(t(n)-t(1))^2} & i \neq 1, n \quad j = n \end{cases} \tag{3.18}$$

where $\frac{\partial s}{\partial t}(i,j)$ *is the entry at the* i^{th} *row and* j^{th} *column of the sensitivity matrix,* $\frac{\partial s}{\partial t}$.

Proof. Solving for a, b using constraints in Definition 3.3.5 yield (3.17). Equation (3.18) follows from differentiation of (3.17). ∎

Using a first order Taylor approximation of s as a function of t,

$$s \approx \frac{\partial s}{\partial t}(t - \mu_t) + \mu_s = \frac{\partial s}{\partial t}U_t c_t + \mu_s \tag{3.19}$$

we find the integral nonlinearity has distribution,

$$s \sim normal(\mu_s, \Sigma_s) \tag{3.20}$$

where $\mu_s = s(t)$, at $t = \mu_t$ using (3.17) and $\Sigma_s = \frac{\partial s}{\partial t}\Sigma_t \frac{\partial s}{\partial t}^T$ at $s = \mu_s$. In general, the rank r_s of Σ_s is much less than full rank since not all the INL errors are linearly independent. In other words, any realization of s must fall into a space spanned by r_s independent basis vectors. A set of such independent basis vectors can be found by the singular value decomposition of Σ_s. Mathematically,

$$\Sigma_s = U_s \Sigma_{cs} U_s^T \tag{3.21}$$

where U_s is a 2^N x r_s transformation matrix, Σ_{cs} is an r_s x r_s diagonal matrix, and r_s is the rank of Σ_{cs}. The columns of U_s are called the *INL signatures* and form a set of orthonormal vectors spanning the *INL space*. For storage and computational efficiency s is expressed as the following equation,

$$s = U_s c_s + \mu_s \tag{3.22}$$

where $c_s \sim normal(0, \Sigma_{cs})$ is a small, zero mean multivariate normal distribution with r_s statistically *independent* variates.

The motivation behind finding the distribution of s is that INL is an important user constraint. By using the distribution of s, bounds on s can be determined quickly and made to fit user constraints during design.

Definition 3.3.6 *The $\pm k\sigma$ INL bounds are*

$$s_{\pm k\sigma}(i) = \mu_s(i) \pm k\sqrt{\Sigma_s(i,i)}. \tag{3.23}$$

These bounds are useful for design since any realization of s will have a high probability of falling within the $k\sigma$ bounds if $k \geq 2$. Furthermore, once the design is done, circuit yield can be estimated by Monte Carlo integration over the distribution s.

In addition, the contribution of INL from each component can be precisely calculated. The INL contribution matrix, p_s, is defined as the cross-covariance between the INL vector s and component variation vector v. An entry at $p_s(i,j)$ represents the contribution from component variation v_j to the INL of code i, s_i. For example,

Definition 3.3.7 $p_s = E[sv^T]$

Lemma 3.3.2 $p_s = \frac{\partial s}{\partial t} \frac{\partial t}{\partial v} \Sigma_v$

Proof. Using (3.19), we have

$$p_s = E[sv^T] = E[(\frac{\partial s}{\partial t}(t - \mu_t) + \mu_s)v^T] \tag{3.24}$$

and using (3.3), we have

$$p_s = E[\frac{\partial s}{\partial t} \frac{\partial t}{\partial v} vv^T] + E[\mu_s v^T] \tag{3.25}$$

$$p_s = \frac{\partial s}{\partial t}\frac{\partial t}{\partial v}\Sigma_v \tag{3.26}$$

since v has zero mean, and $E[vv^T] = \Sigma_v$. ∎

Another useful piece of information to designers is the contribution to INL signatures from each component. Using this information, the importance of each component can be ranked, and the worst component can be pinpointed for redesign. The INL signature contribution matrix, q_s, is defined as the cross-covariance between the vector c_s and component variation vector v. An entry at $q_s(i,j)$ represents the contribution from process variable v_j to the magnitude of signature i, c_i. For example,

Definition 3.3.8 $q_s = E[c_s v^T]$

Lemma 3.3.3 $q_s = U_s^{-1}\frac{\partial s}{\partial t}\frac{\partial t}{\partial v}\Sigma_v$

Proof. Using (3.19), (3.22) and (3.3), we have

$$c_s = U_s^{-1}\frac{\partial s}{\partial t}(t - \mu_t) = U_s^{-1}\frac{\partial s}{\partial t}\frac{\partial t}{\partial v}v \tag{3.27}$$

$$q_s = E[c_s v^T] = E[U_s^{-1}\frac{\partial s}{\partial t}\frac{\partial t}{\partial v}vv^T] = U_s^{-1}\frac{\partial s}{\partial t}\frac{\partial t}{\partial v}\Sigma_v \tag{3.28}$$

using the definition. ∎

Besides INL, designers are also interested in differential nonlinearity, DNL, defined as follows,

Definition 3.3.9 *The differential nonlinearity vector, d, of a D/A converter is an $n-1$-dimensional vector obtained by a first order difference of the integral nonlinearity s.*

$$d = D_+ s \tag{3.29}$$

where D_+ is an $n-1$ by n matrix called the first order difference operator,

$$D_+(i,j) = \begin{cases} -1 & i = j \\ 1 & i = j - 1 \\ 0 & else \end{cases} \tag{3.30}$$

As DNL is a linear transformation of INL, most results related to INL carry over to DNL immediately. For instance, the differential nonlinearity has distribution,

$$d \sim normal(\mu_d, \Sigma_d) \tag{3.31}$$

where $\mu_d = D_+\mu_s$, $\Sigma_d = D_+\Sigma_s D_+^T$. In general, the rank r_d of Σ_d is much less than full rank since not all the DNL errors are independent. In other words, any realization of d must fall into a space spanned by r_d independent basis vectors. A set of such independent basis vectors can be found by the singular value decomposition of Σ_d. Mathematically,

$$\Sigma_d = U_d \Sigma_{cd} U_d^T \tag{3.32}$$

where U_d is an $2^N - 1$ x r_d transformation matrix, Σ_{cd} is an r_d x r_d diagonal matrix, and r_d is the rank of Σ_{cd}. The columns of U_d are called the *DNL signatures* and form a set of orthonormal vectors spanning the *DNL space*. For storage and computational efficiency d is expressed as the following equation,

$$d = U_d c_d + \mu_d \tag{3.33}$$

where $c_d \sim normal(0, \Sigma_{cd})$ is a small, zero mean multivariate normal distribution with r_d statistically *independent* variates.

Definition 3.3.10 *The $\pm k\sigma$ DNL bounds are:*

$$d_{\pm k\sigma}(i) = \mu_d(i) \pm k\sqrt{\Sigma_d(i,i)}. \tag{3.34}$$

These bounds are useful for design since any realization of d will have a high probability of falling within the $k\sigma$ bounds if $k \geq 2$. In addition, the contribution of DNL from each component is defined similarly as the DNL contribution matrix, p_d.

Definition 3.3.11 $p_d = E[dv^T]$

Lemma 3.3.4 $p_d = D_+ \frac{\partial s}{\partial t} \frac{\partial t}{\partial v} \Sigma_v$

Proof.

$$p_d = E[dv^T] = E[D_+ sv^T] = D_+ E[sv^T] \tag{3.35}$$

from (3.29) and (3.26). ∎

The DNL signature contribution matrix, q_d, is defined as

Definition 3.3.12 $q_d = E[c_d v^T]$

Lemma 3.3.5 $q_d = U_d^{-1} D_+ \frac{\partial s}{\partial t} \frac{\partial t}{\partial v} \Sigma_v$

Proof. Using (3.33), (3.29), (3.19) and (3.3), we have

$$c_d = U_d^{-1} D_+ \frac{\partial s}{\partial t}(t - \mu_t) = U_d^{-1} D_+ \frac{\partial s}{\partial t} \frac{\partial t}{\partial v} v \tag{3.36}$$

$$q_d = E[c_d v^T] = E[U_d^{-1} D_+ \frac{\partial s}{\partial t} \frac{\partial t}{\partial v} v v^T] = U_d^{-1} D_+ \frac{\partial s}{\partial t} \frac{\partial t}{\partial v} \Sigma_v \tag{3.37}$$

using the definition. $\blacksquare$

Traditionally, more complex converter specifications such as yield, harmonic distortion and signal-to-noise ratio are obtained from chip data since simulations are computationally infeasible for these system specifications. However, using the proposed model these specifications can be estimated efficiently. For example, yield is typically defined as the fraction of circuits that have INL and DNL within some bounds, s_{max} and d_{max}, respectively. Therefore, yield is obtained by Monte Carlo integration over a region in distributions, s and d. Using (3.19) and (3.29), the regions are

$$|\frac{\partial s}{\partial t} U_t c_t + \mu_s| < s_{max} \tag{3.38}$$

$$|D_+ (\frac{\partial s}{\partial t} U_t c_t + \mu_s)| < d_{max} \tag{3.39}$$

In the Monte Carlo simulations, many realizations of c_t are generated from a random number generator, some of which are labeled success if they fall inside both regions. Yield is estimated from the fraction of success over a finite number of realizations.

3.3.4 Model Validation

A model must be verified against measurements to test its validity. Our model,

$$f(code, V_{in}, \hat{t}), \hat{t} = realization(t) \tag{3.40}$$

describes the behavior of any memoryless A/D due to noise and process variations completely. However, in our more specific Gaussian model, the Gaussian assumption of t must be verified. Therefore, we need to verify process variation effects in (3.9) by verifying t has the multivariate normal distribution for a variety of converter types.

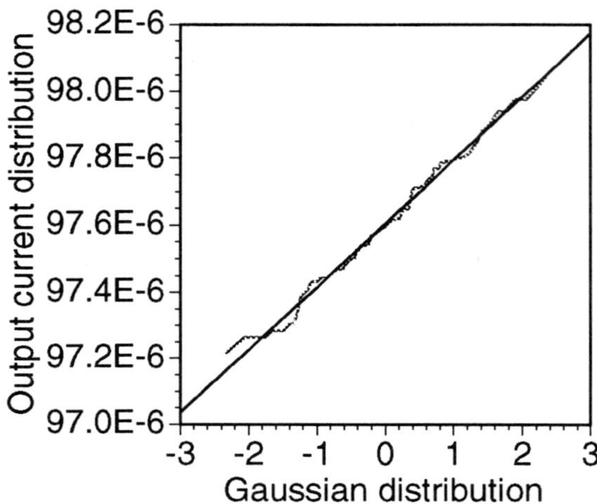

Figure 3.2 Quantile-quantile plot of output distribution vs. normal distribution for code=100

The verifications are done using Monte Carlo simulations with SPICE, as well as real chip data. Assuming typical process variations such as standard deviation of transistor length and width being $0.05\mu m$ and standard deviation of resistor value being one percent, Monte Carlo simulations of a 10-bit current-switched D/A converter consisting of 1068 transistors and 345 parasitic resistors have been performed for randomly selected input codes. The random codes are obtained using a uniformly distributed random number generator on an Hewlett Packard 42S calculator.

The output current distributions for all such simulations match well with the normal distribution, as verified by the quantile-quantile plots in Figure 3.2 and Figure 3.3 for randomly selected input codes 100 and 900, respectively. Each point on the graph represents a comparison between the cumulative distribution function of the output current and that of the normal distribution. Perfect matches fall in a straight line, validating components of the output current vector, t, have normal distributions.

Another implication of the A/D model is that the errors of a converter can be decomposed into its errors signatures. To verify that errors signatures are sufficient to characterize an actual A/D, we obtained the measured error, $\hat{t} - \mu_t$, of a fabricated cyclic A/D [201], and try to fit the measured data, $\hat{t}$, with a linear combination of error signatures using linear regression techniques

$$\hat{t} = \mu_t + U_t \hat{c} + \delta \tag{3.41}$$

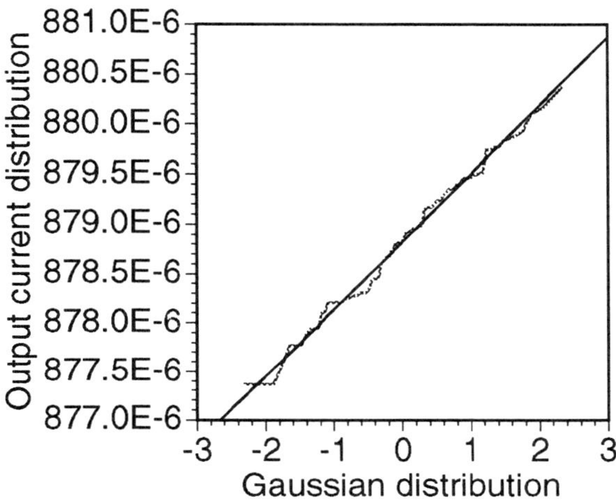

Figure 3.3 Quantile-quantile plot of output distribution vs. normal distribution for code=900

where we solve for $\hat{c}$, an unknown realization of c that minimize the measurement noise $||\delta||^2$, where $|| \cdot ||^2$ denotes the inner product of a vector with itself. Figure 3.4 shows the curve $\hat{t} - \mu_t$ matches well with the curve $U\hat{c}$, proving that the model sufficiently captures the A/D behavior.

3.3.5   Conclusion

We have presented a behavioral representation of Nyquist data converters. The representation captures the behavior of a memoryless Nyquist data converter, including statistical variations. The variations are classified into noise and process variations according to how these non-idealities affect the converter behavior. To describe noise effects, a joint probability density function is used. To describe process variations effects on the converter transfer function, a Gaussian model is used.

Using the behavioral representation, a new strategy to estimate system performance is developed. The performance specifications of a converter, including offset error, full scale gain error, INL, DNL, harmonic distortion and signal-to-noise ratio, are estimated in two steps. First, the converter model parameters are extracted from the circuit. Then, the converter performance is computed using only the model parameters since the model captures the converter behavior. Given that the behavioral model is valid, offset error and gain error are computed exactly, while INL and DNL are

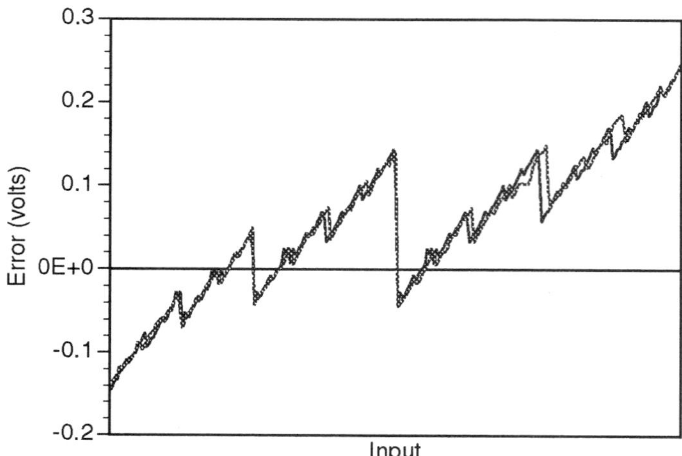

Figure 3.4 Comparison of measured and model generated errors

approximated using first order Taylor approximation. Worst case harmonic distortion and signal-to-noise ratio are approximated using Monte Carlo simulations [176].

From SPICE simulations, the distribution of D/A errors for a 10-bit converter were shown to agree well with the Gaussian distribution; thus, validating our Gaussian error assumption. Furthermore, behavioral simulation results agree well with measurements for an 8-bit cyclic A/D converter [201]; thus, validating the behavioral model.

3.4 NOISE IN MIXED-MODE SAMPLED DATA SYSTEMS

3.4.1 Background

The simulation of mixed analog and digital systems is crucial in system verification as more analog and digital circuits are integrated on the same integrated circuit in data acquisition, automotive, or disk drive electronics applications. For these systems, there is a trend to reduce system power consumption by reducing the supply voltage or minimizing parasitic capacitances. One of the fundamental limit to these power reduction techniques is the electronic noise inherent in the mixed analog/digital sub-systems. While reducing the supply voltage reduces the allowable signal power, the noise power due to physical effects remain the same. Due to the potential decrease in signal-to-noise ratio, it is desirable to simulate the mixed-mode system for noise performances. Unfortunately, there are no noise simulators that can analyze noise effects for mixed-mode systems.

To develop such a simulator, we follow the *behavioral simulation* paradigm in which circuits are modeled mathematically as in [182][98][178][180]. We present a new *noise behavioral model* and a *direct noise analysis* approach for mixed-mode systems. Using the appropriate model, there is no need for circuit or macromodel simulation, but a *direct algebraic* approach where the objects being manipulated are noise characterizations.

In Section 3.4.2, the problem is defined. In Section 3.4.3, the traditional Monte Carlo method for noise simulation is described. Then, new behavioral models (Sections 3.4.4 and 3.4.5) and new approach for noise analysis (Section 3.4.4) are presented with experimental results (Section 3.4.7).

3.4.2 Problem Definition

The class of circuits under consideration includes sampled-data systems that have mixed analog/digital inputs and outputs. This includes, but is not limited to, A/D converters, D/A converters, and receivers of digital signals. The objective of the simulation is to determine the noise effects due to electronic noise. Thermal, flicker, and shot noise in the transistors and thermal noise in resistors contribute to the noise of the circuit. Traditionally, electronic noise is modeled with a Gaussian random process described by its power spectral density [107]. From *the noise sources, the system architecture,* and *the deterministic input,* we seek the distribution of the output. For example, we seek the continuous output distribution of a D/A converter, the bit-error rate of a receiver, or the output code distribution of an A/D converter. To illustrate the problem, Figure 3.5 shows a switched capacitor implementation of a cyclic A/D converter. Due to noise effects, the output of the comparator is wrong with a non-zero probability. Since the N-bit digital output code is a logic function of N successive outputs of the comparator, we seek the probabilities for each of the 2^N possible output codes given a deterministic input.

3.4.3 Previous Work

Traditionally, the SPICE-like simulators [141] analyze analog circuit noise in the frequency domain. SPICE linearizes the circuit at the operating point, adds sinusoidal sources in parallel to the noisy elements, and analyzes the resulting AC equivalent

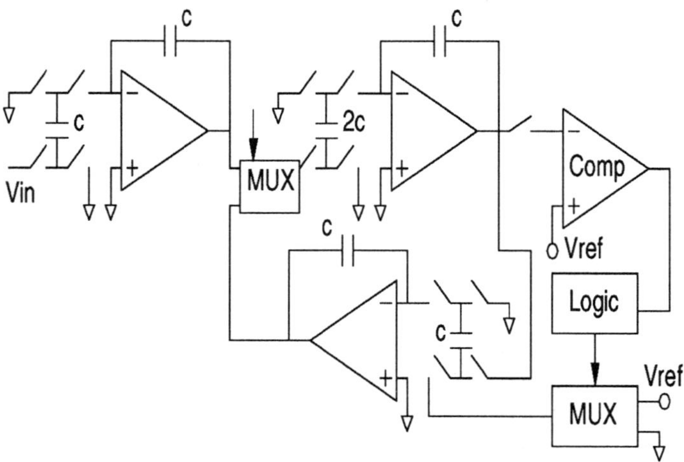

Figure 3.5 Cyclic A/D

circuit. The approach has two problems. Firstly, many components in mixed-mode systems, such as the comparator, have discontinuous transfer functions and cannot be linearized. Secondly, mixed-mode systems are usually non-steady state (e.g. an A/D produces an output code after a *finite* time), so an operating point does not exist for the circuit. Simulators such as [307] analyze noise in switched-capacitor circuits using linear systems theory, but fail to handle noise in mixed-mode switched-capacitor circuits as mixed-mode circuits are in general nonlinear.

Without simulators for noise effects, designers use the Monte Carlo noise simulation technique. In this technique, they model the circuit as a deterministic network of components, and inject random numbers into the signal path. They run the system many times, and hope to get a wrong output code by chance.

Figure 3.6(a) illustrates a sample-and-hold, a major component in a sampled-data system. The electronic noise source, v, has *power spectral density, $S_v(f)$*, where f is the frequency. The component is modeled by a delay with an additive noise, n, as shown in Figure 3.6(b). At each simulation time point, the simulator takes samples of n and add them into the signal before the delay. We compute the distribution of n from the noise power spectral density, $S_v(f)$. Due to the lowpass filter formed by R and C, the voltage on the capacitor when the switch, S1, is closed has power spectral density given by

$$\frac{S_v(f)}{1 + (2\pi f RC)^2} \tag{3.42}$$

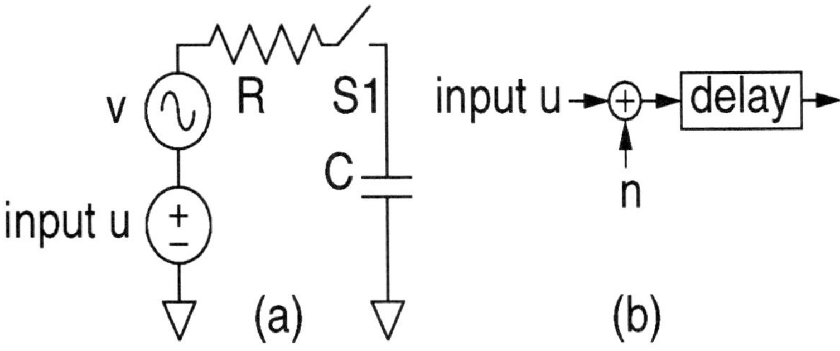

Figure 3.6   Sample-and-hold

As S1 switches, thus sampling the voltage onto the capacitor, we compute the equivalent *sampled noise power spectral density* on the capacitor by shifting and adding the original spectral density. From the sampling theorem, the sampled noise power spectral density, $S_d(f)$, is

$$S_d(f) = \sum_{k=-\infty}^{\infty} \frac{S_v(f - kf_s)}{1 + (2\pi(f - kf_s)RC)^2} \tag{3.43}$$

where f_s is the switching frequency. The inverse Fourier transform of $S_d(f)$ gives the time domain discrete autocovariance function $c_d(t)$ from which the sampled noise n is defined. The sampled noise, n, is a random vector with a multivariate normal distribution,

$$n \sim normal(0, \Sigma_n) \tag{3.44}$$

where $\Sigma_n(i, j) = c_d(iT - jT)$, n is a random vector with m components, mT is the total duration of the simulation, and T is the sampling period. Entry $n(i)$ in random vector n represents a noise sample at time iT. We generate realizations of noise sample $n(i)$ from a random number generator and inject them into the system at time iT. Correlated samples specified in Σ_n can be generated using filtering of uncorrelated random samples.

Despite the simplicity of the Monte Carlo approach, it has problems with computing low error probabilities and machine dependency. For a system with low error probability p_e, the number of system runs required for a confident statistical error estimate is proportional to $\frac{1}{p_e^2}$. For example, to simulate a receiver with bit-error rate of 10^{-8}, we need approximately 10^{16} samples. Furthermore, pseudo-random number generators on computers often do not generate such a large sequence of independent random numbers, but will re-use old random numbers instead. In such instances, the extra simulations do not yield a more accurate statistical estimate.

To compute the probability of rare events, special variance reduction techniques for Monte Carlo simulations such as importance sampling can be used. In general, Monte Carlo simulations can be expressed as the following integral,

$$\int_\Omega g(x)f(x)dx \qquad (3.45)$$

where $g(x)$ is the parameter function, $f(x)$ is the probability density function, and Ω is the sample space. In simple Monte Carlo, the integral is computed by generating samples of x from density function $f(\cdot)$ and summing $g(x)$. An *important sample*, x, provides useful information such as non-zero $g(x)$. Importance sampling [119] forces generation of important samples by using an artificial density function instead of $f(x)$. We can rewrite (3.45) as

$$\int_\Omega g(x)\left[\frac{f(x)}{h(x)}\right]h(x)dx \qquad (3.46)$$

where $h(x)$ is an artificial density function designed to increase the number of important samples, and $\frac{f(x)}{h(x)}$ is the weight of a sample.

Importance sampling is viable provided $h(x)$ is given, but no general algorithm has been proposed for choosing $h(x)$. There are typically many different noise sources, so the number of possible $h(x)$ is large. As a result, automatic selection of a suitable $h(x)$ out of many possibilities is difficult.

To circumvent the difficulty of choosing $h(x)$, Kahn introduced "splitting" techniques [147] as another form of importance sampling. In this technique, he splits an important sample into many neighboring samples with appropriate weights, thus increasing the number of important samples.

3.4.4 Behavioral Model of Noise

The key disadvantage of the Monte Carlo technique is the representation of noise by computer generated samples of noise because many noise samples are required for good accuracy. In contrast, we represent noise by a *random signal*. We sample the random signal n at time iT to obtain a noise sample $n(i)$ in (3.44). Each noise sample, X, is a *random variable* described by its distribution function f_X, or alternatively by all of its *moments*, where the kth moment x_k is defined as

$$x_k = E[X^k] = \int_{-\infty}^{\infty} u^k f_X(u)du \qquad (3.47)$$

Although not all random variables can be described by their moments, we assume that the random variables we consider can be completely described by their moments.

In our noise model, we will approximate a random variable by its first $k + 1$ moments from x_0 to x_k, where k is an integer selected by the user based on accuracy requirements. For example, we represent a Gaussian random variable X with zero mean ($E[X] = 0$) and variance $\sigma^2 = (E[X^2] - (E[X])^2)$ using the first three moments as

$$X = \begin{bmatrix} x_1 \\ x_2 \\ x_3 \end{bmatrix} = \begin{bmatrix} 0 \\ \sigma^2 \\ 0 \end{bmatrix}$$

In mixed-mode sampled-data systems, signals such as reference, supply voltage, and ground are deterministic. We represent a deterministic signal $X = a$ by its first three moments also as

$$X = \begin{bmatrix} a \\ a^2 \\ a^3 \end{bmatrix}$$

since $E[X] = a, E[X^2] = a^2, E[X^3] = a^3$.

The moments are parameters to characterize a distribution function. If all moments are used, then the distribution function is completely characterized. The advantage of using a finite number of lower order moments is computational efficiency. Instead of manipulating a distribution function, we manipulate a set of numbers representing the moments. Furthermore, from the Central Limit Theorem, random variables resulting from many noise sources should converge to a Gaussian distribution which can be characterized by the first two moments, x_1 and x_2. As a result, low order moments are often the principal components of a random variable in practical circuits. The disadvantage of using a finite number of moments is that effects of higher order moments are not considered.

Although the moments describe the distribution of a random variable, they do not provide information about the correlation between each random variable. In general, noise in a system are correlated because the noise sources themselves may be correlated or the system may contain architectures with *reconvergent fan-out*. For example, when the signal path diverges as shown in Figure 3.7, the resulting signals A and B are correlated since they originate from a single noise source. In the presense of correlated noise, *joint moments* as well as moments are necessary to describe the noise statistics completely. As a first step, we will use only moments and focus on systems with noise that are *independent* for the following reasons.

- Noise sources from physically different components should be independent of each other.

- In most sampled-data systems, white noise dominates non-white noise. Thus, successive samples in time of a noise source are approximately independent.

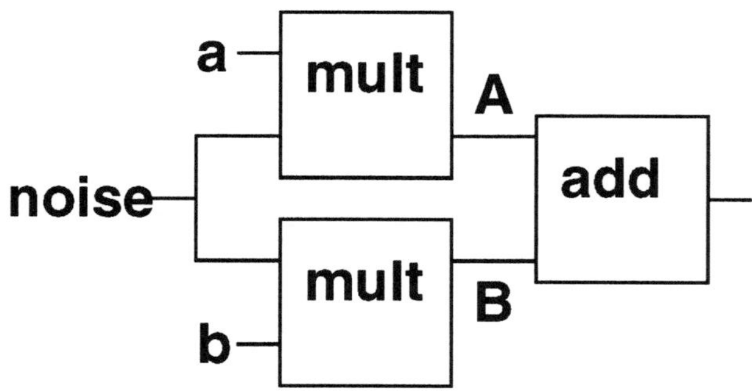

Figure 3.7 Reconvergent fan-out causes correlated noise

- We focus on systems such as converters and finite impulse response filters that lack reconvergent fan-out.

3.4.5 Behavioral Models of Components

As we generalize analog signals from a deterministic representation to a stochastic representation in Section 3.4.4, we must generalize the system components to handle such random signals as well. Components in a mixed-mode sampled-data system fall into three classes; *linear, mildly nonlinear,* and *strongly nonlinear.*

Linear Components

The output of a *linear component* is a linear combination of the inputs. The output moments can be computed using binomial expansion and the independence assumption. For example, an adder sums two random variables X and Y to produce an output Z. We find the moments of Z from the moments of X and Y using

$$z_k = E[Z^k] = E[(X + Y)^k] = \sum_{j=0}^{k} \binom{k}{j} E\left[X^{k-j}Y^j\right] \tag{3.48}$$

using the binomial expansion of $(X + Y)^k$. From the independence of X and Y assumed in Section 3.4.4, we then have

$$z_k = \sum_{j=0}^{k} \binom{k}{j} EX^{k-j}EY^j = \sum_{j=0}^{k} \binom{k}{j} x_{k-j}y_j \tag{3.49}$$

Mildly Nonlinear Components

The output of a *mildly nonlinear* component is a polynomial function of the inputs. The output moments can be computed from input moments using binomial expansion and the independence assumption. For example, a multiplier multiplies two inputs X and Y to produce Z.

$$Z = XY \tag{3.50}$$

We find the moments of Z from the moments of X and Y using

$$z_k = E[(XY)^k] = E[X^k Y^k] = E[X^k]E[Y^k] = x_k y_k \tag{3.51}$$

Strongly Nonlinear Components

For our purposes, the output of a *strongly nonlinear* component is a piecewise polynomial function (with m pieces) of the inputs. The domain is partitioned into m disjoint pieces, where r_i is the domain for the ith polynomial function. The *scenario* for $X \in r_i$ corresponds to the input, X, falling in the domain of the ith polynomial function. For this scenario, the output, Z, is given by the conditional distribution

$$Z = Z|X \in r_i \tag{3.52}$$

and $X \in r_i$ implies that X should be replaced by

$$X = X|X \in r_i \tag{3.53}$$

The expected value of the output due to all possible scenarios is the sum of (3.52) over all possible scenarios,

$$Z = \sum_{i=1}^{m} (Z|X \in r_i)Pr(X \in r_i) \tag{3.54}$$

In Section 3.4.6, we propose to compute the sum using a new simulation algorithm.

In mixed-mode sampled-data systems, the comparator is a strongly nonlinear component which has a piecewise constant transfer function as shown in Figure 3.8. The comparator has as inputs a random variable X and a threshold value t, and produces the probability p_H for the scenario of output high (H), the probability p_L for the scenario of output low (L), the conditional distribution of X given the discrete output Y being low ($f_{X|Y}(x|L)$ in (3.53)), and the conditional distribution of X given Y being high ($f_{X|Y}(x|H)$ in (3.53)). Mathematically, the probability for output above t is defined as

$$p_H = \int_t^{\infty} f_X(u)du \tag{3.55}$$

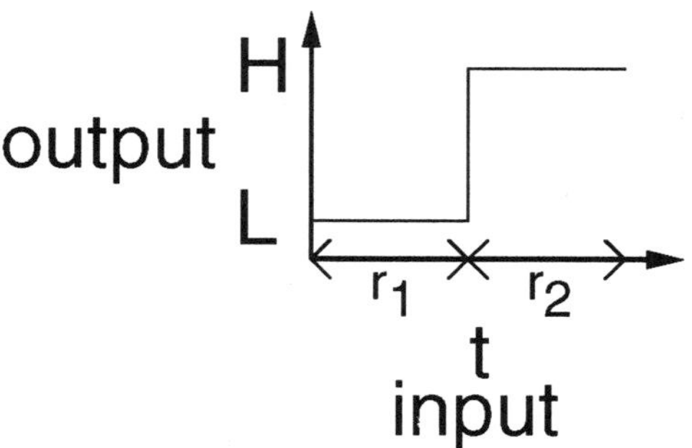

Figure 3.8   Comparator transfer function

where f_X is constructed from the given moments x_n [176]. Conversely, the probability for output low is defined as

$$p_L = \int_{-\infty}^{t} f_X(u)du = 1 - p_H \qquad (3.56)$$

The conditional distribution of X given that $Y = y$ is

$$f_{X|Y}(x|y) = \frac{P\{Y = y|X = x\}}{P\{Y = y\}} f_X(x) \qquad (3.57)$$

Also, we have $P\{Y = H\} = p_H$ and

$$P\{Y = H|X = x\} = \left\{ \begin{array}{ll} 1 & x \geq t \\ 0 & x < t \end{array} \right.$$

Applying (3.57) in our comparator, we find the conditional distribution of X given $Y = H$ is

$$f_{X|Y}(x|H) = \left\{ \begin{array}{ll} \frac{f_X(x)}{p_H} & x \geq t \\ 0 & x < t \end{array} \right. \qquad (3.58)$$

Conversely, the conditional distribution of X given $Y = L$ is

$$f_{X|Y}(x|L) = \left\{ \begin{array}{ll} 0 & x \geq t \\ \frac{f_X(x)}{p_L} & x < t \end{array} \right. \qquad (3.59)$$

Once the conditional distributions are determined, they are converted to a set of $k + 1$ moments [176].

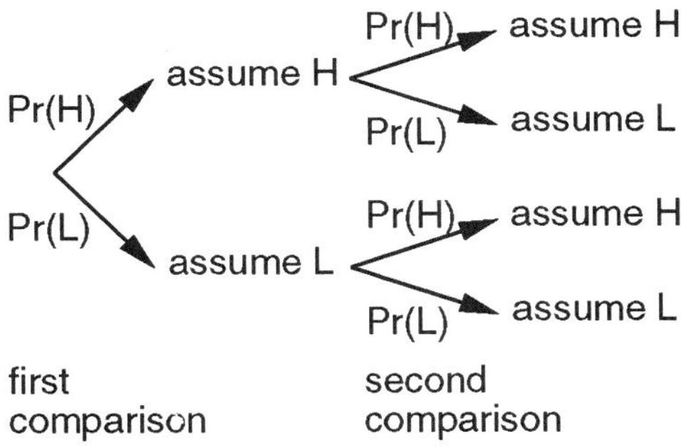

Figure 3.9 Scenario tree

3.4.6 System Simulation Algorithm

After describing the random signal representation, we focus in this section on the algorithm used to simulate the noise in sampled-data systems. The algorithm is event-driven as described in detail below, yet it must handle noise represented as random signals in mixed-mode, sampled-data systems. In such systems, the strongly nonlinear components are special. For example, the comparator is a special interface since its input is analog while its output is digital (e.g. H or L). Due to noise, the analog signals are assumed random; therefore, both comparator output states are possible. The technique used in our algorithm is to compute the probability of each possible outcome, and continue the simulation recursively for each possible *scenarios*. Schematically, the simulation follows a tree of scenarios based on comparator outputs as shown in Figure 3.9. To compute the probability of a particular scenario occurring, we take the product of the probabilities on the unique path from the tree root to the scenario.

The simulation algorithm is as follows:

(a) Extract noise parameters from circuits using (3.42) and (3.43). Construct moments of distributions for noise sources.

(b) Topologically sort network to find the evaluation order of components. The procedure always succeeds because all feedback paths are broken by at least one delay in sampled-data systems.

(c) Evaluate each component in order, return if end of list.

(d) When evaluating a comparator, check distribution at comparator input X, find probabilities for output Y = H and L, execute recursively (e) and (f)

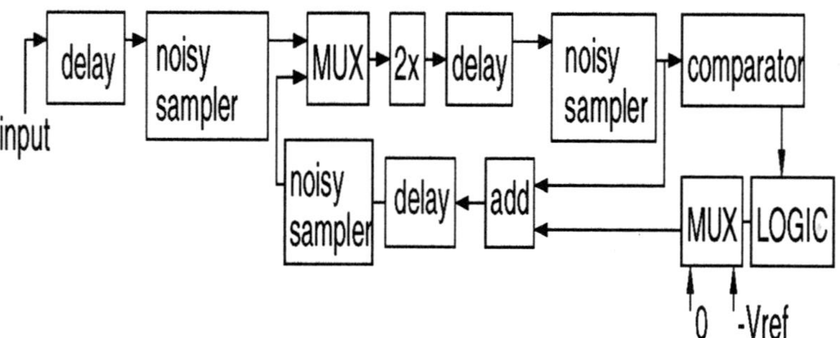

Figure 3.10 Noise model for cyclic converter

(e) Assume the scenario with output Y=H, find conditional distribution of X | Y=H, replace X with X | Y=H, continue simulation in (c)

(f) Assume the scenario with output Y=L, find conditional distribution of X | Y=L, replace X with X | Y=L, continue simulation in (c)

3.4.7   Experimental Results

A 12-bit cyclic A/D (schematic in Figure 3.5 and model for noise calculations in Figure 3.10) has been modeled in the C++ language for both Monte Carlo simulations and direct noise calculations. The gain parameter of a sample-and-hold can be computed from a circuit level implementation using SPICE transient analysis given the fixed system clock period. In this experiment, we assumed the gain to be one for $1x$ S/H and two for $2x$ S/H. The comparator threshold can be computed using SPICE also, but we assumed it to be $1V$. Input voltage is half of the full range input. The effective noise of a sample-and-hold can be computed using (3.43) and (3.44). However, for this experiment we assume that the effective noise is given and is white with standard deviation of $0.365mV$, which may be considered high. However, a high noise value is used to prevent excessive CPU time in Monte Carlo simulations, while a lower noise value would not affect our method. Figure 3.11 compares the predicted output code distribution from a reference Monte Carlo simulation (10^7 trials) with the output code distributions from direct noise calculations ($k = 2$) and ($k = 3$), where k is the highest moment of interest. Notice that the results are comparable, but the CPU times are much less (9620s vs. 0.1s and 3.7s). Next, in Figure 3.12 we compare the Monte Carlo method with the direct method for accuracy as a function of CPU time. The number of trials in Monte Carlo simulations is varied from one hundred to seven hundred thousand. Assuming the reference simulation gives the true solution, the

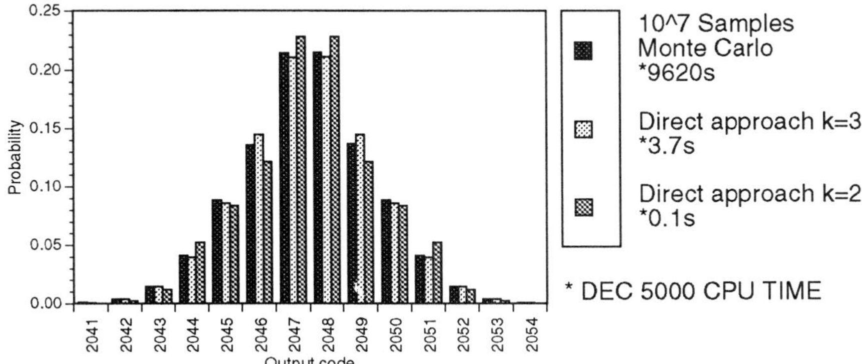

Figure 3.11   Twelve-bit cyclic A/D code distribution

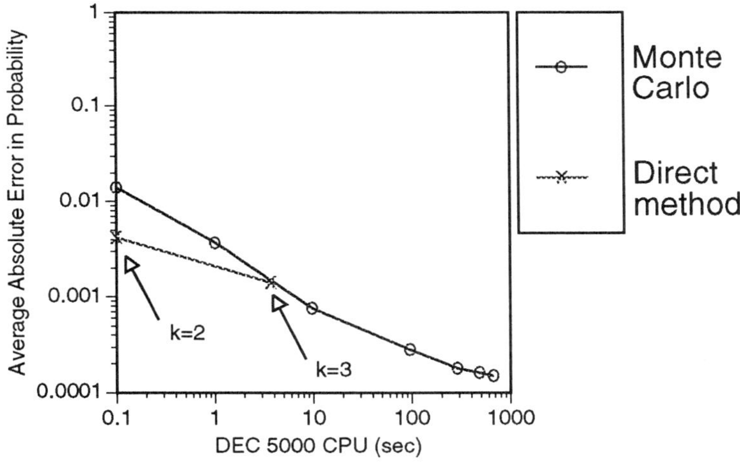

Figure 3.12   Error vs. CPU time

absolute errors in the probabilities are computed for all output codes. The average of the estimated absolute error is then plotted against CPU time. The error curve for the Monte Carlo method suggests that its accuracy is inversely proportional to CPU time. The error curve for the direct method shows that error decreases with higher order approximation. Besides, the direct method has smaller error than the Monte Carlo method for comparable CPU times.

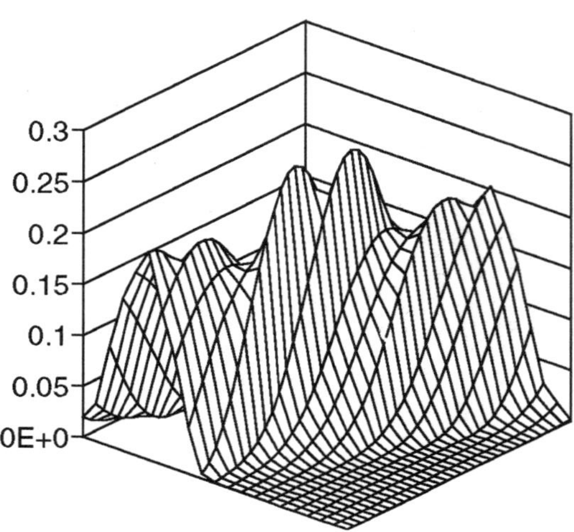

Figure 3.13   Joint probability density function for high component noise

To investigate noise effects for different inputs, we computed noise effects for a voltage range of several LSBs of input voltage centered around half the reference. First, we divide the input range into 30 uniformly spaced inputs. Then, we compute the output code distribution for each input. The result, obtained using 1.3 DEC 5000 CPU seconds, is the joint probability density function of the input and output shown in Figure 3.13.

Notice that the probabilities for wrong codes are as high as that of the correct code, rendering the A/D unusable. In top-down design methodologies, designers can try various noise parameters to determine the proper component noise level from system level constraints. For example, if the noise level in the sample-and-hold is reduced by ten times, then the A/D noise performance is much better as shown in Figure 3.14. Consequently, designers can use behavioral simulation for noise to determine the optimal component noise level based on noise performance constraints.

3.4.8   Conclusion

We have presented a "direct" noise analysis approach for mixed mode systems, and compared our approach with the traditional Monte Carlo approach. The approach is approximate and computes noise effects by performing arithmetic on a finite number

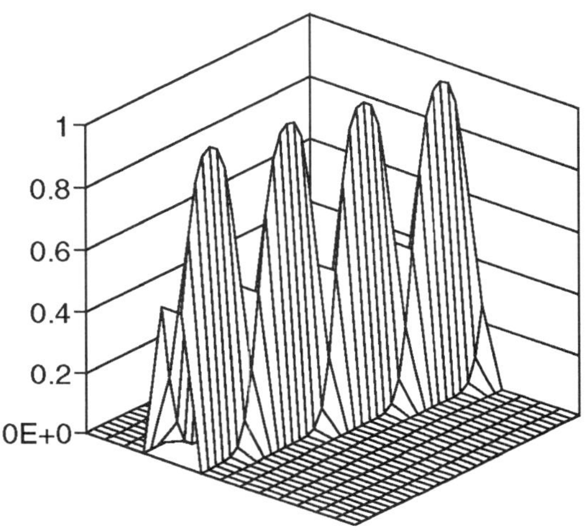

Figure 3.14   Joint probability density function for low component noise

of moments of distribution functions that characterize electronic noise.   One key advantage of this approach is its ability to compute low error probabilities [176].

3.5   VCO AND DETECTORS IN PHASE-LOCK SYSTEMS

In this section we present behavioral representations for VCOs and detectors that are essential circuit components in any phase-lock system.   The representations include sufficient second-order effects for realism, are *independent* of the circuit component architectures, and are general for use in many diverse phase-lock system modeling applications.   Behavioral models for sinusoidal and square wave VCOs are presented in Section 3.5.1 and 3.5.2, respectively.   In Section 3.5.1, results of parameter extraction for a VCO are compared with SPICE and macromodeling results.   A behavioral model for detectors and parameter extraction techniques are presented in Section 3.5.3.

3.5.1   Sinusoidal VCO

A sinusoidal VCO ideally accepts as input a voltage, v, and outputs a sinusoidal signal whose frequency is linearly proportional to the input voltage.   In a practical VCO, the frequency is nonlinearly related to the input voltage, the output contains harmonic

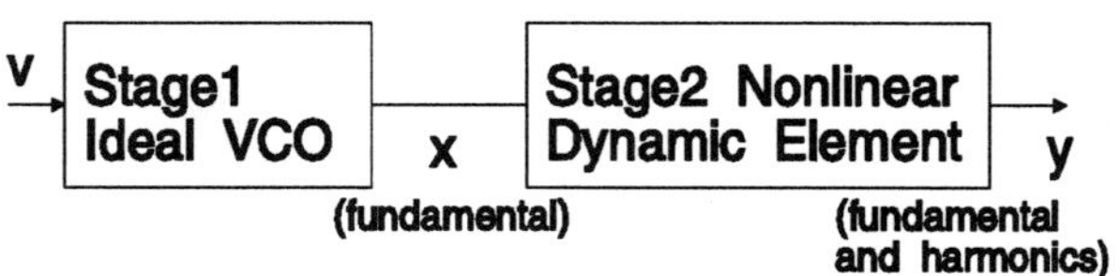

Figure 3.15   Two stage VCO model

components (distortion), and the phase has random variations (phase noise). Previous work [284][300] includes macromodels for SPICE which do not model distortion or noise effects. In our model, non-ideal behaviors can be captured by a two stage model shown in Figure 3.15, where stage 1 is an ideal VCO to generate a signal, x, followed by a nonlinear dynamic stage 2 whose output, y, contains fundamental and harmonic frequencies. Given the input v, the output of stage 1, x, is given by

$$f = g(v) \tag{3.60}$$

$$\Phi(t) = 2\pi \int_0^t f(\tau)d\tau + \Psi(t) \tag{3.61}$$

$$x(t) = cos(\Phi(t)) \tag{3.62}$$

where g is the nonlinear relationship between the frequency f and v, Φ is the instantaneous phase, and Ψ is any stationary random process to represent phase noise (usually Gaussian with mean ψ_0 and variance σ^2). Since f and Ψ are in general time-varying, x is not sinusoidal. When such a signal is fed into the nonlinear dynamic stage 2, the output y is given by

$$y = \sum_{n=1}^{\infty} \int \cdots \int h_n(\tau_1, \ldots, \tau_n) x(t - \tau_1) \ldots x(t - \tau_n) d\tau_1 \ldots d\tau_n \tag{3.63}$$

where h_n are the Volterra kernels [314] of stage 2. To solve (3.63), we make the *quasi-static* approximation.

Quasi-Static Approximation

In a practical sinusoidal VCO, the phase noise Ψ is small, the control range is small, and the control voltage varies much more slowly than the frequency f. Therefore, we propose to determine VCO model parameters for a range of *constant* values of v using (3.63). The quasi-static approximation means that *during simulation for time-varying*

$v(t)$, *we use the model parameters corresponding to instantaneous values of* v. If v is a constant, then x is a pure sinusoidal, and (3.63) has solution,

$$y = \sum_{n=0}^{N} |A_n(\omega)| cos(n\Phi(t) + \angle A_n(\omega)) \tag{3.64}$$

where N is the order of approximation, $\omega = \frac{d\Phi}{dt}$ is the constant frequency, and A_n are complex coefficients given by

$$A_0(\omega) = \frac{1}{2}H_2(\omega, -\omega)e^{j\Psi}, A_1(\omega) = (H_1(\omega) + \frac{3}{4}H_3(\omega, \omega, -\omega))e^{j\Psi},$$

$$A_2(\omega) = \frac{1}{2}H_2(\omega, \omega)e^{j\Psi}, A_3(\omega) = \frac{1}{4}H_3(\omega, \omega, \omega)e^{j\Psi}, etc$$

Summarizing, the VCO model consists of stage 1 described by (3.60), (3.61), and (3.62), and stage 2 described by (3.64). The model parameters are the function g, the phase noise Ψ, and the complex coefficients A_n.

VCO Parameter Extraction

Parameters g, Ψ, and coefficients A_n can be estimated from VCO outputs $y(t, v)$ for different fixed values of v obtained from SPICE simulations or laboratory measurements. The following steps are followed: (a) For a constant input v, estimate from output y the fundamental frequency, f, using the sinusoidal minimum error method [21] or the complex demodulation method [27]. In the latter case, the idea is to multiply $y(t, v)$ with $e^{j2\pi \hat{f} t}$, followed by low pass filtering to get $z(t)$, where $\hat{f}$ is our guess of f. If $\hat{f} = f$, the phase of $z(t)$ is constant; otherwise, it is slowly time-varying for $\hat{f} \approx f$. Using Newton-Raphson [246], solve for f such that the phase of $z(t)$ is constant. (b) With f known, the magnitude and phase of the fundamental and harmonics can be similarly calculated using demodulation. Specifically, magnitude is the average of $2|z(t)|$ and phase is the average of $-\angle z(t)$. (c) Repeat (a) and (b) for different v to estimate the nonlinear function g for (3.60), as well as for more data points on $A_n(\omega)$ in (3.64). (d) Estimate σ for (3.61) from laboratory measurements or by hand analysis [2]. (Notice that SPICE does not simulate noise in the time domain.) (e) Assign value for ψ_0, which is the only initial condition.

Example VCO Parameter Extraction

Shown in Figure 3.16 is a schematic of a Pierce oscillator with varactor tuning that consists of circuit elements $M1 = 15\mu m/1\mu m$, $M2 = 35\mu m/1\mu m$, $L = 129.75nH$,

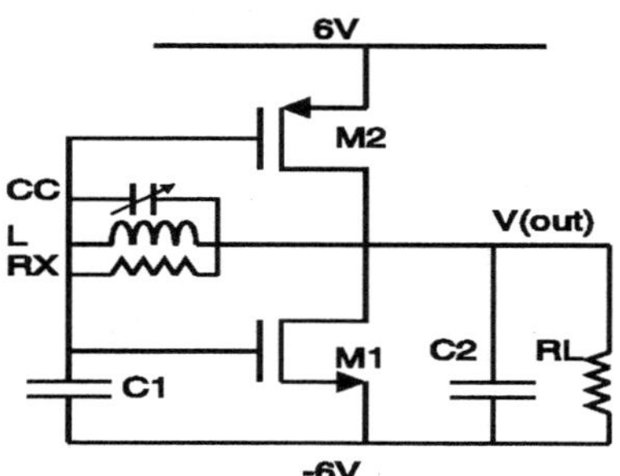

Figure 3.16 Pierce oscillator

v	0	1	2
n	(gain,phase)	(gain,phase)	(gain,phase)
1	(2.88V, -3.72)	(2.88V, -4.10)	(2.88V, -4.47)
2	(1.63mV, -0.89)	(2.00mV, -1.29)	(2.02mV, -1.77)
3	(78.2mV, -3.02)	(78.0mV, -4.17)	(78.2mV, -5.27)
4	(30.2μ V, -1.90)	(45.6μ V, -0.92)	(36.3μ V,-1.08)
5	(6.65mV, -1.37)	(6.63mV, -3.29)	(6.69mV, -5.12)

Figure 3.17 Behavioral parameters for Pierce oscillator

$C1 = C2 = 20pF$, $RL = 100\Omega$, $RX = 1M\Omega$, and voltage-controlled capacitor $CC = 2pF/V$. Its behavioral model parameters (Figure 3.17) were extracted from three SPICE transient simulations for three different control voltage v. Figure 3.18 shows the SPICE output for $v = 0$ and the residual error in the 5th order (N=5) behavioral approximation used in this case. In contrast, the SPICE macromodels in [284] do not model harmonics, so the lower error bound is given by the first order residual error shown in Figure 3.19. Thus, the behavioral model is *more accurate* than macromodels, as well as *faster* since evaluating the behavioral model only involves evaluating the function in (3.64).

Error Analysis

To estimate the error of the quasi-static approximation when the control input is not constant, we let the frequency, f, be modulated by a sinusoidal signal with small

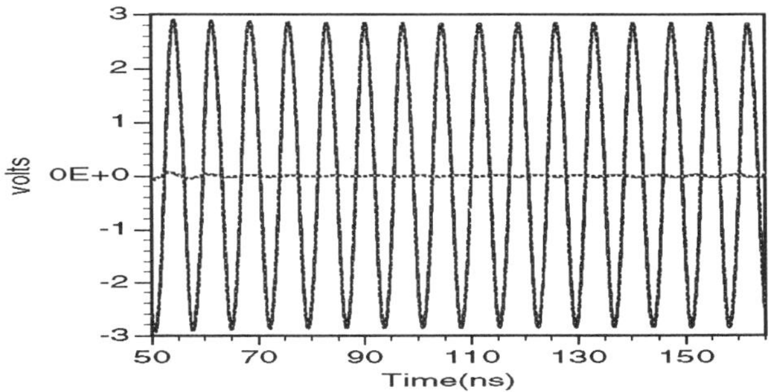

Figure 3.18 SPICE simulations and residual error in behavioral approximation

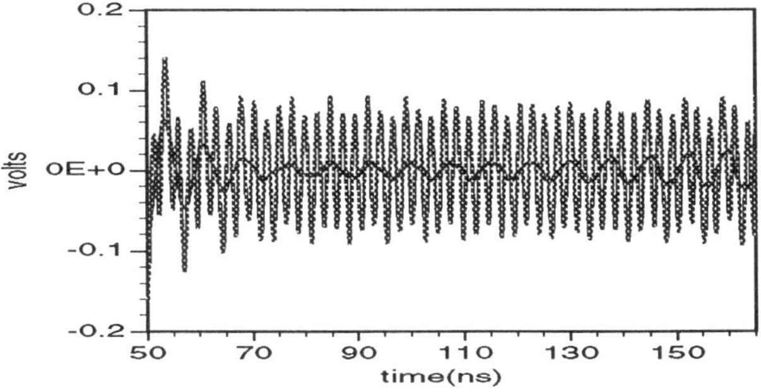

Figure 3.19 Residual errors in first and fifth order approximation

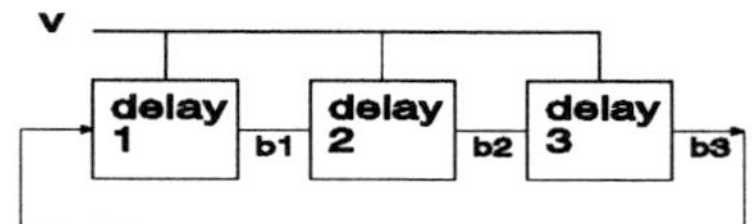

Figure 3.20   Square wave VCO with $m = 3$ delays

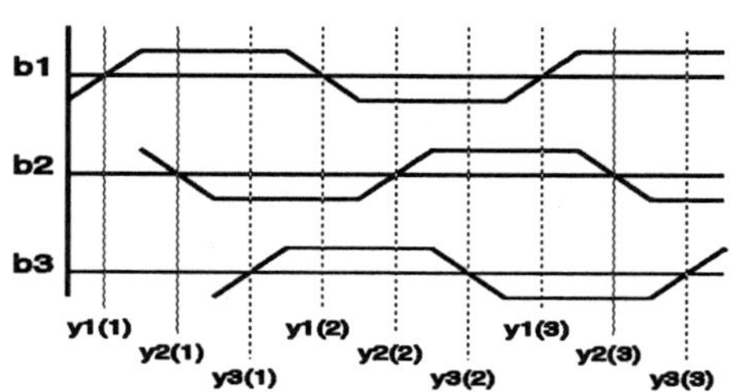

Figure 3.21   VCO square waveforms

amplitude (narrow band frequency modulation). For instance,

$$f = g(v) = f_c + \beta f_m \cos(2\pi f_m t)$$

where β is the modulation index, and f_m is the modulation frequency. Assuming the phase noise is negligible, it can be shown that the error of the quasi-static approximation is bounded by an error term proportional to β. As a result, when the modulation index β is sufficiently small, the error becomes negligible.

3.5.2   Square Wave VCO

In addition to oscillators with sinusoidal outputs, there are oscillators with square wave outputs. In contrast to the sinusoidal VCO where the waveform is represented by (3.64), the waveform of a square wave is represented by the zero crossings (low-to-high and high-to-low transition times). In general, a VCO consists of m delay elements connected in a ring structure (Figure 3.20). Each of the m outputs are phased shifted by $\frac{1}{2mf}$ seconds, where f is the frequency of the VCO. We represent the jth transition time of output i with $y_i(j)$ as shown in Figure 3.21. Similar to the sinusoidal VCO, the frequency is controlled by the input v using equation

$$f_i = g_i(v) \tag{3.65}$$

where g_i represents control relationship for element i, and f_i represents the effective frequency of delay element i. Unfortunately, v varies considerably in a square wave

VCO; therefore, the quasi-static approximation is not appropriate in this case. The transition times are determined by the delay between adjacent delay elements. For example, $y_i(j)$ for $i \neq 1$ is equal to $y_{i-1}(j)$ plus the delay of element i. Because of this relationship, it is convenient to use a recursive definition for $y_i(j)$,

$$\pi = 2\pi m \int_{y_{i-1}(j)}^{y_i(j)} f_i(\tau)d\tau + \Psi_i(y_i(j)) - \Psi_i(y_{i-1}(j)), i \neq 1 \qquad (3.66)$$

$$\pi = 2\pi m \int_{y_m(j-1)}^{y_i(j)} f_i(\tau)d\tau + \Psi_i(y_i(j)) - \Psi_i(y_m(j-1)), i = 1 \qquad (3.67)$$

where Ψ_i is the random phase noise of element i. Summarizing, the model consists of (3.65), (3.66), and (3.67). Model parameters are the number of delay elements m, the voltage-control relationship g_i, the phase noise Ψ_i, and the initial condition $y_1(1)$.

3.5.3 Detectors

A detector compares two signals (input waveform and VCO waveform). Traditionally, circuit designers represent a detector by its phase characteristic which plots the average product of the input waveform and the VCO waveform as a function of the phase difference between the two waveforms. One problem with this modeling approach is that the phase characteristic is defined only when the input waveform and the VCO waveform have the same frequency. As a result, this approach cannot be used during phase-locked loop acquisition. Another problem is that the phase characteristic depends on the shape of the waveforms, so the characteristic is not defined until input signals are applied. As a result, the phase characteristic is not appropriate as a behavioral representation.

Detectors are classified into two broad categories [90]: multipliers (zero memory circuits) and sequential circuits (with memory). In this paper, sequential circuits are represented by state machines with analog inputs, a and b, analog outputs $g(a, b)$, and clock triggered by zero crossings of a and b. Moreover, the same representation can be used for multipliers if they are considered as state machines with a single state and no clock. In general, all detectors can be conveniently represented by a state transition table such as the one shown in Figure 3.22 for a phase-frequency detector. Signals a, b, and c are the input, VCO, and output waveforms, respectively. The first column is the state number; the second column is the next state if a crosses zero; the third column is the next state if b crosses zero; t_d are gate delays of next state logic; the last column is the output $g(a, b)$ for each state. State transition tables for common detectors are shown in Figure 3.23. Since all known detectors can be written in this representation, this representation is general and independent of the circuit architecture.

state	a crosses (next, delay)	b crosses (next, delay)	c g(a,b)
0	$(1, t_d)$	$(2, t_d)$	0.0
1	$(1, t_d)$	$(0, t_d)$	Iup
2	$(0, t_d)$	$(2, t_d)$	Idown

Figure 3.22   Behavioral model for phase frequency detector

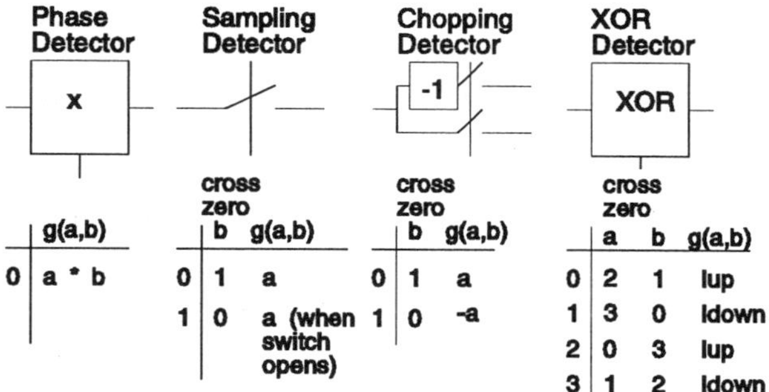

Figure 3.23   Common detectors

Detector Parameter Extraction

For a given detector type, the next state logic is fixed. Non-idealities are the transition delays, t_d, and deviations of the output function. t_d is computed by measuring the total gate delays in the next state logic. Typically, $g(a, b)$ is very simple; for instance, $g(a, b) = c_0 + c_1 a + c_2 b + c_3 ab$. Its non-idealities include constant offset error in c_0 and multiplier gain error in c_3. All four coefficients $c_0 \ldots c_3$ can be extracted using four DC SPICE simulations with different values of a and b, and solving for the coefficients in a system of four equations.

3.6 DELAY- AND PHASE-LOCKED SYSTEMS

3.6.1 Introduction

PLLs and delay-locked loops (DLLs) are an important part of the class of mixed-mode nonlinear dynamic systems which includes PLLs, DLLs, oversampling data converters, switching power supplies. PLLs and DLLs have many applications; they are used in receivers, clock generators [322], clock recovery [171], data synchronization [312], etc.

In general, mixed-mode nonlinear dynamic circuits are stiff systems. Traditional simulators such as SPICE spend large amounts of CPU time on computing the accurate waveforms of the digital circuits, yet the waveforms of the much slower analog circuits determine the system behavior. Consequently, traditional simulators take too much time to compute the system behavior by simulating the whole system over the interval required by the analog circuits.

A transistor-level simulation of a complete phase-locked loop circuit, even just to simulate the deterministic behavior, is prohibitively expensive if not impossible because of the inherently stiff nature of the circuit. Designers have been using macromodeling or "behavioral" simulation techniques to simulate mostly the deterministic behavior of PLL circuits. We would like to be able to simulate not only the deterministic behavior (such as acquisition and tracking) but also the timing jitter/phase noise performance of a PLL frequency synthesizer used in an RF transceiver. Hence, a "behavioral level" modeling and simulation strategy for nonlinear noise analysis of PLLs is needed.

Key specifications for PLLs are acquisition time, capture range, lock range and phase noise/timing jitter. Timing jitter is a major and very important PLL specification. For instance, this jitter can cause data errors in communication systems, create inaccurate readings in instrumentation systems such as spectrum analyzers. Prediction of timing jitter through simulations is crucial at the early stages of the top-down design. Timing jitter simulation imposes tight restrictions on the numerical simulation algorithms to be used. Numerical noise created by the simulation algorithm should be negligible when compared with the timing jitter to be computed for accurate results.

3.6.2 Numerical Algorithms

Behavioral models of circuit components which operate only on timing information, such as VCOs, phase detectors (PDs), and phase/frequency detectors (PFDs), have already been described [180]. With these models, a wide variety of phase/delay-locked systems can be simulated.

Simulation of systems, which are composed of the interconnection of such behavioral models, can be done by using a modified integration method. Digital circuits are simulated with an event-driven simulator, while the analog circuits are simulated with a differential equation solver. By interfacing the analog simulator with the digital one using a protocol, the two simulators are synchronized. At the interface, integration must pause at breakpoints which correspond to the exact times when the digital signals switch. Because switching times are generally not known ahead of time, they must be solved for during the integration. For instance, the time-derivative of instantaneous phase in a square wave VCO model is expressed as

$$\dot{\Phi}(t) = 2\pi\, f(v(t)) \tag{3.68}$$

where $v(t)$ is the VCO control voltage, and $f(v)$ is, in general, a nonlinear function relating the effective frequency of VCO to the control voltage [180]. The waveform at the output of the VCO is represented by zero crossings, which occur when $\Phi(t)$ takes values which are integer multiples of π. At a time point during the integration of (3.68), if a zero crossing occurs in between the previous time point and current time point, the integration algorithm should track back and hit the point where the zero crossing occurs, as illustrated in Figure 3.24. There are two schemes to solve for the timing of switching events:

- *Interpolation method.* Switching times are estimated based on a polynomial interpolation using the information from current and previous time points. For instance, with a first-order interpolation method, let a switching event be detected between the time points t and Δt, where Δt is the current time step size. Then,

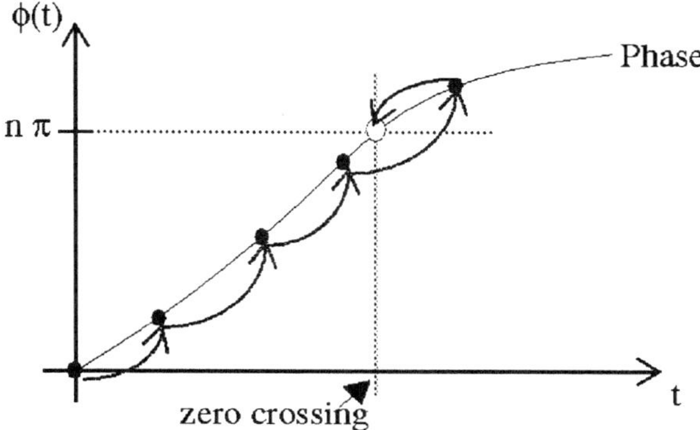

Figure 3.24 Integration Algorithm with Zero Crossings

the time for the switching event is estimated by

$$t_s = t + \left(\frac{\Phi(t_s) - \Phi(t)}{\Phi(t + \Delta t) - \Phi(t)} \right) \Delta t$$

$$\Phi(t_s) = n\pi \qquad n = 1, 2, 3, \ldots$$

(3.69)

- *Iterative method.* Switching times are calculated using a bisection search (Figure 3.25). If at time $t + \Delta t$, the phase $\Phi(t)$ is larger than $n\pi$ while at t it was smaller, then there has been a switching event in between. To calculate the switching time, we divide the present time step by 2. Then we evaluate $\Phi(t + \Delta t/2)$. If this is still to the right of the switching time, we repeat the bisection procedure. If it is to the left, then we return the control to the integration algorithm. To prevent the algorithm from taking too much of a time step, we limit the maximum step size to $1/4$ of the last time step.

We have implemented an integration algorithm, including these schemes as options, to simulate phase/delay-locked systems. The integration algorithm is a variable time-step, variable order one based on the implicit backward differentiation formula [263][248]. A behavioral model for a charge-pump PLL (Figure 3.26) [91], which consists of a ring-oscillator VCO, a PFD, a charge-pump and a second-order loop filter, was built to analyze the behavior of the numerical algorithms. The algorithms were compared by doing timing jitter simulations of this PLL. Accuracy, in this context, is used to mean the accuracy of the solutions for the timing of switching events. A comparison of the interpolation method with the iterative method was made. There are basic differences in the behavior of these methods. The accuracy of the iterative method can be set to

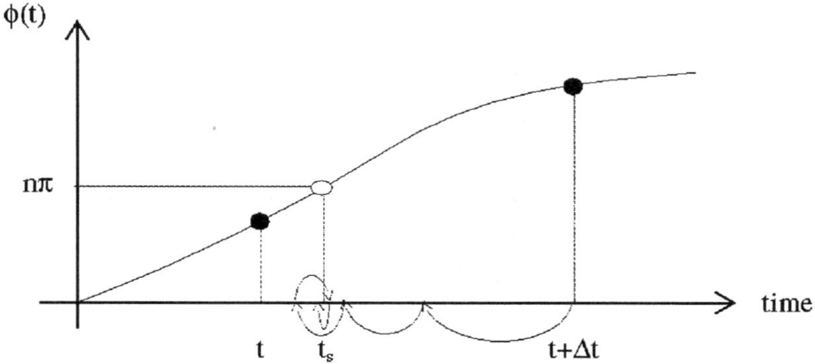

Figure 3.25   Iterative Search for Switching Times

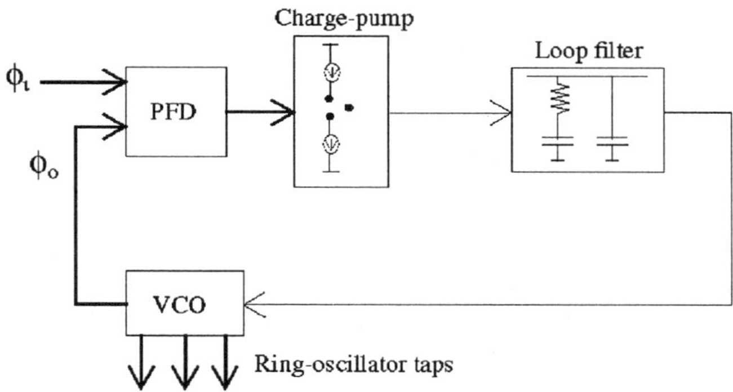

Figure 3.26   Charge-Pump PLL

any desired value. However, there is no control on the accuracy of the interpolation method. During an unlocked state, while PLL is acquiring the input, the accuracy of the interpolation method decreases considerably. When PLL is "in lock," accuracy is inversely proportional to the noise level in the circuit. This is, in fact, a desired property, because more accuracy is needed for smaller noise levels. On the other hand, the accuracy of the iterative method stays at the desired/set level regardless of the state or noise level of the circuit. Figure 3.27 shows the response of the above PLL to a frequency step at its input in the presence of noise. Even after acquisition, deviations around the mean value occur because of the noise in the circuit. Figure 3.27 also shows the error in the solution of the timing of switching events for the interpolation method and the iterative method. The error for the iterative method in Figure 3.27 is not visible on the plot that uses the same scale as the plot for the interpolation method, because the error for the iterative method was several orders of magnitude smaller. The details can be seen in the blow-up plot. For the PLL considered above, the accuracy of the interpolation method was found to be sufficient for doing timing jitter simulations when the circuit is "in lock." The results obtained by the interpolation method were verified by using the iterative one. Although this was the case for the PLL considered, there is no guarantee on the accuracy of the interpolation method in general. On the other hand, the accuracy of the iterative method does not depend on the type, state or noise level of the circuit. The accuracy can also be set to a desired level where numerical noise will not interfere with the results.

3.6.3   Timing Jitter Simulation

Timing jitter is a very important specification for phase/delay-locked systems. Design for low timing jitter is probably the most challenging part of the design of such systems. The difficulty of design for low timing jitter is amplified by the lack of efficient simulation tools for its prediction. Prediction of timing jitter through simulations is crucial at the early stages of the top-down design. Any modeling or design methodology, and simulation tools, which do not address this problem are not suitable for high performance phase/delay-locked system design.

The models we use in behavioral simulation include statistical second-order effects [180]. Analysis of a phase/delay-locked system in the presence of noise, when the system is not "in lock," is a hard problem. Fortunately a detailed observation of the system when it is in the unlocked state is almost never required, as long as the loop is guaranteed to lock on to the input in the specified amount of time. This can be assured by simulations. At this point, it is assumed that the system is "in lock" for timing jitter simulation. The expression "in lock" is used in a time-averaged sense here, not in the instantaneous sense. For instance, for a PLL (say with a ring-oscillator VCO, charge-

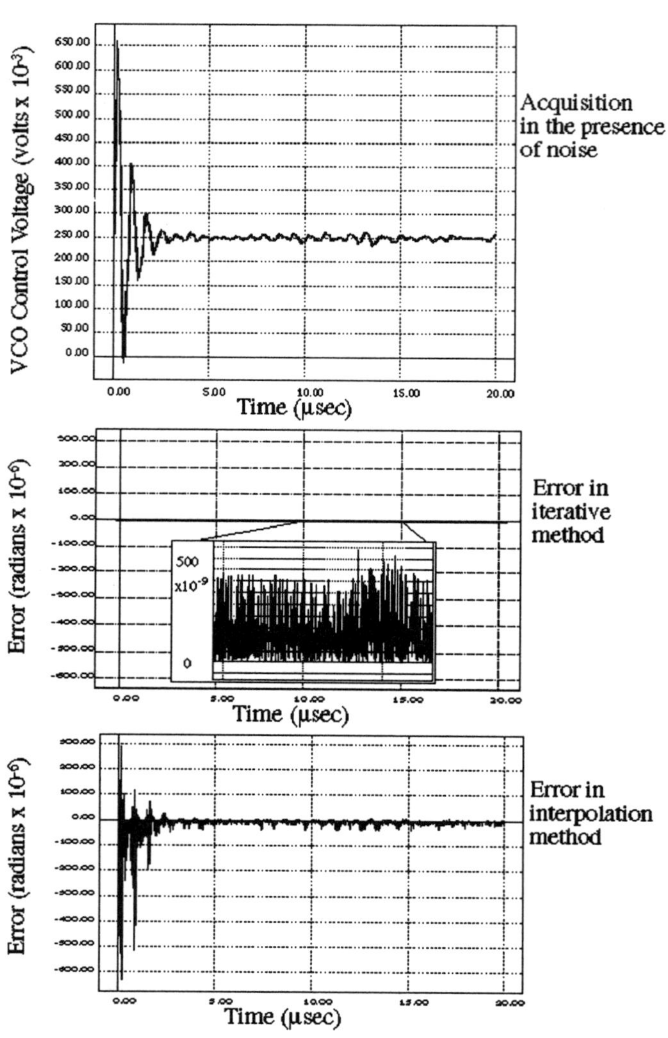

Figure 3.27 Acquisition in the Presence of Noise

pump and PFD), instantaneous frequency of the VCO will have variations around a mean value because of the noise in the circuit. The mean value of the VCO frequency will be equal to the input frequency. It is assumed that timing jitter performance of the circuit is above a level where it does not lose lock (in a time-averaged sense). This is a reasonable assumption, because if it does not hold, that means that we are confronted with a bad design which has deterministic performance problems. It is important to understand that the "in lock" assumption does not mean that we are using a linear model for the circuit. The linearity assumption might be valid for very low noise levels, but higher noise levels will excite the nonlinearities. For behavioral timing jitter simulation, we use the full nonlinear models of the components, which were described. The method that is used for timing jitter simulations is Monte Carlo in time domain. The CPU time disadvantage of the Monte Carlo method disappears because of the high efficiency of behavioral simulation. As discussed before, having a numerical algorithm with controllable numerical noise makes it possible to predict timing jitter accurately. In behavioral timing jitter simulation, circuit with noise sources, which is "in lock," is simulated over a period of time. Then, the characteristics of jitter at the output is calculated by a statistical analysis of the data obtained over that time period.

3.6.4 Behavioral Simulation in Bottom-Up Verification of a Phase-Locked Loop

The traditional approach to the bottom-up verification of phase/delay-locked system designs is to use a transistor-level simulator such as SPICE. Transistor-level simulation of a phase/delay-locked system takes too much time to be practical, because the system is stiff. Macromodeling these systems in a circuit simulator was proposed to circumvent the efficiency problems of transistor-level simulation [300][284]. This approach also results in impractical simulation times, because many circuit elements are needed for accurate models. Still, the accuracy attained by the macromodeling approach is not as high as needed for verification of designs. On the other hand, behavioral simulation achieves the goal of verifying complex system behavior efficiently. This is made possible by accurate statistical behavioral models which include analog second-order effects. Evaluations of behavioral models are fast which results in an efficient system simulation. Bottom-up verification of a phase-locked system using behavioral simulation is done in two steps:

- *Set up the behavioral models for the components.* The model parameters are extracted using SPICE from the transistor-level description of components.

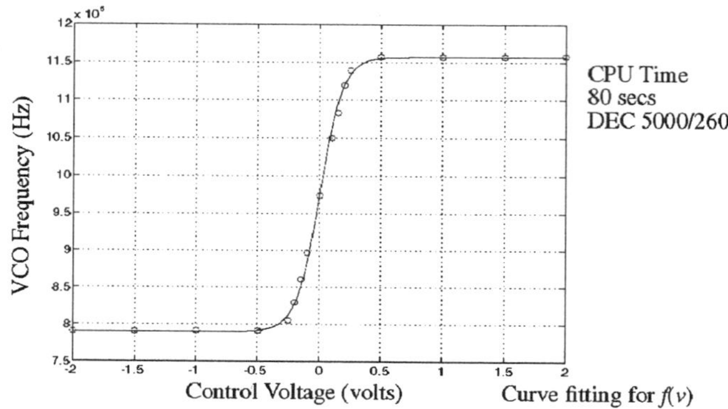

Figure 3.28   Extraction with SPICE

- *Simulate the system in time-domain using the behavioral simulator to calculate the performance measures.* Behavioral simulation is much faster than circuit simulation.

To illustrate this procedure, component behavioral models for a bipolar PLL [106] were set up, and behavioral simulation was used to analyze the acquisition characteristics. Figure 3.28 shows SPICE domain extraction of the relation between the effective frequency of VCO and the control voltage ($f(v)$ in (3.68)). Other model parameters are extracted in a similar way from the transistor-level description. Model extraction is done only once for a circuit. Then created models are used in many behavioral simulations to analyze the acquisition characteristics, stability, timing jitter and others. Figure 3.29 shows the response of the modeled PLL to a frequency step at its input (both SPICE and behavioral simulation).

3.6.5   Behavioral Simulation in Top-Down Design of a Multi-Phase Clock Generator

The role of behavioral simulation in top-down design of a phase/delay-locked system will be illustrated with a multi-phase clock generator. A PLL with a ring-oscillator VCO or a DLL can be used to generate multi-phase clocks (Figure 3.30). These two architectures were compared for their timing jitter performance using behavioral simulation. Figure 3.31 shows the relationship between clock jitter and percentage delay cell jitter for a given design of both architectures. Ring-oscillator VCO for PLL,

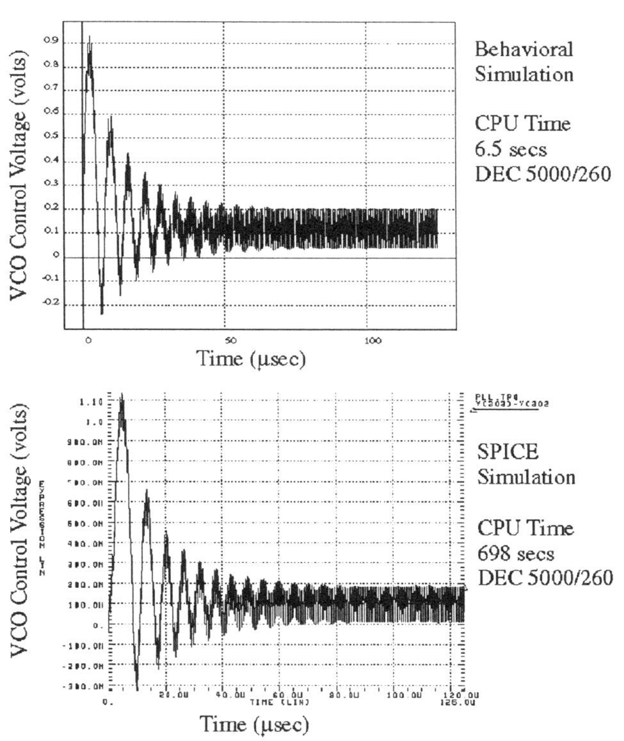

Figure 3.29   Behavioral and SPICE Simulation of Acquisition

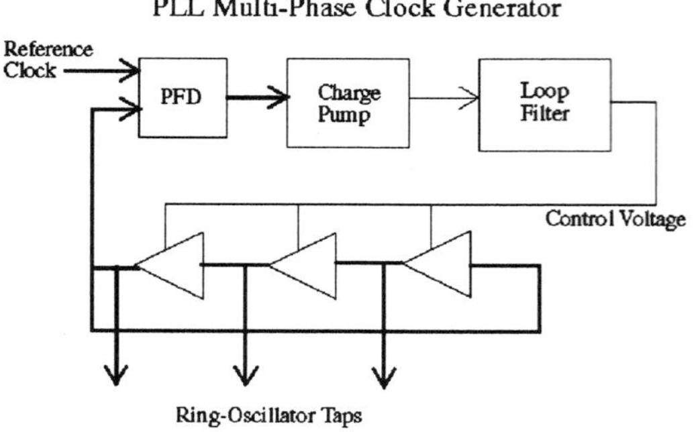

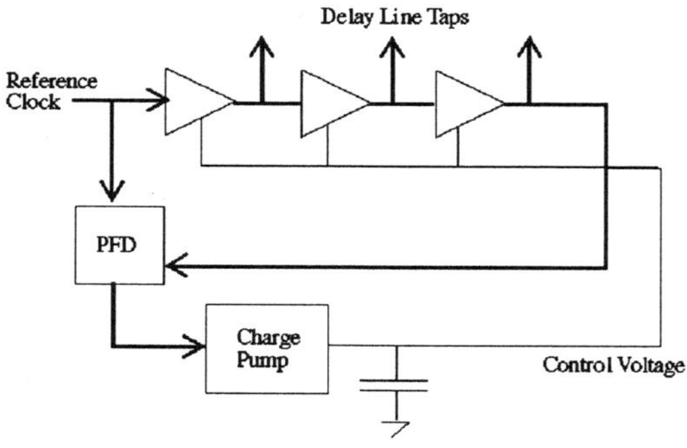

Figure 3.30   PLL and DLL Multi-Phase Clock Generator

as well as the delay line of DLL, has 5 delay cells. Reference clock frequency is 50 MHz. From Figure 3.31, we conclude that DLL has better timing jitter performance, when compared with a PLL, for fixed percentage delay cell jitter. Then, the relationship between clock jitter and percentage delay cell jitter is used to predict amount of jitter allowable an a delay cell, given a clock jitter allowance. In this way, the clock jitter constraint is mapped on to the delay cell jitter constraint.

3.6.6   Phase Noise in Phase-Locked Loop Frequency Synthesizers

Up to this point, we have concentrated on the time-domain characterization of noise effects (timing jitter) in phase/delay-locked systems. Time-domain characterization is suitable for PLLs used in applications such as clock generation, clock recovery and

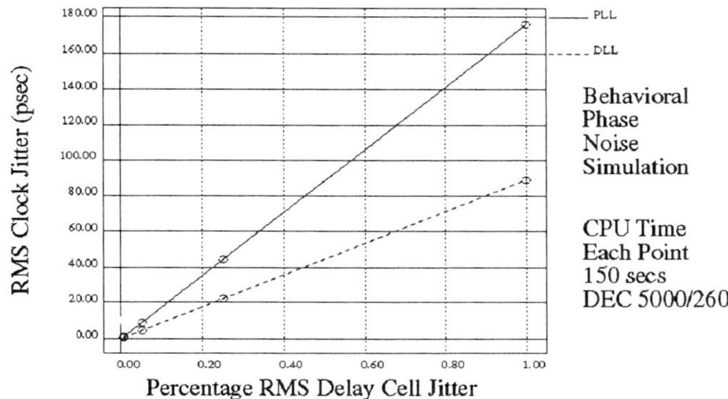

Figure 3.31 Clock Jitter for PLL and DLL Clock Generator

data synchronization. On the other hand, noise effects in frequency-synthesizer PLLs used in RF transceiver applications are characterized in the frequency domain (phase noise).

In particular, we would like to be able to estimate the phase noise spectrum of the LO signal generated by a frequency synthesizer. The phase noise of the LO signal is a result of the interactions of the components of a frequency synthesizer (noisy VCO, noisy crystal reference, loop filter and other components such as phase detectors and frequency dividers) governed by the nonlinear dynamics of a feedback loop.

The deterministic behavior of a VCO is modeled by a nonlinear map relating the instantaneous frequency to the control signal. The noise performance of the VCO is modeled with its open-loop phase noise spectrum. In [67], we present a transistor-level numerical method to calculate the phase noise spectrum for open-loop VCOs. [67] also presents the application of this method to several oscillator circuits and arrives at a random walk phase noise model (considering shot and thermal noise sources only) for the open-loop VCOs. A random walk phase noise model is equivalent to an *uncorrelated* cycle-to-cycle jitter model for a VCO [202]. Similarly, the reference oscillator is modeled by its oscillation frequency and phase noise spectrum. All the phase and phase/frequency detectors, frequency dividers are modeled by state machines capturing possible nonideal behavior (e.g. the dead zone in PFDs).

In phase noise spectrum estimation of the local oscillator (LO) signal generated by a frequency synthesizer, we use the mixed-signal numerical integration algorithm presented in Section 3.6.2 to simulate the PLL in time-domain. The sampled phase noise for the VCO and the reference signal are introduced into the simulation using

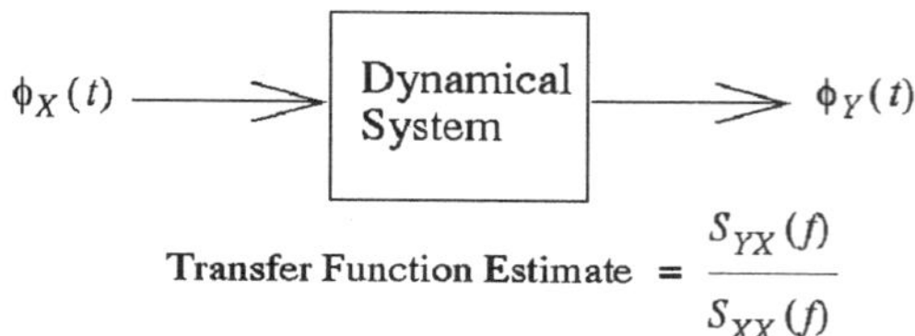

$$\text{Transfer Function Estimate} = \frac{S_{YX}(f)}{S_{XX}(f)}$$

Figure 3.32   Transfer Function Estimate

random number generators. This is done in such a way that rapidly fluctuating random quantities enter the numerical integration only as threshold values to be crossed by a state variable (i.e. the state variables modeling the phase of the VCO or the reference). This avoids the need to set the maximum time step that can be taken by the numerical integration algorithm to be smaller than the correlation time of a noise source introduced directly in the differential equation. In the case of white noise, for which the correlation time is ideally zero, the maximum time step has to be set to a very small value making the simulation highly inefficient. In our case, the time step selection is not affected by the correlation time of the phase noise sources. The crossing times of the random threshold values correspond to the transitions of the VCO or the reference, and they are calculated during numerical integration.

The calculated transition times of the VCO (closed-loop in a PLL) are saved during simulation. Note that, even though the phase noise processes for the reference and open-loop VCO are nonstationary, the phase noise process for the closed-loop LO output is wide-sense stationary as a result of the phase corrections done by the feedback loop. The closed-loop, self referenced measurement scheme [202] is used with the saved transition times of the VCO to extract a sample path of the sampled phase noise process for the LO signal. Then, the power spectral density for phase noise is estimated with the time-averaged periodogram method using FFTs [92]. Similarly, an estimate of the "transfer function" between the reference phase noise and the LO phase noise as well as the one between the open-loop VCO phase noise and the closed-loop LO phase noise can be calculated. This is accomplished using the time-averaged periodogram method to calculate the spectral density of the input process and the cross spectral density between the output and the input processes. Then the estimate for the transfer function is calculated by dividing the cross spectral density by the input spectral density (Figure 3.32). In general, the output LO phase noise and the phase noise of the reference and the open-loop VCO can not be related by a linear time invariant (LTI) transformation because of the nonlinearities in the models. In that sense, the transfer function estimate calculated using the above method is a "heuristic" one considering the transformation to be LTI. Actually, a useful measure of the degree to which two wide-sense stationary processes X and Y are approximately related by

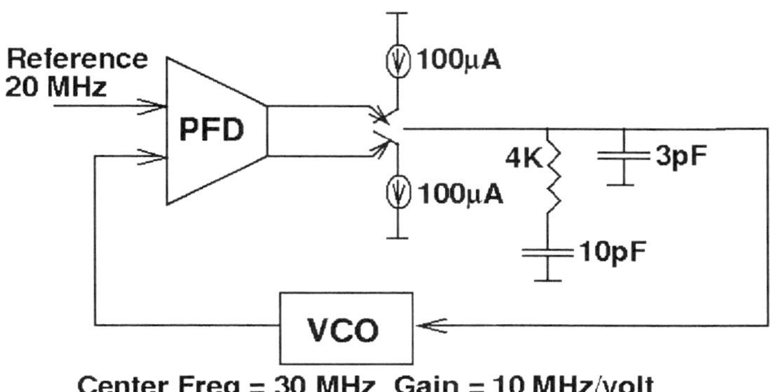

Figure 3.33   Charge-Pump PLL

an LTI transformation is the coherence function [92] defined by

$$\rho(f) = \frac{S_{XY}(f)}{\sqrt{S_X(f)\,S_Y(f)}} \qquad 0 \le |\rho(f)| \le 1. \tag{3.70}$$

$|\rho(f)| \equiv 1$ if and only if $X(t)$ and $Y(t)$ are exactly related by an LTI transformation [92]. The coherence function can be calculated also using the time-averaged periodogram method.

Examples for Phase Noise Simulation of PLLs

Now, we illustrate use of the behavioral phase noise simulation method described above with phase noise simulations of a simple PLL circuit shown in Figure 3.33. In the first simulation example, we have used an ideal PFD model, a white phase noise model for the reference and a random walk phase noise (uncorrelated cycle-to-cycle jitter) model for the open-loop VCO. The upper plot in Figure 3.34 shows the phase noise spectrums for the open-loop VCO, reference and the closed-loop VCO. Bottom plot is an histogram of the sampled phase noise for the closed-loop VCO output. As expected, at low frequencies (below the loop bandwidth) phase noise is dominated by the reference phase noise, and at high frequencies by the open-loop VCO phase noise. Practical PFDs with tri-state outputs often have a "dead zone" region where the output is insensitive to small phase differences between the two inputs [245]. To see the effect of the "dead zone" region on the resultant phase noise spectrum, we have simulated the PLL circuit first with a noiseless reference and VCO and a PFD which can not generate pulses at its outputs with a duration time less than 2 ns. We have also performed a simulation with a white phase noise reference, a random walk phase noise

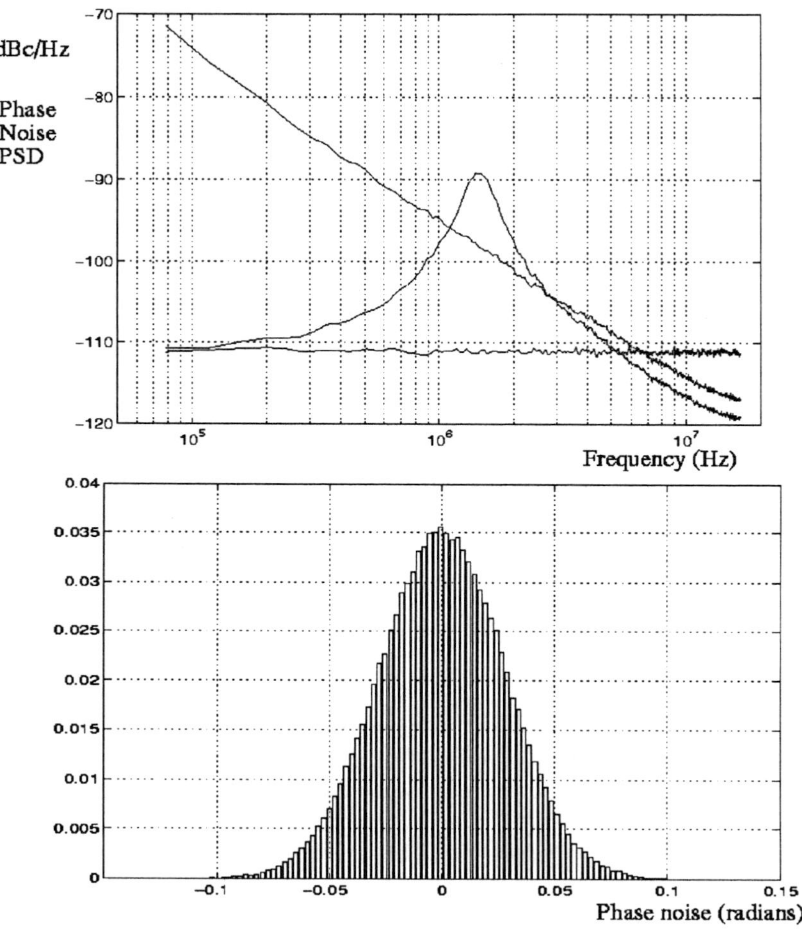

Figure 3.34   Ideal PFD with Noisy Reference and VCO

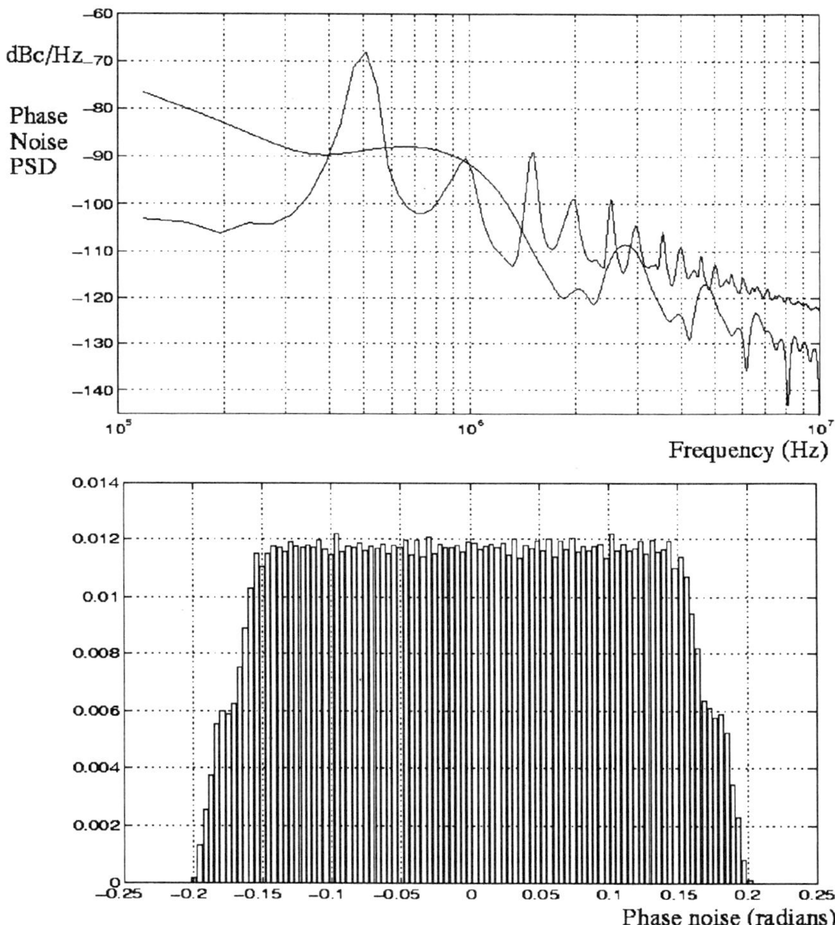

Figure 3.35   PFD with a Dead Zone

VCO and the same PFD with the dead zone. The upper plot in Figure 3.35 shows the spectrum of phase noise for the closed-loop VCO for both cases. The bottom plot is the histogram of the sampled phase noise for the case with a noisy reference and VCO and a PFD with a dead zone. To eliminate the phase noise performance degradation caused by the dead zone of the PFD, an "alive zone" PFD [245] can be used. The alive zone phase comparator introduces a reset delay into a standard edge-sensitive PFD and creates minimum duration pulses at both of its outputs. These occur at each phase comparison and activate the fourth state of the charge pump which has both positive

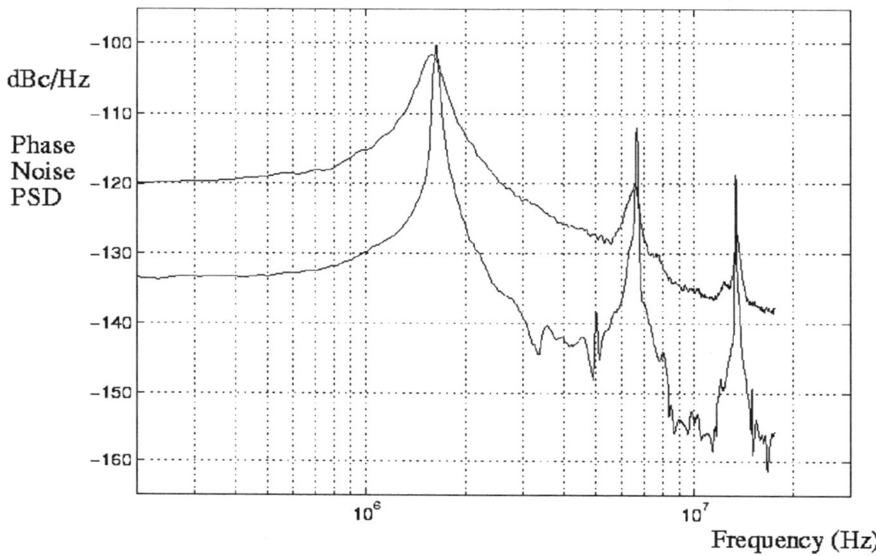

Figure 3.36   Alive Zone PFD with Current Mismatch

and negative currents on at the same time, a state not possible for a standard PFD [245]. The positive and negative charge pumps both deliver a charge greater than zero at each phase comparison. With zero phase error, ideally the net charge pumped sums to zero. But any mismatches between the positive and negative currents will cause phase noise performance degradation. To see the effect of the current mismatch, we have simulated the PLL circuit first with a noiseless reference and VCO, an alive zone PFD with a minimum pulse duration of 3 ns, and a charge-pump with 10% current mismatch. We have also performed a simulation with a white phase noise reference, a random walk phase noise VCO, the same alive zone PFD with a 10% current mismatch. Figure 3.36 shows the phase noise spectrum of the closed-loop VCO for both cases.

4

ARCHITECTURAL MAPPING AND OPTIMIZATION

4.1 ARCHITECTURAL MAPPING

Once a behavioral simulator has been developed, the next step in the development of the design methodology is in its application to architecture mapping. One key contribution of this work is to define a mechanism by which hierarchy can be exploited successfully. The example in Figure 4.1 will be used to illustrate this.

Performance specifications for INL, signal-to-noise ratio (SNR), and timing jitter are specified. Assuming that an architecture has been selected, component parametric values in the circuit must be found, i.e. transistor widths and lengths (W_x, L_x), resistance values (R), capacitance values (C), and inductance values (L). Without

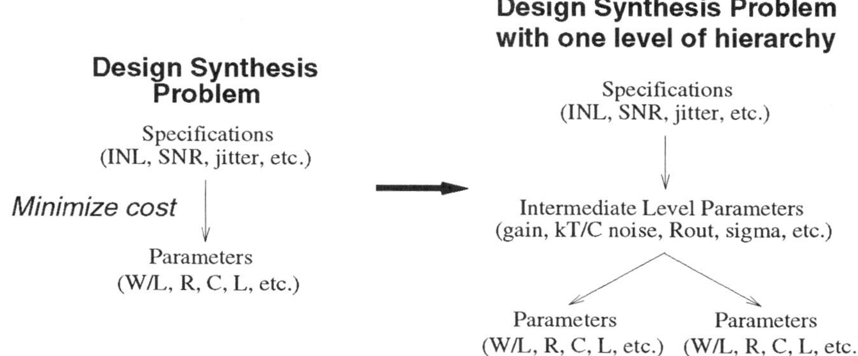

Figure 4.1 Design Synthesis Problem with and without Hierarchy

hierarchy, the problem can be formulated as

$$minimize \qquad \text{cost}(W_x, L_x, R, C, L)$$
$$s.t. \quad \text{INL}(W_x, L_x, R, C, L) \le \text{specified value for INL}$$
$$\text{SNR}(W_x, L_x, R, C, L) \ge \text{specified value for SNR}$$
$$\text{jitter}(W_x, L_x, R, C, L) \le \text{specified value for jitter}$$
$$\text{process constraints}$$

where W_x, L_x, R, C, and L are vectors representing all of the transistors and passive components.

Suppose, however, that this problem contains too many design variables and cannot be solved as formulated. It is transformed from the non-hierarchical problem on the left in Figure 4.1 to the hierarchical problem on the right. The performance specifications are first divided into a set of *intermediate level parameters*. These must characterize all of the second-order performance specifications of the low-level blocks which cause performance degradation in the overall system. Satisfying this condition is often difficult. In general, they have to be selected on a design-by-design basis. If they are chosen improperly, i.e. a cause of system performance degradation is missing, when the chip is fabricated and tested, it could fail to meet specifications. In this example, the intermediate level parameters are the gain and kT/C noise for an operational amplifier and output resistance (R_{out}) and a matching term, σ, for a current source.

We use behavioral models to calculate INL, SNR, and jitter as functions of gain, kT/C noise, R_{out}, and σ. Since behavioral models are *architecturally independent*, they give no data on the design cost as a function of intermediate level parameter values. Rather than minimizing cost, since this cannot be calculated, we choose to maximize the design *flexibility* of the lower-level components. For an individual variable, the goal is to choose reasonable parametric values. For example, specifying that the kT/C noise should be 0 yields an infeasible design, since T must be greater than 0. We build a *flexibility function* to reflect this information. For multiple variables, the flexibility function is set up to balance design tradeoffs. For example, suppose lowering the gain, lowers the constraint on kT/C noise. Maximizing the flexibility will prevent the constraint on gain from being too low, allowing for a reasonable kT/C noise requirement.

An example flexibility function is

$$flex_P(x) = -ax^2 - \frac{b}{x} + c \qquad (4.1)$$

For variables where *decreasing* values ease design difficulty, $b = 0$. For variables where *increasing* values ease design difficulty, $a = 0$. Two points are required to solve

Difficulty	$b = 0$ (parabolic)		$a = 0$ (hyperbolic)	
	Parameter	Flexibility	Parameter	Flexibility
moderate	6	0	6	0
difficult	10	-10	2	-10

Table 4.1 Points on the Flexibility Curve

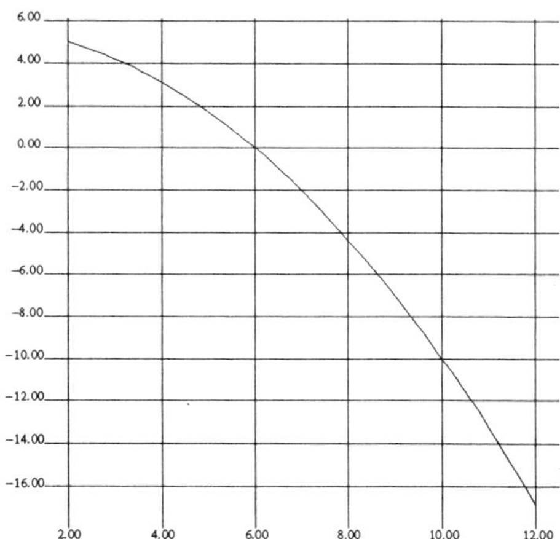

Figure 4.2 Flexibility Function—Parabolic

for the remaining coefficients. We use as a heuristic, a design point of "moderate" difficulty, assigning it *flexibility* = 0, and a "difficult" design point, assigning it *flexibility* = -10. Table 4.1 shows an example selection. For the two examples, the variable values for the moderate and difficult cases are listed in the parameter column, and the chosen values for flexibility are listed in the flexibility column.

Solving for the coefficients for the $b=0$ case, we obtain

$$flex_P(x) = -\frac{5}{32}x^2 - \frac{45}{8}$$

This function is graphed in Figure 4.2. The parameter value is on the x-axis. The flexibility value is on the y-axis. Marginal flexibility information is included in these curves. For variable values of high design difficulty, increasing the variable value incurs a high penalty in flexibility, while for variable values of low design difficulty, decreasing the variable value does not aid greatly in flexibility. The intuition behind this is as follows. Suppose gain is the design variable. A large value for gain is difficult

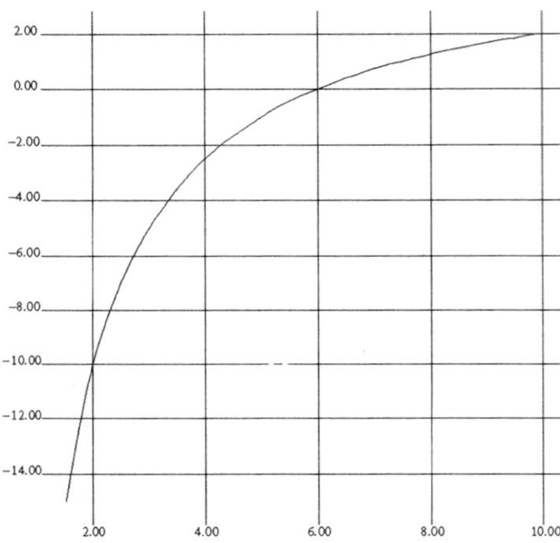

Figure 4.3   Flexibility Function—Hyperbolic

to achieve. The design becomes increasingly difficult as the value of gain increases. However, small values of gain are easy to design. Asking for a reduced value in gain usually offers little advantage. This is reflected by the small change in the flexibility function around points of low gain.

Solving for the coefficients for the $a{=}0$ case, we obtain

$$flex_P(x) = -\frac{30}{x} + 5$$

This is graphed in Figure 4.3. Once again marginal information is included. In this case for variable values of high design difficulty, *decreasing* the variable value incurs a high penalty in flexibility, while for variable values of low design difficulty, *increasing* the variable value does not aid greatly in flexibility. kT/C noise is an example.

Once flexibility functions have been developed for each of the parameters, they are summed to create the overall flexibility function. The high-level problem in Figure 4.1 can now be formulated as

$$
\begin{aligned}
maximize \quad & f_{gain}(gain) + f_{noise_{kT/C}}(noise_{KT/C}) + f_{R_{out}}(R_{out}) + f_\sigma(\sigma) \\
s.t. \quad & INL(gain, noise_{kT/C}, R_{out}, \sigma) \leq \text{specified value for INL} \\
& SNR(gain, noise_{kT/C}, R_{out}, \sigma) \geq \text{specified value for SNR} \\
& jitter(gain, noise_{kT/C}, R_{out}, \sigma) \leq \text{specified value for jitter}
\end{aligned}
$$

Since flexibility functions are heuristically determined, overall optimality is lost for the design. However, if a solution is found, *feasibility* is insured. If the design is

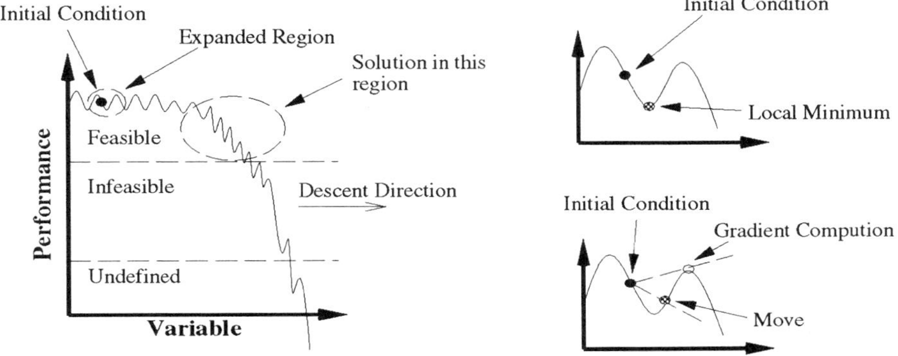

Figure 4.4   Optimization Function Surface

found to be too suboptimal when completed, the flexibility functions can be rebuilt to reflect data from previous design passes for future iterations. [273] attempts to *a priori* build extremely accurate contours (flexibility functions) to also guarantee an optimal result. However, this is extremely expensive, since not only does the entire design space need to be simulated, but also all possible combinations of architectures must be explored *before* any designs can be obtained. Our method has the advantage that a design solution is *immediately* found, so that the designer has a feasible solution at the end of each iteration.

The low-level optimization problems are directly formulated as

$$
\begin{aligned}
minimize \quad & \mathrm{cost}(W_x, L_x, R, C, L) \\
s.t. \quad & gain(W_x, L_x, R, C, L) \geq gain \text{ from high level} \\
& noise_{kT/C}(W_x, L_x, R, C, L) \geq noise_{kT/C} \text{ from high level} \\
& \text{process constraints}
\end{aligned}
$$

and

$$
\begin{aligned}
minimize \quad & \mathrm{cost}(W_x, L_x, R, C, L) \\
s.t. \quad & R_{out}(W_x, L_x, R, C, L) \leq R_{out} \text{ from high level} \\
& \sigma(W_x, L_x, R, C, L) \leq \sigma \text{ from high level} \\
& \text{process constraints}
\end{aligned}
$$

Ideally, to solve these problems, we would like to use an "off-the-shelf" *optimization program* such as MINOS [214] which allows for nonlinear objective and nonlinear constraint functions. However, often the functions used are simulations which are solved using iterative techniques. These result in solutions with a large amount of numerical noise (sometimes as high as 10%), which usually caused the general purpose optimization programs to fail. Three common modes of failure are illustrated in Figure 4.4. The graph on the left shows the performance versus the input. The initial condition is the point on the left. The descent direction is to the right. The feasible region is where the performance is high. In this example, the solution is in the descent direction somewhere near the boundary between feasible and infeasible region.[1] The following are the modes of failure:

1. Finding a false local minimum—The region near the initial condition is expanded and shown in the upper-right-hand corner of the figure. In this situation, because of numerical noise, a local minimum appears to exist near the initial condition. The optimizer may fall into this valley. The optimizer will terminate, and return this as the solution.

2. Iterate excessively because of gradient computation error—The area near the initial condition is expanded again and shown in the lower-right-hand corner of the figure. The gradient computation with the step size taken is shown. It indicates that the performance is increasing as the variable increases. The optimizer may make a move in that direction. The move step, however, may be smaller than the one used in the gradient computation. So, upon moving, the performance decreases. The optimizer will detect this and shrink the step sizes until a consistent solution is found. However, since it is in the noise, only random chance will allow a consistent solution to be found. Thus, in general, the step size will continue to shrink until the iteration count is exceeded, and the program will terminate with an error.

3. Wander into an undefined region—Often in the circuit design space, there are regions that are undefined, e.g. no DC convergent regions in SPICE. The graph on the left indicates an undefined region below the infeasible region. Some algorithms such as the projected augmented Lagrangian algorithm used in MINOS, allow an intermediate point to wander into the infeasible region. If it should wander into the undefined region, the simulator will not return a result, and the optimization process will terminate, or may continue but with unexpected results.

In our design examples, we tried MINOS first. It usually succeeded when we either had analytic functions or when the standard error of the simulator was low (typically less

---

[1]This is not the only variable in the problem. If there were only one variable, the solution would be exactly at the boundary.

than 10^{-5}). When MINOS failed to find a solution directly, we tried to either modify the problem formulation given to MINOS or to use other optimization algorithms.

Once a solution has been found, we advocate that the designer examine the optimizer's solution. Even if there are no numerical noise problems, gradient based optimizers generally only return a local optimum, unless techniques have been implemented to search for other solutions, or the space is known to be convex. The solution, then, may be very sensitive to the initial condition. Much of the optimality obtainable depends on the designer choosing a reasonable starting point. A commonly used technique to avoid local optimum solutions is simulated annealing. However, because of function evaluation through simulation, simulated annealing techniques are usually too CPU expensive.

4.2 OPTIMIZATION TOOLS

As seen in the previous section the use of optimizers plays an important role. They are used not only in high-level architectural mapping, but also in many other phases of design. This section describes the basic optimization problem formulation, discusses a generic optimization platform, OPTZ, that has been developed, and presents some examples to illustrate its use.

4.2.1 General Problem

C++ classes have been implemented to allow users to specify optimization problems of the following form:

$$
\begin{aligned}
\text{Minimize or maximize} \quad & F(x) + c^T x + d^T y \\
\text{subject to} \quad & f(x) + A_1 y \leq, =, \geq b_1 \\
& A_2 x + A_3 y \leq, =, \geq b_2 \\
& l_x \leq x \leq u_x \\
& l_y \leq y \leq u_y
\end{aligned}
$$

x is the vector of variables which will be used in nonlinear functions. y is the vector of the remaining variables. $F(x)$ contains the nonlinear portion of the objective function. $f(x)$ is a vector of nonlinear functions. l_x, l_y, u_x, and u_y are the lower and upper bounds on x and y.

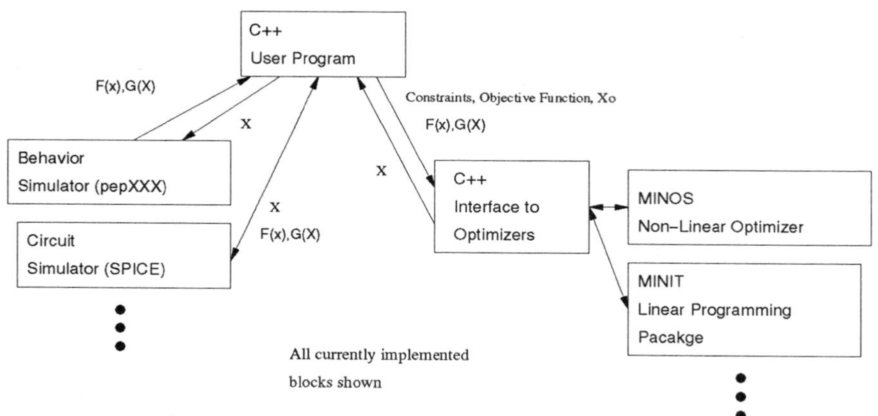

Figure 4.5   Optimizer Interface

4.2.2   Optimization Platform

An overview of the optimization process is shown in Figure 4.5. (1) The problem statement is made, and results are requested. (2) The optimizer returns with a vector, x, and asks for function evaluation. OPTZ forwards x to the function evaluator which returns $F(x)$ and, if they are available, the gradients with respect to x, $G(x)$. Most useful to us are function evaluators which deal with integrated circuit analysis such as behavioral simulators and circuit simulators. (3) The optimizer uses $F(x)$ and $G(x)$ and computes the next x vector if necessary. (4) This process continues until the optimizer finishes. The result and status are returned.

Currently, two optimizers are supported by the interface. MINOS [216] is the first. It uses a variety of techniques to solve nonlinear programming problems. For example, if the constraints are linear, but the objective function is nonlinear, it uses a *reduced-gradient algorithm* with a *quasi-Newton algorithm*. If the constraints and the objective functions are nonlinear, then it uses a *projected augmented Lagrangian algorithm*. MINOS provides a nice user interface for the following reasons: (1) it automatically uses finite difference to compute derivatives if derivatives are not provided; (2) it can handle an arbitrary number of provided derivatives; (3) it contains many options to control its operation; and (4) it is well documented. The second optimizer that is connected to our interface is MINIT [184], a linear programming package.

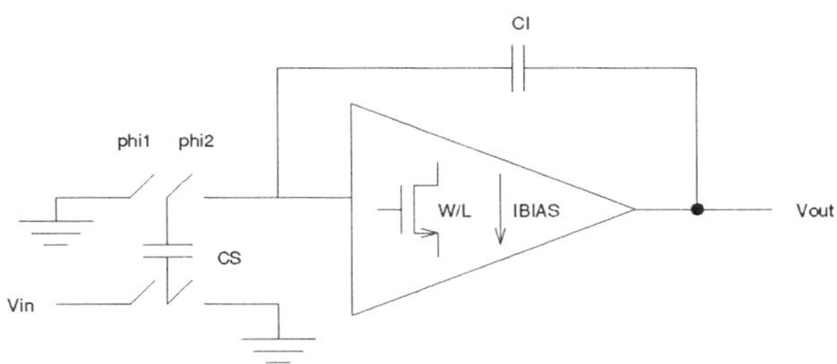

Figure 4.6   Sample-and-Hold Schematic

P mW	C_S,C_I pF	W_I μm	L_I μm	I_{BIAS} mA	RES	Area mm^2	Power mW	CPU min.	SIM
2	0.264	1632	2.02	0.194	12.00	3.28	1.94	16	5744
5	0.250	622	2.25	0.338	12.00	1.49	3.38	42	8078
1.5	*	*	*	*	*	*	*	*	*

Table 4.2   Sample-and-Hold Optimization Results

4.2.3   Examples

Several examples of circuit optimizations will be presented. Both behavioral and SPICE simulations were used to evaluate circuit performance. The purpose of this section is to not only demonstrate the generality of OPTZ but also to provide insight on the CPU requirements for these problems.

4.2.4   Sample and Hold

This example is for the sample-and-hold (S/H) shown in Figure 4.6. The variables are: (1) C_S (C_I is set equal to C_S), bound to be between 0.25 pF and 10 pF; (2) the width and the length of the input source coupled pair of the operational amplifier, bound to be between 2 μm and 3000 μm; and (3) the bias current in the input stage of the operational amplifier, bounded to be between 100 μA and 100 mA. The performance constraint is that the circuit must have 12 bits of resolution, and there may be a constraint on power (P). Initially, C_S is set to 1 pF. The width and the length of the input transistor are set at 100 μm and 20 μm respectively. And the bias current is set to 0.5 mA. The objective is to minimize area. Table 4.2 shows the optimization results. The 1.5 mW case was infeasible.

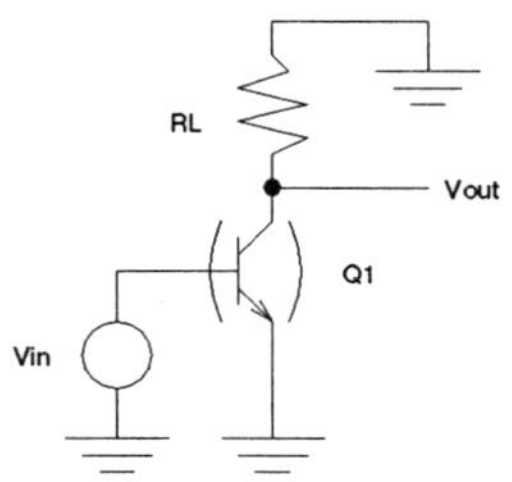

Figure 4.7　Common Emitter Amplifier

R_L $k\Omega$	f_{-3db} MHz	CPU min.	SIM
31.83	1.00	35	171

Table 4.3　Common Emitter Amplifier Optimization Results

4.2.5　Common Emitter Amplifier

Gain bandwidth trade-off is illustrated by the small-signal common emitter (CE) amplifier in Figure 4.7. C_π and C_μ are are both set to 5 pF. There is only one variable, R_L, in this circuit. It is bound to be greater than 100 Ω and is initially set at 10 $k\Omega$. The frequency response (f_{-3db}) is constrained to be greater than 1.0 MHz.

The results are shown in Table 4.3. They are optimal. Figure 4.8 shows the path of the optimizer (selected R_L values) with respect to iteration number.

4.2.6　Transimpedance Amplifier

A more complicated example, an input stage for a wide band, low-noise monolithic fiber-optic preamplifier, is shown in Figure 4.9. In this design, there are six variables, six nonlinear constraints, and a nonlinear objective function. Four types of SPICE analyses are employed—"OP," "DC," "AC," and "NOISE."

The problem statement is as follows:

$$\begin{aligned} minimize \quad &I_{\text{NOISE}}(R_F, W_{M1}, R_{L1}, R_{L3}, R_{E2}, R_{L3}) \times 10^8 + \\ &\text{POWER}(R_F, W_{M1}, R_{L1}, R_{L3}, R_{E2}, R_{L3}) \times 10^3 - \\ &f_{-3dB}(R_F, W_{M1}, R_{L1}, R_{L3}, R_{E2}, R_{L3}) \times 10^7 \end{aligned}$$

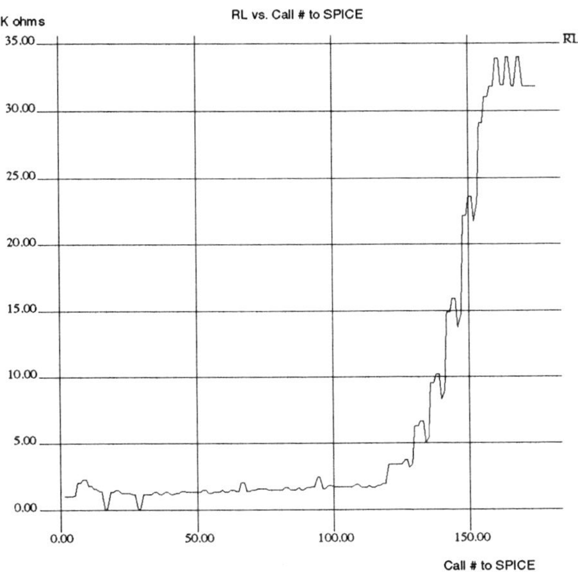

Figure 4.8   Common Emitter Optimization

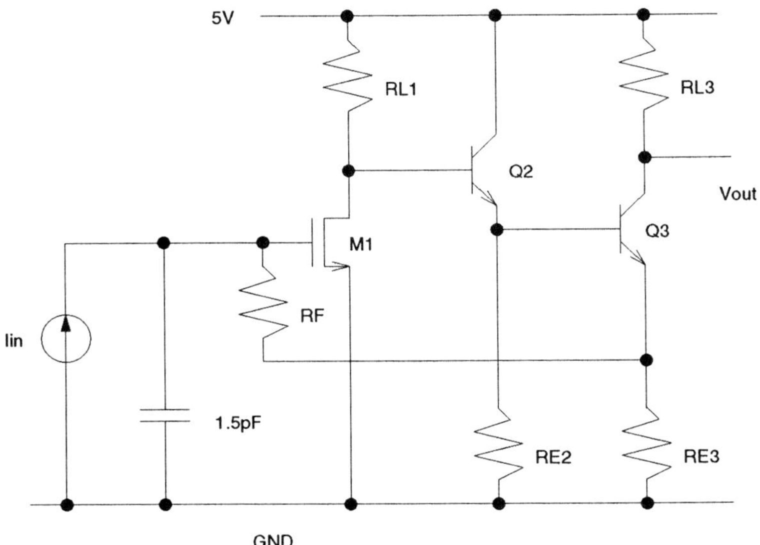

Figure 4.9   Transimpedance Optimization

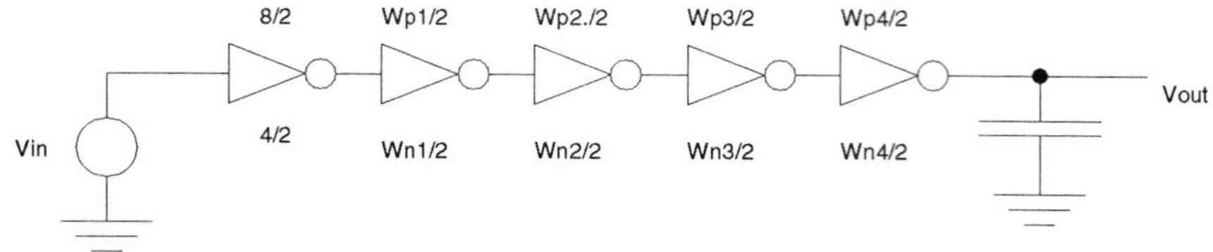

Figure 4.10   Inverter Chain Optimization

$$\text{s.t} \quad \text{GAIN}(R_F, W_{M1}, R_{L1}, R_{L3}, R_{E2}, R_{L3}) \geq 5000 \ \Omega$$
$$\text{RIPPLE}(R_F, W_{M1}, R_{L1}, R_{L3}, R_{E2}, R_{L3}) \leq 0.01$$
$$f_{-3dB}(R_F, W_{M1}, R_{L1}, R_{L3}, R_{E2}, R_{L3}) \geq 150 \ \text{MHz}$$
$$\text{DC}_{\text{OUT}}(R_F, W_{M1}, R_{L1}, R_{L3}, R_{E2}, R_{L3}) \geq 3.5 \ \text{V}$$
$$\text{POWER}(R_F, W_{M1}, R_{L1}, R_{L3}, R_{E2}, R_{L3}) \leq 20 \ \text{mW}$$
$$\text{MAXIN}(R_F, W_{M1}, R_{L1}, R_{L3}, R_{E2}, R_{L3}) \geq 140 \ \mu\text{A}$$
$$R_F \geq 5 \ \text{k}\Omega$$
$$W_{M1} \geq 3 \ \mu\text{m}$$
$$R_{L1} \geq 1 \ \text{k}\Omega$$
$$R_{L3} \geq 100 \ \Omega$$
$$R_{E2} \geq 100 \ \Omega$$
$$R_{E3} \geq 100 \ \Omega$$

I_{NOISE}, POWER, f_{-3dB}, GAIN, RIPPLE, DC$_{\text{OUT}}$, POWER, and MAXIN are computed by SPICE. MAXIN is defined as the point where the gain is nonlinear by 5% to the linear gain. The initial conditions were: $R_F = 6.5$ kΩ, $W_{M1} = 150$ μm, $R_{L1} = 5.2$ kΩ, $R_{E2} = 1.7$ kΩ, $R_{L3} = 500$ Ω, and $R_{E3} = 500$ Ω. The chosen values were: $R_F = 6.13$ kΩ, $W_{M1} = 144$ μm, $R_{L1} = 6.72$ kΩ, $R_{E2} = 1.37$ kΩ, $R_{L3} = 1.09$ kΩ, and $R_{E3} = 1.54$ kΩ. It required 3.8 hours of CPU time to run, and required 429 SPICE simulations. The results obtained were: GAIN = 5094 Ω, RIPPLE = 0.01, $f_{-3db} = 232$ MHz, DC$_{\text{OUT}} = 4.32$ V, POWER = 11.37 mW, $I_{\text{NOISE}} = 32.81$ nA, MAXIN = 140 μm.

4.2.7   Inverter Chain

Figure 4.10 shows the classic inverter chain problem. The exact specifications for the problem were obtained from ECSTASY [281]. The lower bound and the initial

Optimizer	W_{n1} μm	W_{n2} μm	W_{n3} μm	W_{n4} μm	Delay ns	CPU min.	SIM
ECSTASY	14.9	56.1	231	1325	5.1	n/a	8
MINOS	15.1	65.1	516	2530	4.7	17.5	148

Table 4.4   Inverter Chain Optimization Results

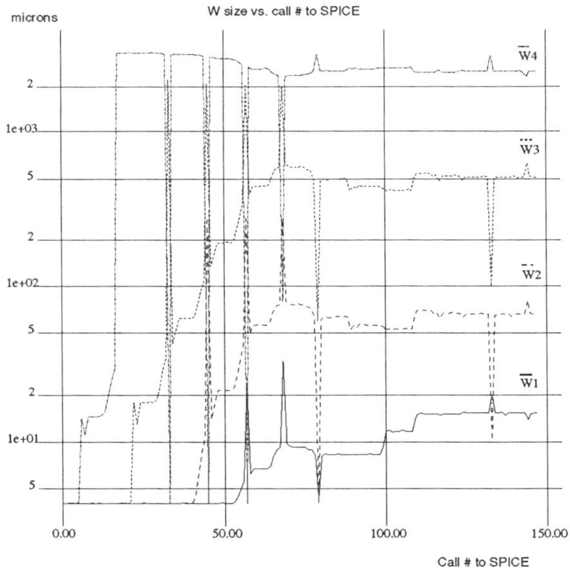

Figure 4.11   Inverter Chain Optimization (Widths)

condition for the widths of the transistors was 4 μm. The widths of the PMOS devices were set to be twice the size of the NMOS devices. The objective was to minimize the delay time. Table 4.4 shows the MINOS results compared to ECSTASY's. The Ws for the ECSTASY run and the SIM value was obtained from [281]. The delay value was obtained by simulating the result.

We present here another illustration of the optimizer's path. Figure 4.11 shows the optimizer's selection of width versus iteration number with a corresponding graph in Figure 4.12 showing the value of the objective function versus iteration number.

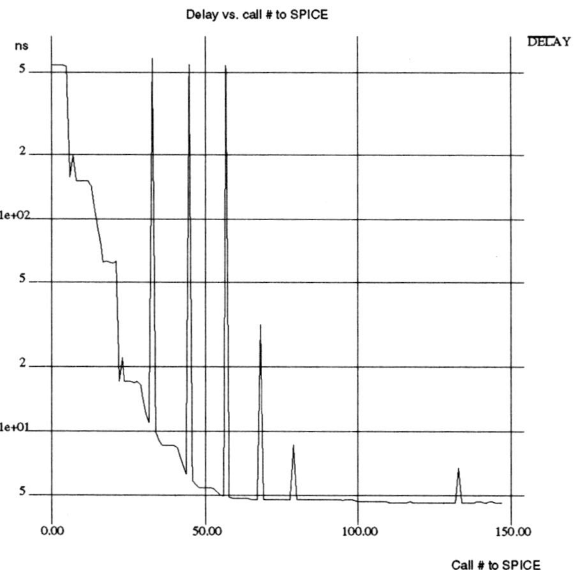

Figure 4.12   Inverter Chain Optimization (Delay)

5

CONSTRAINT-DRIVEN LAYOUT SYNTHESIS

5.1 INTRODUCTION

Research on CAD systems for reliable physical assembly of analog circuits has progressed at a considerably slower pace than that for digital counterparts. Part of the reason has been the intrinsic difficulty of defining and controlling performance in analog circuits. High performance can be achieved by taking advantage of the physical characteristics of integrated devices and of the correlation between electrical parameters and their variations due to statistical fluctuations of the manufacturing process. Device matchings, parasitics, thermal and substrate effects must all be taken into account. The nominal values of performance functions are subject to degradation due to a large number of parasitics which are generally difficult to estimate accurately before the actual layout is completed.

Another reason might be the present difficulty to identify a level of abstraction where generic models such as the ones developed for digital synthesis can be derived. All these concerns need be addressed in each phase of the design equally carefully, since severe performance degradation, even if localized only in some components, often jeopardizes the functionality of the whole system.

5.1.1 Bottom-Up vs. Top-Down Approaches

The approach generally adopted by designers consists of building complex layouts bottom-up, starting from the simplest components of the systems and estimating component specifications using rough approximations mostly derived from experience. This approach often results in a series of time-consuming design loops, hence multiple re-designs are needed for the whole system.

It is our belief that the design loops could be drastically reduced if a top-down approach were used. In a top-down approach the order of the synthesis phases is reversed. First, top-level specifications are rigorously mapped onto constraints on the physical details of the layout, in such a way that the satisfaction of the low-level constraint implies satisfaction of the overall system specifications. Then, the entire physical assembly, partitioned in its basic steps, module generation, placement, routing and compaction, is performed enforcing all physical constraints. A bottom-up verification step based on extraction concludes the assembly.

There are several advantages to this approach. First, a tight control of performance can be maintained in each phase of the physical assembly independently, hence specification violations can be identified early, thus enhancing the robustness of the process. Second, all physical constraints are derived so as to minimize the effort that each layout tool requires for its enforcement, hence improving its efficiency. Finally, due to the generality of the constraint generation process, the scheme can be easily extended to encompass a wide variety of non-idealities usually encountered in layout.

The scheme, called *constraint-driven layout design*, was originally formulated for a class of layout problems and then generalized by us to the extent of physical assembly for analog and mixed-signal ICs. This chapter presents the generalized constraint-driven layout design methodology and the techniques used for each phase of the physical assembly.

5.1.2   Generalized Constraint-Driven Layout Design

The design flow of the constraint-driven physical assembly system is illustrated in Figure 5.1. First, *high-level specifications* are translated into a set of bounds on low-level *physical constraints*. *A priori* parasitic estimates are used to determine feasible bounds. Among all possible sets of bounds, the one maximizing the *flexibility* of the tool to be used is chosen. Flexibility is a function which measures how easily the tool is able to meet the given set of constraints.

Then, at each step, the existence of a feasible configuration is tested and feedback paths are provided to resolve situations of infeasibility. These situations can occur as the result of partially inaccurate estimations on parasitics and other circuit non-idealities at early layout phases due to incomplete information about the physical implementation. In these cases, mechanisms are provided for the re-generation of constraints by correcting *a priori* estimations.

Each layout phase is organized as illustrated in Figure 5.2. The design task is con-

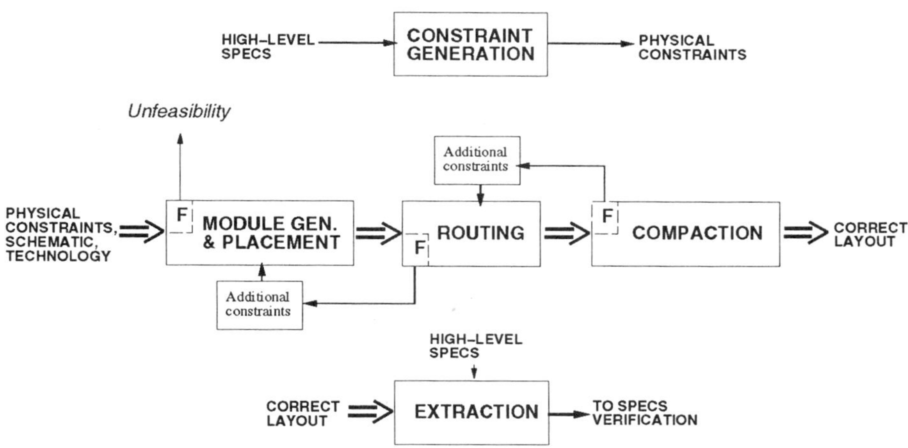

Figure 5.1 Constraint-driven layout design: Traditional design partition into tasks has been modified by adding information paths between layout phases.

strained by a set of input specifications, which are either high-level performance specifications or additional design constraints introduced by other layout phases. Constraints are translated into a set of bounds on parasitics by a *constraint generator*, based on estimates of the feasible values of each parasitic. These bounds drive each tool independently. The resulting layout is then analyzed to check whether performance specifications have actually been met. If some constraints have been violated, the values of the extracted parasitics can provide more accurate estimates to the constraint generator. The constraint generator also executes the feasibility check. In fact, low-level bounds must be feasible, i.e. they must lie between the minimum and maximum possible values estimated for the parameters. Such early detection of infeasibility provides an efficient control of design iterations, thus minimizing overall computation time. Feedback control paths provide previous design phases with information on those critical parasitics for which it was not possible to determine feasible bounds with the current configuration.

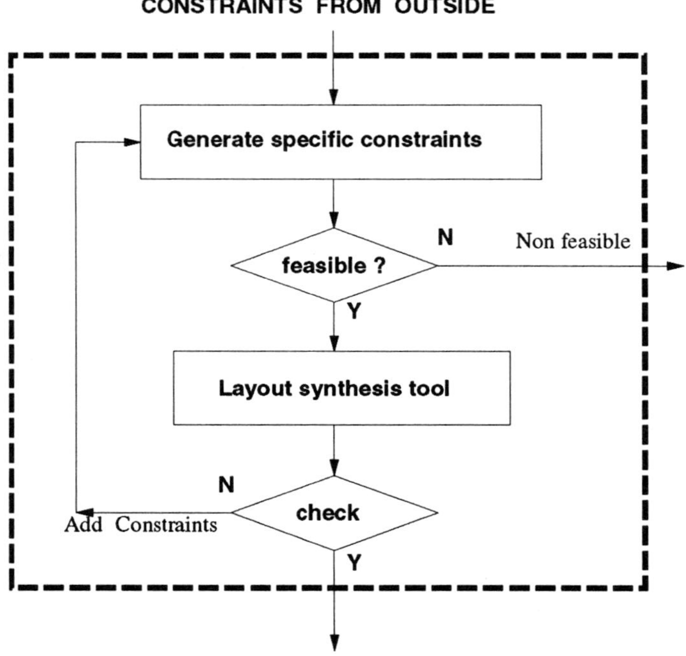

Figure 5.2 The organization of each layout phase. The internal feedback path provides information to the constraint generator. External feedback paths provide information on the reasons of failure to meet performance specifications.

5.2 CONSTRAINT GENERATION

5.2.1 Problem Formulation

For a given circuit C, let us define *performance* $\mathbf{K}$ as the finite array of all measures that evaluate a parametric behavior for C and N_K its size. Performance specifications are expressed as the maximum allowed performance degradation from nominal, due to process variance and the parasitics caused by the realization of the layout details. Both absolute parasitic values and mismatch play a role in the deviation of performance measures from nominal.

Let us define V as the finite set of all possible operating points for C. Assume that $\mathbf{K}$ can vary around an operating point $v \in V$ and let $\mathbf{K_v}$ be the performance at v, the *nominal performance*. Assume that all parasitics significantly affecting performance are known. Let us define $\mathbf{\Delta K} = \mathbf{K} - \mathbf{K_v}$ as the degradation of performance $\mathbf{K}$ from its nominal due to all known parasitics. A specification on $\mathbf{K}$ is defined to be a constraint on the maximum allowed performance degradation $\mathbf{\Delta K}$,

$$\mathbf{\Delta K} \leq \overline{\mathbf{\Delta K}}. \tag{5.1}$$

The *constraint generation problem* consists of finding bounds on a subset of parasitics, whose enforcement guarantees the satisfaction of all performance specifications.

Problem 1 *Given a circuit C with performance $\mathbf{K}$ and a finite set of parasitic components, find bounds on all parasitics, such that (5.1) holds.*

The solution of this problem is nontrivial for two reasons. First, performance $\mathbf{K}$ is generally an array of nonlinear functions of parasitics, often not representable in a compact form. In addition, the number of parasitics is generally much larger than the size of the performance array. Hence a naive approach based on solving (5.1) with respect to each parasitic is not feasible. In the remainder of this chapter techniques are presented that address this problem in an efficient and rigorous fashion.

5.2.2 Mapping Specifications onto Layout Constraints

Our approach to the constraint generation problem is based on the work described in [51], [50] and [42]. Let us assume that performance $\mathbf{K}$ is continuously differentiable around operating point v. Then, performance degradation $\mathbf{\Delta K}$ can be represented in terms of its sensitivity with respect to all relevant parasitics. We denote by N_p the number of layout parasitics, by $\mathbf{p} = [p_1 \ \ldots \ p_{N_p}]^T$ the array of all such parameters, and by $\mathbf{p}^{(0)} = [p_1^{(0)} \ \ldots \ p_{N_p}^{(0)}]^T$ the array of their *nominal* values. Each performance K_i is a nonlinear continuously differentiable function of all parasitics $K_i = K_i(\mathbf{p})$ and the array of the N_k performance functions will be indicated as $\mathbf{K} = \mathbf{K}(\mathbf{p}) = [K_1(\mathbf{p}) \ \ldots \ K_{N_k}(\mathbf{p})]^T$. If all parasitics are subject to variations with respect to their nominal values, let $\Delta\mathbf{K}(\mathbf{p}) = \mathbf{K}(\mathbf{p}) - \mathbf{K}(\mathbf{p}^{(0)})$ be the corresponding degradation of $\mathbf{K}$ due to such variations.

A generalized expression for the computation of sensitivities from a set of arbitrary performance functions has been derived in [228][280]. With this formulation, all performance functions can be represented in a compact and rigorous way, as long as they are continuous and sufficiently regular in an interval around their nominal value. The sensitivity of K_i with respect to p_j is defined as[1]

$$S_{i,j} = \left.\frac{\partial K_i(\mathbf{p})}{\partial p_j}\right|_{\mathbf{p}^{(0)}}. \tag{5.2}$$

The matrix of all sensitivities is

$$\mathbf{S} = \left[\begin{array}{ccc} S_{1,1} & \ldots & S_{1,N_p} \\ \ldots & \ldots & \ldots \\ S_{N_k,1} & \ldots & S_{N_k,N_p} \end{array}\right].$$

Sensitivities are computed for each performance function, with respect to each parameter that may be introduced or modified in the layout phase, i.e. parasitics and geometric parameters. Several techniques have been developed for efficient numerical calculation of sensitivities in time and frequency domain [69][48][228]. Performance degradations are approximated by linearized expressions using sensitivities [51]. These approximations are acceptable if degradations are small compared to the nominal values. The array of all degradations of performance functions due to parasitic variations is

$$\Delta\mathbf{K}(\mathbf{p}) \approx \mathbf{S}\left[\mathbf{p} - \mathbf{p}^{(0)}\right]. \tag{5.3}$$

Before the definition of layout details, one cannot take advantage of the possible cancellation effects due to positive and negative sensitivities for different parasitics. Hence, each performance constraint is modeled only with respect to the parasitics whose sensitivity is either positive or negative, depending on the sign of the constraint itself. Assuming that the performance model of (5.3) is used, (5.1) becomes

$$\Delta\mathbf{K}(\mathbf{p}) - \overline{\Delta\mathbf{K}^+} \leq 0 \tag{5.4}$$

$$\Delta\mathbf{K}(\mathbf{p}) + \overline{\Delta\mathbf{K}^-} \geq 0 \tag{5.5}$$

---

[1] Here and in what follows the non-normalized notation, first used in [50], is used for sensitivities, without loss of generality.

where $\overline{\Delta \mathbf{K}^+}$ and $\overline{\Delta \mathbf{K}^-}$ are the vectors of constraints, in absolute value, on the degradation of performance functions $\mathbf{K}(\mathbf{p})$ in the positive and negative direction respectively. They can be different and one of them can eventually be infinite. By substituting the linearized expression (5.3) into inequalities (5.4) and (5.5), the general problem can be rewritten as

$$\mathbf{S}^+ \left[\mathbf{p} - \mathbf{p}^{(0)} \right] - \overline{\Delta \mathbf{K}^+} \leq 0 \tag{5.6}$$

$$\mathbf{S}^- \left[\mathbf{p} - \mathbf{p}^{(0)} \right] - \overline{\Delta \mathbf{K}^-} \leq 0 \tag{5.7}$$

where $\mathbf{S}^+$ is the matrix of the worst-case positive sensitivities and $\mathbf{S}^-$ is the matrix of the absolute values of the worst-case negative sensitivities

$$\begin{aligned} \mathbf{S}^+{}_{i,j} &= \max\left(0, S_{i,j}\right) \\ \mathbf{S}^-{}_{i,j} &= \max\left(0, -S_{i,j}\right) \end{aligned}$$

In the remainder of this book the '+' and '−' signs have been omitted in the notations of sensitivities and constraints. Expressions (5.6) and (5.7) are given for positive and negative directions, and the general problem formulation becomes

$$\mathbf{S} \left[\mathbf{p} - \mathbf{p}^{(0)} \right] - \overline{\Delta \mathbf{K}} \leq 0. \tag{5.8}$$

We want to determine an array of *bounds* $\mathbf{p}^{(\mathbf{b})} = [p_1^{(b)} \; \ldots \; p_{N_p}^{(b)}]^T$ for all parasitics, such that inequality (5.8) holds as long as each parasitic remains below its bound, i.e.

$$\mathbf{S} \left[\mathbf{p}^{(\mathbf{b})} - \mathbf{p}^{(0)} \right] - \overline{\Delta \mathbf{K}} = 0. \tag{5.9}$$

It is necessary that all bounds be *feasible* and *meaningful,* i.e. all layout structures associated with the bounds must be physically realizable. Let $p_j^{(min)}$ and $p_j^{(max)}$ be respectively the minimum and maximum possible values which can be assumed by parasitic p_j, and let $\mathbf{p}^{(\mathbf{min})} = [p_1^{(min)} \; \ldots \; p_{N_p}^{(min)}]^T$ and $\mathbf{p}^{(\mathbf{max})} = [p_1^{(max)} \; \ldots \; p_{N_p}^{(max)}]^T$. The array of bounds $\mathbf{p}^{(\mathbf{b})}$ must satisfy the following inequalities

$$\begin{cases} \mathbf{p}^{(\mathbf{b})} - \mathbf{p}^{(\mathbf{min})} \geq 0 \\ \mathbf{p}^{(\mathbf{b})} - \mathbf{p}^{(\mathbf{max})} \leq 0 \end{cases} \tag{5.10}$$

The constraint-generation can be reformulated as the problem of finding a solution to (5.9), subject to (5.10).

5.2.3 Methods for the Evaluation of Sensitivities

In general, the constraint generation problem requires sensitivities to be computed for a general performance K_i with respect to a set of parasitics p_j or a set of design parameters π_j. Suppose K_i is an explicit differentiable function of p_j or π_j, or it can be modeled as such. Then, the sensitivity of K_i, defined in (5.2), can be derived using a number of techniques, both numerical, analytical or a combination of the two [229][130][48].

If on the contrary, K_i is not continuous or strongly nonlinear, two strategies can be adopted. The first consists of approximating performance K_i as a Taylor series, thus making the derivation of constraints a complex task. The second option consists of calculating *worst-case sensitivities* for K_i.

To illustrate the method, consider the array $\mathbf{\Pi}$ of all parameters π_j and $K_i = K_i(\mathbf{\Pi})$. Let us split vector $\mathbf{\Pi}$ in subvectors $\mathbf{\Pi}'$ and $\mathbf{\Pi}''$. The two vectors include the parasitics that show a linear and a nonlinear behavior, respectively. $\mathbf{\Pi}'$ is defined as the vector of all parameters such that

$$\left| (\mathbf{S}_{i,\mathbf{\Pi}'})^T \Delta\mathbf{\Pi}' - \Delta K_i \right| < \epsilon, \tag{5.11}$$

with $0 < \Delta\mathbf{\Pi}' < \delta$, for some $\epsilon, \delta > 0$. The problem of finding a worst-case sensitivity $\overline{\mathbf{S}}_{i,\mathbf{\Pi}'}$ is equivalent to that of solving

$$\begin{aligned} maximize \quad & \mathbf{S}_{i,\mathbf{\Pi}'} \\ s.t. \quad & \mathbf{\Pi}'' \in I \end{aligned} \tag{5.12}$$

where I is the feasibility interval of $\mathbf{\Pi}''$.

5.2.4 Constraint Generation Engine

Absolute Parasitic Constraints

In general, an infinite number of solutions exist for the constraint-generation problem. PARCAR [50] is a constraint-generator, namely a tool able to find a solution to the constraint-generation problem under particular assumptions. Among all solutions, PARCAR chooses the one maximizing the layout tool *flexibility*, which is a measure of how easily the tool is able to meet the constraints. To illustrate this concept, consider the following example. Suppose that the bound for a given parasitic p_j is close to its lower limit $p_j^{(min)}$, then the implementation of a layout geometry associated with p_j will be arduous and a small number of solutions for its realization will be available.

This might result into a significant loss of flexibility due to the limitations that will be necessarily imposed onto the remaining layout to be implemented. If, on the contrary, the bound is close to $p_j^{(max)}$, the effort required is lower, and the constraint easier to meet. Therefore a flexibility function defined as

$$F = 1 - \frac{\|\mathbf{p}^{(\mathbf{max})} - \mathbf{p}^{(\mathbf{b})}\|_2}{\|\mathbf{p}^{(\mathbf{max})} - \mathbf{p}^{(\mathbf{min})}\|_2}.$$

reflects the advantage of choosing a certain set of bounds. A discussion of this definition and of the quadratic norm choice can be found in [50]. In PARCAR a geometric norm is used and the constraint-generation problem is solved by minimizing a quadratic function (the geometric norm) subject to linear constraints (5.9) and (5.10), using a standard quadratic programming (QP) package.

The quality of the result depends on the estimates of parasitic limits $\mathbf{p}^{(\mathbf{min})}$ and $\mathbf{p}^{(\mathbf{max})}$, which become more and more accurate as layout details are defined during the design. The values of $\mathbf{p}^{(\mathbf{min})}$ and $\mathbf{p}^{(\mathbf{max})}$ are generally not known *a priori*. However, it is possible to compute suitable estimates, depending on the layout algorithm used. For example, the minimum value of the cross-coupling capacitance between unrouted nets can be set either to zero, or to the crossover capacitance due to unavoidable crossings. The latter estimate, however, is possible only if the router is able to detect unavoidable net crossings. This is the case for a channel router, where wire paths have been predefined in the global routing phase. With maze routing, on the contrary, the minimum value is always set to zero.

A substantial speed-up of the QP solver is achieved by removing from the problem those parasitics whose cumulative contribution to performance degradation is negligible. A threshold value $\alpha < 1$ is defined (in PARCAR we set $\alpha = 0.01$). For each performance function K_i, all parasitic effects on performance are sorted by increasing value. The first n_i parasitics in the sorted list which satisfy

$$\sum_{j=1}^{n_i} S_{i,j} p_j^{(max)} \leq \alpha \overline{\Delta K_i}, \tag{5.13}$$

are considered non-critical with respect to the threshold α. To compensate for this simplification in the constraint-generation problem, (5.9) is modified by replacing $\overline{\Delta \mathbf{K}}$ with $(1 - \alpha) \overline{\Delta \mathbf{K}}$. Notice that the sorting order may be different for different performance functions. Let P_i denote the set of n_i non-critical parasitics sorted according to performance K_i. When all performance functions are considered simultaneously, the set P of all non-critical parasitics is

$$P = \bigcap_{i=1}^{N_k} P_i.$$

The set P, determined in this way, is eliminated from further analysis. Different sorted lists are maintained for each kind of parasitics and elimination is carried out separately. This simplification can be very effective, since in most cases it permits the elimination of a significant portion (80-90%, typically) of all parasitics.

Constraints on Mismatch

The importance of *device matching* in integrated circuits has been shown not only for active but also for passive elements [6][240]. Designers generally impose matching constraints on circuit devices to ensure that voltage and current mismatches in these devices are bounded. These constraints can then be mapped directly onto constraints on the geometry of the physical implementations [240]. For this reason device matching constraints are usually qualitative in nature, mostly dictated by the expertise of the designer.

Matching enforcement of device parameters or interconnect parasitics is often referred to as the maximization in correlation of electrical parameters associated with given circuit components. In general the task is accomplished by minimizing the physical distance between the components or by means of topologies specifically designed to overcome the effect of technological gradients and random mask errors.

With tighter specifications and more complex circuits however the concept of matching as the solution of a maximization problem becomes a problem since many, possibly conflicting matching specifications might be required. This problem was first addressed by us in [42], where the rather imprecise definition of matching was replaced with a rigorous one, which allows advantageous trade-offs to be implemented. The approach consists of two phases: sensitivity characterization and constrained optimization.

Consider two parasitics p_1 and p_2. Within the limits of linear approximation (5.3), their contribution to the degradation of performance, K_i is

$$\Delta K_i\big|_{1,2} = S_{i,1}p_1 + S_{i,2}p_2 = 2S_{i,p}p + \frac{S_{i,\Delta}}{2}\Delta p, \qquad (5.14)$$

where

$$
\begin{aligned}
p &= \frac{p_1+p_2}{2} & S_{i,p} &= S_{i,1} + S_{i,2} \\
\Delta p &= p_1 - p_2 & S_{i,\Delta} &= \frac{S_{i,1}-S_{i,2}}{2}
\end{aligned}
\qquad (5.15)
$$

It is evident that if

$$\left|\frac{S_{i,\Delta}}{S_{i,p}}\right| \gg 1 \qquad (5.16)$$

the contribution of p_1 and p_2 to the degradation of K_i can be significantly reduced by increasing the correlation between the two parasitics, i.e. by enforcing matching between them. Inequality (5.16) determines quantitatively the benefit deriving from matching enforcement. For an arbitrary parasitic pair (ℓ, j), their mismatch and average sensitivities are computed. If relation (5.16) holds, the mismatch $\Delta p_{\ell j}$ and the average value $p_{\ell j}$ replace p_ℓ and p_j in the list of parasitics. In our approach, the magnitude requested to ratio $\left| \frac{S_{i,\Delta}}{S_{i,p}} \right|$ is user-defined. In our tests, we have obtained good results by requiring the ratio to be at least 10, and this value has been used in all the examples of this book.

From the information on the range of each parasitic, the range of variation of the average and mismatch values is computed as

$$\frac{p_\ell^{(min)} + p_j^{(min)}}{2} \leq p_{\ell j} \leq \frac{p_\ell^{(max)} + p_j^{(max)}}{2},$$

$$p_\ell^{(min)} + p_j^{(max)} \leq \Delta p_{\ell j}^{(mismatch)} \leq p_\ell^{(max)} + p_j^{(min)}, \quad \forall\, \ell \neq j. \tag{5.17}$$

One can recognize that the constraint generation problem for parasitic average $p_{\ell j}$ and a simple parasitic component are identical. The constraint generation for parasitic mismatch on the other hand is handled in a slight different manner. Suppose a solution for $\Delta p_{\ell j}$ has a negative value, then this term could possibly generate cancellation in the approximation formulae for performance degradation ΔK_i of (5.3) and (5.6), thus creating false results. To eliminate this problem, the variable $\Delta p_{\ell j}$ is split into two, $\Delta p_{\ell j}{}^+$ and $\Delta p_{\ell j}{}^-$, which are consequently added as a contribution to the approximate negative and positive components of the performance degradation. $\Delta p_{\ell j}{}^-$ and $\Delta p_{\ell j}{}^+$ represent the lower- and upper-bound of the allowed mismatch.

This approach is often expensive computationally, since it involves the generation of a large number of mismatches, and the constraint generation problem has to be solved on a large set of parameters. A more efficient approach is to use the matching requirement expressed in (5.16). For every pair of parasitics, their mismatch and average sensitivities are computed and compared against each other. If relation (5.16) holds, the original parasitics are discarded and substituted by their average value and the mismatch. Otherwise they are kept and mismatches between them are not considered. With this approach, the computational cost of the constraint generation problem is affected very slightly, only one being the parameter added for each group of three or more matched parasitics.

Matching constraints are often expressed as a difference between parameter ratios rather than of simple parameters. This is often the case when trying to establish specific geometric constraints from topological constraints during the technology mapping process. Given two independent parameters x and y and a performance $K_i(x/y)$ then

$$S_{i,(x/y)} = \frac{\partial K_i}{\partial (x/y)} = \frac{y^2}{y-x}(S_{i,x} + S_{i,y}), \tag{5.18}$$

and thus, the sensitivity of performance K_i with respect to the mismatch difference $\Delta\frac{x}{y} = \frac{x_1}{y_1} - \frac{x_2}{y_2}$ can be written as

$$S_{i,\Delta\frac{x}{y}} = \left(\frac{S_{i,\frac{x_1}{y_1}} - S_{i,\frac{x_2}{y_2}}}{2}\right) = \frac{y_1^2}{2(y_1 - x_1)}(S_{i,x_1} + S_{i,y_1}) - \frac{y_2^2}{2(y_2 - x_2)}(S_{i,x_2} + S_{i,y_2}). \tag{5.19}$$

Throughout this analysis it was always assumed that x_i and y_i are independent variables. This is usually the case in layout since orthogonal geometries, such as channel width and length in MOS transistors, are generally influenced by independent sources.

Generating Matching and Symmetry Constraints

Assume for simplicity but without loss of generality that all active devices can be represented by a two port. Furthermore, assume that a model relating the output to the input port is available and that only one performance K_i is considered. Let V_{D_ℓ} and I_{D_ℓ} be the quantities characterizing respectively input and output of device D_ℓ. Let $I_{D_\ell} = f(\mathbf{\Pi_0} + \mathbf{\Pi}, V_{D_I})$, where $\mathbf{\Pi}$ is the vector of all deviations from a nominal value $\mathbf{\Pi_0}$, of all technological parameters affecting the device. Suppose the set of sensitivities $\{S_{i,\Pi_{\ell m}}\}$ of performance K_i with respect to all vector elements of $\mathbf{\Pi}$ associated with device D_ℓ is available. Then, in first approximation, the degradation $\Delta K_{i,D_\ell}$ of performance K_i with respect to technological deviations in device D_ℓ can be expressed as

$$\Delta K_{D_\ell} = \sum_m S_{i,\Pi_{\ell m}} \Pi_m. \tag{5.20}$$

where, for reasons that we will be clear later, the sign of sensitivities has been dropped.

Consider now a pair of devices D_ℓ and D_j, then the degradation due to the parameter mismatch of the devices can be computed as

$$\Delta K_{i,D_\ell - D_j} = \sum_m S_{i,\Delta\Pi_m} \Delta\Pi_m. \tag{5.21}$$

where $S_{i,\Delta\Pi_m}$ is the sensitivity of K_i with respect to parameter Π_m, computed using (5.15) and $\Delta\Pi_m$ are the components of vector difference $\mathbf{\Pi}(\mathbf{D}_\ell) - \mathbf{\Pi}(\mathbf{D_j})$.

Now, assuming that the components $\Delta\Pi_m$ are independent random variables with zero mean, the variance of the degradation of performance K_i with respect to the variances of the mismatches of all technological parameters relevant to the pair of devices D_ℓ and D_j, is computed as

$$\sigma^2(\Delta K_{i,D_\ell - D_j}) = \sum_m |S_{i,\Delta\Pi_m}|^2 \, \sigma^2(\Delta\Pi_m). \qquad (5.22)$$

Consider for instance a pair of matched MOS transistors. The variance of the degradation due to technological mismatches can be expressed as

$$
\begin{aligned}
\sigma^2(\Delta K_{i,m_1 - m_2}) \;=\; & S_{i,\Delta W}{}^2 \, \sigma^2(\Delta W) + S_{i,\Delta L}{}^2 \, \sigma^2(\Delta L) \qquad (5.23) \\
& + S_{i,\Delta C_{ox}}{}^2 \, \sigma^2(\Delta C_{ox}) + S_{i,\Delta\mu_n}{}^2 \, \sigma^2(\Delta\mu_n) \\
& + S_{i,\Delta V_{TO}}{}^2 \, \sigma^2(\Delta V_{TO})
\end{aligned}
$$

where W, L, C_{ox}, μ_n and V_{TO} are respectively channel width, channel length, gate oxide capacitance, mobility and threshold voltage of the transistors.

In [240] a direct relation has been shown between these variances and the relative orientation and distance between device pairs. This information can be used to translate the maximum allowed performance degradation into the physical separation and relative orientation between pairs of devices. In order to do this, estimations on the minimum and maximum attainable variances for a particular process are needed, to determine the upper- and lower-bound of the performance degradation for *each* pair of devices. At this point the sum of all degradations due to each pair of devices in the circuit should be added to (5.4) and the constrained optimization solved to find the actual variances of each parameter, and consequently numerical values for the physical quantities. Notice that, for consistency, the standard deviation should be added to (5.5), thus making the optimization harder and more time consuming. In order to avoid this additional complexity, the expression for the standard deviation is linearized after substituting the single parameter variances with analytical models in the geometric quantities of interest. Thus a linear expression for the standard deviation of the degradation referred to the pair is obtained as

$$\sqrt{\sigma^2(\Delta K_{i,m_\ell - m_j})} \simeq A_{\ell j} d_{\ell j} + B_{\ell j} r_{\ell j}. \qquad (5.24)$$

where A and B are quantities which depend upon the performance sensitivity of K_i with respect to the pair's technology parameters. $d_{\ell j}$ and $r_{\ell j}$ represent the distance and relative rotation of devices m_ℓ and m_j. Also in this case a simplification mechanism similar to the one proposed in [50] is used for computing the criticality of the mismatch contributions.

The requirement of statistical independence for $\Delta\Pi_m$ can be relaxed if it can be assumed that: (1) the variables are Gaussian; (2) the variance-covariance matrix $\mathbf{A}$

symbol	*meaning*
R_{S_i}	degeneration resistance at the source of transistor M_i
$R_{S\_i,j}$	mismatch between R_{S_i} and R_{S_j}
C_i	substrate capacitance of net i
$C_{i,j}$	cross-coupling capacitance between nets i and j
V_{t_i}	voltage threshold of transistor M_i.
$V_{t\_i,j}$	mismatch between V_{t_i} and V_{t_j}
V_{dd}	Supply voltage
ω_0	Unity-gain bandwidth
A_v	Low-frequency gain
V_{off}	Systematic offset
ϕ_M	Phase margin
τ_D	Switching delay

Table 5.1   Notation for parasitics and performance functions

associated with them is known. In this case due to the positive-definitiveness and symmetric nature of $\mathbf{A}$, a method based on the $\mathbf{LDM}^T$ factorization can be used to translate the original variable set onto one where all the variables are uncorrelated and, since Gaussian, also statistically independent. The method is discussed in detail in Section 5.3.

In the remainder of this book, we shell refer to parasitics, parasitic mismatches and technology mismatches related to devices and interconnect according to Table 5.1.

Symmetry is often used in the layout of analog integrated circuits to minimize the effects of mismatched parasitics on certain performance measures such as offset voltage, Common Mode Rejection Ratio and noise. Symmetric placement and routing forces the parasitics of differential signal paths to be matched, thus reducing non-uniform and unbalanced signals. Deriving topological symmetry constraints in an automatic fashion has been traditionally associated with pattern recognition or expert systems.

An alternative approach, fully quantitative, based on rigorous matching analysis and graph-searching techniques was proposed by us in [42]. Symmetry is recognized as a particular case of matching between devices or interconnections belonging to distinct differential signal paths, which become effective when the circuit is operated in differential mode. A graph-based search algorithm, described in detail in [42], has been designed for the automated detection of all critical symmetry constraints. First, a graph is built, with a node for each circuit net, and an edge for each device, to represent the circuit connectivity. Then all virtual grounds are detected by comparing the common- and differential-mode gains of all nets. The search algorithm recognizes all the sub-graphs whose structure has the following characteristics:

1. symmetric topology,

2. matching constraints between symmetric graph elements,

3. the two halves of the structure are connected with one another by one or more real or virtual ground nets.

Each of these sub-graphs is a differential structure, and the symmetry constraints are all the matching constraints recognized at Step 2. For a thorough discussion of the modes of the search graph as well as details and various algorithmic aspects of the method we refer to in [42].

Example

As a practical example, consider the clocked comparator COMPL, whose schematic is shown in Figure 5.3. This comparator has been used as a benchmark in several recent works on analog CAD [80][12], due to its relevant performance sensitivity to layout details. Consider the following stray resistances (see Table 5.1 for the notations) and the corresponding sensitivities of systematic offset V_{off} with respect to each of them

$$
\mathbf{p} = \begin{bmatrix} R_{S\_1} \\ R_{S\_2} \\ R_{S\_3} \\ R_{S\_4} \\ R_{S\_6} \\ R_{S\_7} \\ R_{S\_20} \\ R_{S\_21} \\ R_{S\_22} \\ R_{S\_23} \end{bmatrix}
\qquad
\mathbf{S} = \begin{bmatrix} 56.53 \\ -56.53 \\ 0.202 \\ -0.202 \\ 11.83 \\ -11.83 \\ 16.76 \\ -16.76 \\ -16.72 \\ 16.72 \end{bmatrix}^T
$$

Offset sensitivities to resistances are expressed in $\mu V/\Omega$. They were computed by SPICE-3 [48] with a precision within the third digit. Therefore for each of the pairs $R_{S\_1,2}, R_{S\_3,4}, R_{S\_6,7}, R_{S\_20,21}, R_{S\_22,23}, R_{S\_21,23}, R_{S\_20,22}, R_{S\_21,22}$ the ratio (5.16) is $\left| \frac{S_{i,\Delta}}{S_{i,p}} \right| \geq 10^3$, i.e. the resistive mismatch is at least 10^3 times more important for offset than the absolute values of these resistances. By simplification

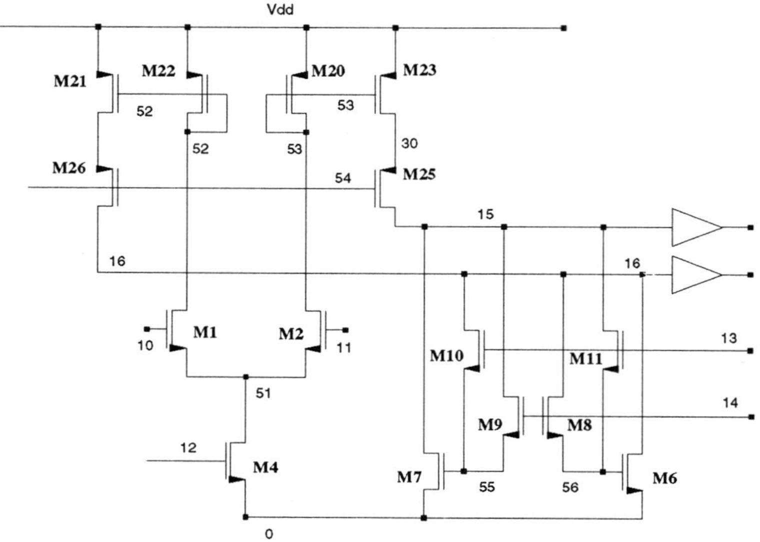

Figure 5.3   Schematic of COMPL

(5.14), offset sensitivities with respect to mismatches become

$$
\mathbf{p} = \begin{bmatrix} R_{S\_1,2} \\ R_{S\_20,23} \\ R_{S\_21,23} \\ R_{S\_20,22} \\ R_{S\_21,22} \\ R_{S\_6,7} \\ R_{S\_3,4} \end{bmatrix}
\qquad
\mathbf{S} = \begin{bmatrix} 56.53 \\ 16.76 \\ 16.74 \\ 16.74 \\ 16.72 \\ 11.83 \\ 0.201 \end{bmatrix}^{T}
$$

The cumulative effect of all average values on performance degradation is negligible according to (5.13), and therefore they are all eliminated from $\mathbf{p}$.

The symmetry-constraint graph-search algorithm detected the following symmetric net pairs

$$(52, 53), (15, 16), (10, 11), (13, 14), (55, 56)$$

and the following device pairs

$$(M_1, M_2), (M_{20}, M_{22}), (M_{21}, M_{23}),$$
$$(M_{25}, M_{26}), (M_6, M_7), (M_{10}, M_{11}), (M_8, M_9).$$

Performance constraints are enforced on the maximum switching delay τ_D and on systematic offset V_{off}

$$\tau_D \leq 7 \text{ ns}$$
$$|V_{off}| \leq 1 \text{ mV} \tag{5.25}$$

In the first steps of layout, we assume that the nominal value of all parasitics is 0, i.e. $\mathbf{p}^{(0)} = [0 \ldots 0]^T$. Simulation yields a nominal value of the switching delay $\tau_D^{(0)} = 4$ ns and null offset. Therefore

$$\mathbf{K} = \begin{bmatrix} \tau_D \\ V_{off} \\ -V_{off} \end{bmatrix} \qquad \mathbf{K}(\mathbf{p}^{(0)}) = \begin{bmatrix} 4.0 \text{ ns} \\ 0.0 \text{ mV} \\ 0.0 \text{ mV} \end{bmatrix} \qquad \overline{\mathbf{\Delta K}} = \begin{bmatrix} 3.0 \text{ ns} \\ 1 \text{ mV} \\ 1 \text{ mV} \end{bmatrix}$$

As expected, sensitivity analysis shows that delay is sensitive to stray capacitances, while resistances and mismatch affect only offset

$$\mathbf{p} = \begin{bmatrix} C_{15} \\ C_{16} \\ C_{55} \\ C_{56} \\ R_{S\_1,2} \\ R_{S\_20,23} \\ R_{S\_21,23} \\ R_{S\_20,22} \\ R_{S\_21,22} \\ R_{S\_6,7} \\ R_{S\_3,4} \end{bmatrix} \qquad \mathbf{S} = \begin{bmatrix} 36 \text{ ps/fF} & 0.0 & 0.0 \\ 36 \text{ ps/fF} & 0.0 & 0.0 \\ 47 \text{ ps/fF} & 0.0 & 0.0 \\ 47 \text{ ps/fF} & 0.0 & 0.0 \\ 0.0 & 0.056 \text{ mV/}\Omega & 0.056 \text{ mV/}\Omega \\ 0.0 & 0.016 \text{ mV/}\Omega & 0.016 \text{ mV/}\Omega \\ 0.0 & 0.016 \text{ mV/}\Omega & 0.016 \text{ mV/}\Omega \\ 0.0 & 0.016 \text{ mV/}\Omega & 0.016 \text{ mV/}\Omega \\ 0.0 & 0.016 \text{ mV/}\Omega & 0.016 \text{ mV/}\Omega \\ 0.0 & 0.011 \text{ mV/}\Omega & 0.011 \text{ mV/}\Omega \\ 0.0 & 0.201 \text{ }\mu\text{V/}\Omega & 0.201 \text{ }\mu\text{V/}\Omega \end{bmatrix}^T$$

Because of symmetries, and since the nominal value of mismatch is 0, offset sensitivities in the positive and negative direction are equal. We use the following conservative minimum and maximum parasitic estimates

$$\min C = 1 \text{ fF}$$
$$\max C = 100 \text{ fF}$$
$$\min R = 0 \ \Omega$$
$$\max R = 50 \ \Omega$$

With these estimates, PARCAR computed the following set of parasitic bounds

$$\mathbf{p}^{(\mathbf{b})} = \begin{bmatrix} 71.96 \text{ fF} \\ 71.96 \text{ fF} \\ 78.52 \text{ fF} \\ 78.52 \text{ fF} \\ 1.0 \ \Omega \\ 7.4 \ \Omega \\ 7.4 \ \Omega \\ 7.4 \ \Omega \\ 7.5 \ \Omega \\ 19.9 \ \Omega \\ 49.5 \ \Omega \end{bmatrix}$$

Here the relation between sensitivity and tightness of bounds is evident. Only a few parameters critically affect the performance of this circuit and therefore need to be bounded tightly. In practice, only the mismatch between the source resistances in the differential pair and the mismatch between the two current mirrors (M_{20}, M_{23}) and (M_{21}, M_{22}) are responsible for offset.

5.3 PLACEMENT WITH ANALOG-SPECIFIC CONSTRAINTS

5.3.1 Placement Engine

PUPPY-A [39] is a macrocell-style placement tool based on simulated annealing (SA) [1]. In PUPPY-A, the cost function is a weighted sum of non-homogeneous parameters controlling parasitics, symmetries and device matching. Let s be a placement configuration, i.e. the set of the positions and rotation angles of all layout modules. The cost function is given by the following expression

$$\begin{aligned} f(\mathbf{s}) = \ & \alpha_{wl} f_{wl}(\mathbf{s}) + \alpha_a f_a(\mathbf{s}) + \alpha_{ov} f_{ov}(\mathbf{s}) + \alpha_{sy} f_{sy}(\mathbf{s}) \\ & + \alpha_{ma} f_{ma}(\mathbf{s}) + \alpha_{we} f_{we}(\mathbf{s}) + \alpha_{co} f_{co}(\mathbf{s}) \end{aligned}$$

where

- $f_{wl}(\mathbf{s})$ is the sum of wire length estimates over all the modules. Two estimation methods are available, one based on semi-perimeter and the other on pseudo-Steiner tree technique.

- $f_a(\mathbf{s})$ is the total area of the circuit. Space for routing is estimated with the *halo* mechanism described in [270]

- $f_{ov}(\mathbf{s})$ is the total overlapping area between cells.

- $f_{we}(\mathbf{s})$ is a measure of the discontinuity of well regions. This parameter is used only in device-level placement. It is given by the sum of the distances between devices that should lay within the same well or substrate region.

- $f_{sy}(\mathbf{s})$ is a measure of the *distance* between placement s and a symmetric configuration, given by the following expression:

$$f_{sy}(\mathbf{s}) = \sum (d(\mathbf{s})_i + \rho_i), \qquad (5.26)$$

where the sum is extended to all symmetric devices. Item $d(\mathbf{s})_i$ is the translation needed to bring the i-th cell to a symmetric position. The value of ρ_i is 0 if mirroring and/or rotation are not needed to enforce symmetry, otherwise it is set to 10.

- $f_{ma}(\mathbf{s})$ is a measure of the mismatch between circuit devices. Its definition is similar to the one of $f_{sy}(\mathbf{s})$:

$$f_{ma}(\mathbf{s}) = \sum (d_i^{(m)} + \rho_i),$$

where the sum is extended to all matched devices. Item $d_i^{(m)}$ is the translation needed to bring the i-th device inside an area of adequate matching characteristics with the other matched devices. This area can be user-specified or automatically computed as explained in Section 5.2. Parameter ρ_i and has the same meaning as in (5.26).

- $f_{co}(\mathbf{s})$ is a penalty function accounting for performance constraint violations. Its computation is key to our performance-driven approach. Estimates $\mathbf{p}^{(\mathbf{min})}$ and $\mathbf{p}^{(\mathbf{max})}$ of minimum and maximum interconnect capacitances and resistances are obtained on the ground of net length estimates and of the available routing layers. Using the linearized expression (5.3), performance degradation can be computed at each annealing iteration, and one of the following cases can apply.

  1. If the maximum degradation is within the specifications, that is

  $$\mathbf{S}\left[\mathbf{p}^{(max)} - \mathbf{p}^{(0)}\right] - \overline{\mathbf{\Delta K}} \leq 0$$

  no cost function penalty is imposed. In fact, in this case constraints (5.9) and (5.10) are met whatever values the parasitics assume.

  2. Otherwise, $f_{co}(\mathbf{s})$, is a function of the constraint violation $\mathbf{\Delta K}(\mathbf{p})$.

In case 2, the penalty term is computed as follows. Let us consider first the performance degradation due to deterministic parasitics which result from the a set of alternative interconnect implementations. Let $\Delta K_i{}^{(min)}$ and $\Delta K_i{}^{(max)}$ be respectively the minimum and maximum values that the degradation of performance K_i can assume with different values of parasitics.

$$f_{co} = \sum_{i=1}^{N_k} C_i$$

where C_i is given by

$$C_i = \begin{cases} 0, & \text{if } \Delta K_i{}^{(max)} \leq \overline{\Delta K_i} \\ \Delta K_i{}^{(max)} - \overline{\Delta K_i}, & \text{if } \Delta K_i{}^{(min)} < \overline{\Delta K_i} \leq \Delta K_i{}^{(max)} \\ (S_r + 1)(\Delta K_i{}^{(max)} - (S_r \rho_c + 1)\overline{\Delta K_i}), \\ \qquad \text{if } \overline{\Delta K_i} \leq \Delta K_i{}^{(min)} \end{cases}$$

ρ_c is the ratio between the maximum and minimum value of the minimum-width unit-length substrate capacitance or resistance of interconnections on the available routing layers. If $S_r \gg 1$ then the values of C_i for feasible and infeasible placements differ by at least one order of magnitude. In our implementation, $S_r = 10$.

Let us consider now the case in which some or all parasitics are non-deterministic. Let us assume that these parasitics can be represented as random variables π_j with a well-defined statistical behavior, i.e. with a bounded, continuous statistical distribution. Moreover, assume that performance is sufficiently linear in the range of values assumed by each variable with high probability[2]. In case π_j are statistically independent, then performance degradation ΔK_i can be represented in terms of its mean μ and variance σ^2 as

$$\begin{aligned} \mu_{\Delta K_i} &\approx \sum_{j=1}^{n_i} S_{i,j} \mu_{\pi_j}, \\ \sigma^2_{\Delta K_i} &\approx \sum_{j=1}^{n_i} |S_{i,j}|^2 \sigma^2_{\pi_j}, \end{aligned} \qquad (5.27)$$

where μ_{π_j} and $\sigma^2_{\pi_j}$ are the mean and variance of parasitic π_j, respectively. Hence, the following set of constraints need be imposed

$$\begin{aligned} \mu_{\Delta K_i} &\leq \overline{\Delta K_i} \\ \sigma^2_{\Delta K_i} &\leq \overline{\sigma^2}_{\Delta K_i}, \end{aligned}$$

where the term $\overline{\sigma^2}_{\Delta K_i}$ represents the specification on the maximum allowable variance on degradation ΔK_i.

---

[2] One can actually set bounds on the minimum linearity over a given range.

In the case of statistical dependence, the set of variables $\{\pi_j\}$ can be translated onto a new set $\{\pi'_j\}$ of uncorrelated variables. If we assume the original variables to have a Gaussian distribution, then the variables in set $\{\pi'_j\}$ are also statistically independent and hence (5.27) can be used to approximate the mean and variance of ΔK_i by replacing the terms π_j and $S_{i,j}$ with π'_j and $S'_{i,j}$, respectively. The term $S'_{i,j}$ is defined as the partial $\partial K_i / \partial \pi'_j$, while the new set of variables $\{\pi'_j\}$ is derived as follows. Let us assume that the original variables π_j be Gaussian. Moreover, assume that variance-covariance matrix $\mathbf{A}$ associated with π_j is known. Then, $\mathbf{A}$ is by construction a positive-definite, symmetric $n_i \times n_i$ square matrix and hence it can decomposed as

$$\mathbf{A} = \mathbf{LDM}^T,$$

where $\mathbf{L}$ and $\mathbf{M}$ are square orthogonal matrices and $\mathbf{D}$ is a diagonal matrix. Then, if every new variable π'_j is represented as a linear combination of the original ones, in a compact form

$$\pi' = \mathbf{L}\pi,$$

where π and π' are $n_i \times 1$ vectors representing the random variables, one can show that, due to the orthogonality of $\mathbf{L}$ and $\mathbf{M}$, $\{\pi'_j\}$ are necessarily uncorrelated and, since Gaussian, statistically independent as well. Hence, $\sigma^2_{\Delta K_i}$ can be computed by replacing the terms $\sigma^2_{\pi_j}$ with $\sigma^2_{\pi'_j}$ and $S_{i,j}$ with $S'_{i,j}$ in (5.27).

- $\alpha_{wl}, \alpha_a, \alpha_{ov}, \alpha_{we}, \alpha_{sy}, \alpha_{ma}$ and α_{co} are non-negative weights. Their initial default values are adjusted dynamically during the algorithm using heuristics so that, at the beginning of the annealing, area and wire length dominate in the expression of the cost function, then their importance decreases progressively, until at low temperatures overlaps, symmetries, and constraint violations become dominating.

5.3.2 Abutment and Control of Junction Capacitances

Device abutment during placement is useful to reduce interconnect and junction capacitances, and to obtain substantial gain in area. It can also be used to merge the diffusion regions of MOS transistors or of other components, such as capacitors, BJTs etc. In PUPPY-A, abutment is obtained in two different ways. The first is by dynamic device abutment (similar to the approach in KOAN [57]), performed by PUPPY-A during the annealing algorithm. In PUPPY-A, dynamic abutment is driven by parasitic constraints, as well as by area and wiring considerations. Instead of randomly choosing the devices to merge, the algorithm operates first on the nets whose parasitics are critical for performance constraints. The second is through the *stack generator* LDO, which efficiently builds stacks containing transistors all with the same width.

LDO [193] implements stacks of folded or interleaved MOS transistors sharing their drain and source diffusions. Dense layouts can be achieved with this approach, the junction capacitances associated with shared diffusions being minimized. Moreover, matching between transistors decomposed into elements stacked together is usually good, in particular if the elements are interleaved. Because of the regularity of these structures, routing is usually dense with this layout style.

The target of the stack generator can be summarized as follows:

1. obtain maximally compact stacks, so that the area occupied by the devices is minimum;

2. keep all critical capacitances at their minimum value by exploiting the abutment of source/drain diffusion areas;

3. provide control over device matching, so that critically matched devices can be decomposed into interleaved elements, and common-centroid structures are obtained when symmetry constraints are enforced;

4. provide control over net length, by conveniently distributing the elements within the stacks.

LDO is based on an algorithm exploiting the equivalence between stack generation and path partitioning in the circuit graph. The algorithm is guaranteed to find all optimum stacked configurations, according to an optimality criterion defined by a cost function, which takes into account parasitic criticality, matching constraints and device area. The stack generation algorithm is based on a two-phase approach, working on the *circuit graph*, i.e. a graph whose nodes are circuit nets, and whose edges are MOS transistors. In the first phase, a dynamic programming procedure generates all possible paths in the circuit graphs, namely in the connected subgraphs whose nodes have no more than two adjacent edges. The second phase explores the compatibility between all paths. By solving a clique problem, an optimum set of paths is selected, which minimizes the cost function and contains all the transistors of the circuit. More details on the algorithm and its implementation can be found in [193].

In LDO, not only are symmetries fully taken into account, but they have proved effective to reduce the computational complexity by limiting the size of the search space, while preserving the admissibility of the algorithm (i.e. the optimum solution is always found). In practice, the higher the number of symmetry constraints, the faster the algorithm runs.

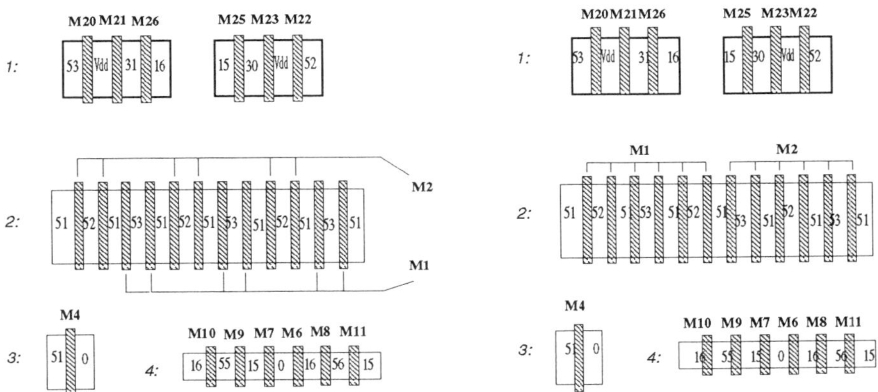

Figure 5.4 Stack implementations of COMPL

By abutting elemental transistors into one stack, their source/drain regions are merged, thus reducing effectively their junction capacitances. The cost function driving LDO tries to minimize the most critical capacitances, according to the tightness of their bounds. Junction capacitances are the only parasitics that can be directly controlled by LDO, because they are directly influenced by the shape and inner organization of the stacks. Routing parasitics, such as interconnect stray resistances and capacitances, can be controlled effectively only after the placement phase. This limitation can be overcome by simultaneously generating and placing the stacks. This has been achieved by means of an annealing move-set extension, to include a move called *alternative solution swap*, which selects randomly a module in the circuit, and swaps it with one of its alternative implementations found by LDO. The criterion whether to accept the move is based on the usual annealing scheme [41].

Example

Consider once again the clocked comparator COMPL. Many possible stack implementations exist for this circuit. Two of such possible solutions are shown in Figure 5.4. All transistors have been grouped in four subcircuits, according to their channel widths, their matching requirements and bulk nets. Only transistors belonging to the same subcircuit can belong to the same stack. The two solutions only differ by the implementation of the stack containing the input differential pair. In the first realization they are interleaved in a common-centroid pattern, which minimizes device mismatch, but usually requires a considerable area overhead, due to the complex routing required. The second solution is symmetric, but without the common-centroid structure. The

nets	without abutment	with abutment	% reduction
15,16	41.2 fF	34.4 fF	16.5%
55,56	13.4 fF	6.6 fF	51%

Table 5.2   Capacitance calculations in stacks produced by LDO

choice between such alternative realizations is left to the user or it can be made automatically during the placement phase on the ground of area and routing considerations. In both solutions, critical nets 55, 56, 15, 16, whose capacitance toward the substrate strongly influences the comparator speed, have been kept in internal positions when possible. Their capacitances are reported in Table 5.2. In both cases stack abutment yielded a reduction of net capacitance. Such a reduction can be exploited to improve the flexibility of the routing stage. For example, consider nets 55 and 56. Abutment allowed each of them to be reduced by more than 6.6 fF, which in our process is the capacitance of a 136 μm-long minimum-width metal-1 wire. Therefore the router is allowed to draw longer wires for the sensitive nets, thus increasing the success rate and the robustness of the entire layout synthesis.

These capacitance values constitute new nominal values and better lower limits, and can be used to compute a new set of bounds. By using these values

$$\begin{aligned}
C_{15}^{(min)} &= C_{15}^{(nom)} = C_{16}^{(min)} = C_{16}^{(nom)} = 34.4 \text{ fF} \\
C_{55}^{(min)} &= C_{55}^{(nom)} = C_{56}^{(min)} = C_{56}^{(nom)} = 6.6 \text{ fF} \\
\max C &= 100 \text{ fF} \\
\min R &= 0 \text{ } \Omega \\
\max R &= 50 \text{ } \Omega
\end{aligned} \tag{5.28}$$

we obtain the following arrays

$$\mathbf{K}(\mathbf{p}^{(0)}) = \begin{bmatrix} 5.5 \text{ ns} \\ 0.0 \text{ mV} \\ 0.0 \text{ mV} \end{bmatrix} \qquad \overline{\mathbf{\Delta K}} = \begin{bmatrix} 1.5 \text{ ns} \\ 1 \text{ mV} \\ 1 \text{ mV} \end{bmatrix}$$

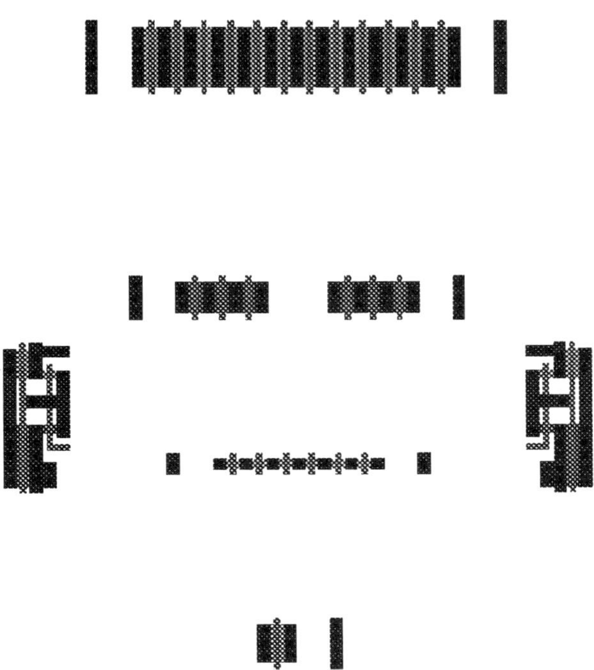

Figure 5.5 Placed layout for COMPL

Here the delay degradation, due to the insertion of junction capacitances, is apparent. The next set of bounds found by PARCAR is the following

$$\mathbf{p}^{(b)} = \begin{bmatrix} 67.1 \text{ fF} \\ 67.1 \text{ fF} \\ 48.9 \text{ fF} \\ 48.9 \text{ fF} \\ 1.0 \ \Omega \\ 7.4 \ \Omega \\ 7.4 \ \Omega \\ 7.4 \ \Omega \\ 7.5 \ \Omega \\ 19.9 \ \Omega \\ 49.5 \ \Omega \end{bmatrix} \tag{5.29}$$

Notice that all bounds on critical capacitances have been lowered, because the degradation allowed to delay is smaller than in the previous step. In fact half of the degradation allowed at the beginning of the layout design has been introduced by junction capacitances alone, and the remaining half will be available to the remaining tools (i.e. placement and routing tools). The placement of Figure 5.5 was obtained with the set of bounds (5.29). After placement, estimates of the minimum values of all critical parasitics can be drawn, taking into account the junction capacitances of all terminals and the estimated minimum length of interconnections between terminals.

$$\mathbf{p}^{(0)} = \mathbf{p}^{(min)} = \begin{bmatrix} 10.1 \text{ fF} \\ 10.1 \text{ fF} \\ 51.0 \text{ fF} \\ 51.0 \text{ fF} \\ 0.0\ \Omega \\ 0.0\ \Omega \\ 0.0\ \Omega \\ 0.0\ \Omega \\ 0.0\ \Omega \\ 0.0\ \Omega \\ 0.0\ \Omega \end{bmatrix} \tag{5.30}$$

5.4 ROUTING WITH ANALOG-SPECIFIC CONSTRAINTS

5.4.1 Channel Routing

In the channel router ART [49], the two-layer gridless channel routing problem [31] is represented by a *vertical-constraint graph (VCG)* whose nodes correspond to the horizontal segments of a net or subnet. An undirected edge links two nodes if the associated segments have a common horizontal span. A directed edge links two nodes if one segment has to be placed above the other because of pin constraints. The weight of an edge is the minimum distance between the center lines of two adjacent segments. Hence the channel routing problem is formulated as the problem of directing all the undirected edges so as to minimize the longest directed path in the VCG. The length of such path corresponds to the channel width. Overconstraints can be solved by assigning to each net more than one node in the VCG. However, with this approach we introduce additional capacitive couplings due to the wire jogs. In the current implementation of ART, one VCG node is supported for each net.

In ART, all parasitic bounds are mapped into constraints for the VCG. Within a channel, nets provide two different contributions to cross-coupling capacitance: crossover

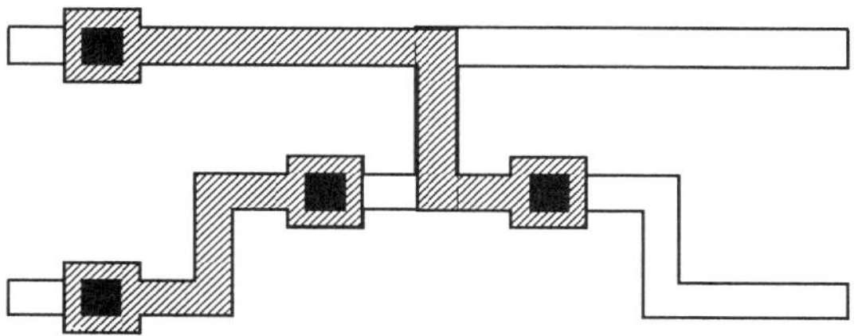

Figure 5.6 Symmetry net-crossing in ART

capacitances between overlapped orthogonal wire segments, and capacitances between segments running parallel to each other. Both depend on the distribution of terminals along the channel edges. Unavoidable crossovers can be determined directly on the ground of the terminal positions. If such a crossover is detected, it introduces a lower bound for the cross-coupling capacitance between the corresponding nets. The coupling between horizontal adjacent edges is controlled by their minimum separation, and therefore it is proportional to the weight of the corresponding edge in the VCG. This contribution can be theoretically reduced at will by inserting sufficient space, or by exploiting the shielding effect due to other wires. Shielding nets can be inserted on purpose, if the presence of a further wire segment in the channel is more convenient in terms of area than extra spacing. In ART, this is automatically carried out by adding a new node and edges to the VCG.

Perfect mirror symmetry can be achieved when symmetric nets are restricted to different sides of the symmetry axis (i.e. they do not cross). If the horizontal spans of a pair of symmetric nets intersect the symmetry axis, perfect mirror symmetry cannot be achieved. However, good parasitic matching can be obtained between the nets with the technique illustrated in Figure 5.6. A "connector" allows two symmetric segments to cross over the axis. Resistances and capacitances of the two nets are matched, because for each one the connector introduces the same interconnect length, the same number of corners on each layer, and the same number of vias. Only coupling capacitances with other nets running close to the connector will suffer slight asymmetries.

5.4.2 Area Routing

ROAD [194] is a maze router based on the A* algorithm [55], using a relative grid with dynamic allocation. For each net, the path found by the maze router is the one of minimum length. If a cost function is defined on the edges of the grid, the path found is the one minimizing the integral of the cost function. In ROAD, the cost function is a weighted sum of several non-homogeneous items. Let $\mathcal{N}$ be the set of all nets. On a given grid edge x with length $L(x)$, on layer l, the cost function for a net $N \in \mathcal{N}$ has the following form

$$F(x) = L(x) \cdot \left(1 + \frac{Cr(x)}{Cr_0} + w_R \frac{R_u(l)}{R_0} + w_{C_u} \frac{C_u(l)}{C_0} + \right.$$
$$\left. + \frac{1}{C_P} \sum_{n \in \mathcal{N} - \{N\}} w_{C_n} C_n(x) \right) \qquad (5.31)$$

where

- $Cr(x)$ is a measure of local area crowding. It is computed in a simplified form, by giving overcongested areas steep cost function "hills," which prevent future wires from crossing these areas. Area crowding $Cr(x)$ is given by

$$Cr(x) = \begin{cases} 0, & \text{if } R \leq 1 \\ Cr_{max} & \text{if } R > 1 \end{cases} \qquad (5.32)$$

  where Cr_{max} is a large constant (the height of the "hills"), and R is the ratio between the needed room for the new wire which has to be built, and the room available on the sides of edge x

$$R = \frac{\text{needed room}}{\text{available room}}.$$

- $R_u(l)$ is the resistance of a minimum-width unit-length wire segment on layer l.

- $C_u(l)$ is the capacitance to bulk of a minimum-width unit-length wire segment on layer l. The model is described in Appendix.

- $C_n(x)$ is the capacitance between a unit-length wire segment located across edge x and the wire implementing net n. The model is described in Appendix.

- w_{R_u}, w_{C_u} and w_{C_n} are weights regulating the relative importance of each item.

- Cr_0, R_0, C_0, C_P are reference parameters providing dimensional homogeneity to the addenda and a meaning to their comparison.

Weights provide an efficient way to limit the magnitude of critical parasitics. Performance sensitivities to parasitics are used to generate the weights for the cost function driving the area router. The contribution of a parasitic to performance degradation is proportional to the sensitivity and inversely proportional to the maximum variation range allowed to that performance. The weight w_j associated to parasitic p_j is defined as

$$w_j = \sum_{i=1}^{N_k} \left(\frac{S_{i,j}^-}{\Delta K_i^-} + \frac{S_{i,j}^+}{\Delta K_i^+} \right) P_0,$$

where P_0 is a normalization factor, such that, if sensitivities are not all zero, at least one weight is set to 1, and the others are all between 0 and 1. The dimensional unit of P_0 ($\Omega, V, F, \ldots$) depends on the parasitic type. With this definition, each item in (5.30) can be interpreted as the contribution to performance degradation due to one of the parasitics introduced by the wire segment routed along edge x of the grid.

The routing schedule is determined with a set of heuristic rules set up and tuned with experimental tests. The higher the number of constraints on a net, the higher is its priority. We define a number of *properties* that a net can have, for instance symmetry, belonging to the class of supply nets, of clocks etc. The priority of net n is given by the expression

$$Pr(n) = \sum_{j} a_j \mathrm{Prop}_j(n) + \sum_{i=1}^{N_p} a_{p_i} w_{p_i}$$

where $\mathrm{Prop}_j(n)$ is 1 if net n has the j-th property, and 0 otherwise. Parameters w_{p_i} are the same parasitic weights used to define the cost function (5.30), while a_j and a_{p_i} are *priority weights* expressing the importance of each property. Priority weights are assigned in such a way that maximum priority is given to symmetric nets, followed by supply nets and then by nets with tight electrical requirements. The most difficult nets are routed first, and the unrouted ones are less and less critical as the circuit crowding increases. If two nets have the same priority, the shorter one is routed first.

After performing the weight-driven routing, parasitics are extracted and performance degradation is estimated and compared with its specifications. If constraints (5.8) are not met, the weights of the most sensitive parasitics are raised and routing is repeated. When the weights of all sensitive parasitics hit their maximum value (that is 1), iterations stop. This means that even considering maximum criticality for the sensitive parasitics, routing is not possible on the given placement, without constraint violations. In this case, the circuit placement needs to be generated again, using a wider range of variation for the detected sensitive parasitics.

Symmetry Management

ROAD is able to find symmetric paths for differential signals with symmetric placements, even in the presence of a non-symmetric distribution of terminals. The algorithm, described in detail in [194], is illustrated in Figure 5.7. Let us assume without

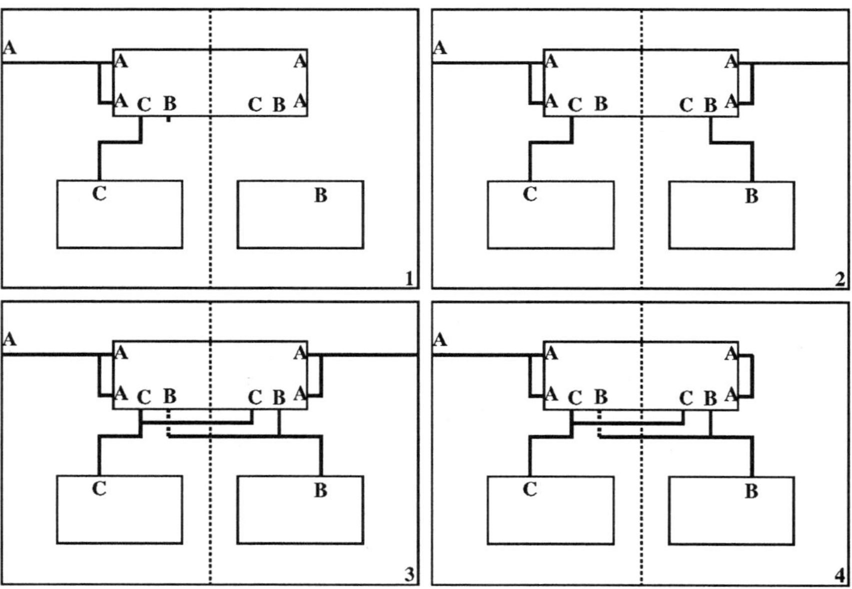

Figure 5.7 Symmetry enforcement in ROAD

loss of generality that the symmetry axis is vertical and it splits the circuit into a left half and a right half. If the placement is not perfectly symmetric, we consider the outline determined by the union of real obstacles and *virtual obstacles* obtained by mirroring each obstacle with respect to the symmetry axis. First every net is built considering only the terminals located on the left side of the symmetry axis or on the border of the wiring space, that is not contained by any virtual image of an obstacle. The wire segments defined in this way are called *left-side segments.* Then, each left-side segment is mirrored with respect to the symmetry axis. Next, the routing is extended to cover the portions of area occupied by virtual obstacles, but not by real obstacles. The segments whose existence is not required for the full net connectivity are pruned. Only those branches of non-symmetric nets should be pruned, that do not cross or run close to symmetric nets.

Electrostatic Shields

Decoupling based only on wire spacing can increase excessively area, and this can be avoided by inserting wire stubs, connected to a virtual ground, shielding critically coupled wires. Shields are built after all wires have been routed. Given two wires to be decoupled, first a grid node between each pair of parallel segments of the two wires is found, or generated by dynamic grid allocation. On each of these node, area congestion is computed with expression (5.32), and a terminal is defined wherever congestion is sufficiently low, i.e. where $Cr(x) < Cr_{max}$. Next, a new wire is routed through all these terminals, and connected (with null weight on resistive constraints) to the proper ground node. If local congestion does not allow to create a suitable shield, then we pass to the re-route phase as described above.

Example

Consider the clocked comparator COMPL. After the placement step, the nominal values of all parasitics have been updated as shown in (5.30). With these new nominal values, the least capacitive interconnect among the terminals of the critical nets would give (by simulation) a total delay of 5.9ns

$$\mathbf{K}(\mathbf{p}^{(0)}) = \begin{bmatrix} 5.9\,\text{ns} \\ 0.0\,\text{mV} \\ 0.0\,\text{mV} \end{bmatrix} \qquad \overline{\mathbf{\Delta K}} = \begin{bmatrix} 1.1\,\text{ns} \\ 1\,\text{mV} \\ 1\,\text{mV} \end{bmatrix}$$

With the high-level constraints (5.25), now only 1.5 ns of delay degradation are allowed to the router, of which 0.4 ns have been recognized as being unavoidable with this placement. Therefore tight bounds will have to be enforced by the router. In fact PARCAR now requires the following bounds

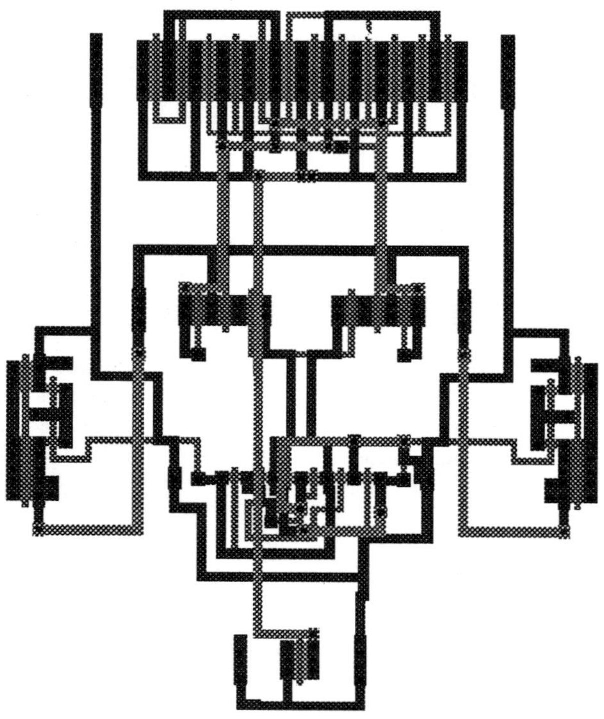

Figure 5.8 Routed layout of COMPL

$$\mathbf{p}^{(\mathbf{b})} = \begin{bmatrix} 76.5 \text{ fF} \\ 76.5 \text{ fF} \\ 39.3 \text{ fF} \\ 39.3 \text{ fF} \\ 1.0 \ \Omega \\ 7.4 \ \Omega \\ 7.4 \ \Omega \\ 7.4 \ \Omega \\ 7.5 \ \Omega \\ 19.9 \ \Omega \\ 49.5 \ \Omega \end{bmatrix}$$

Comparing the capacitive bounds with the junction capacitances (5.28) computed after module generation, it is evident that 42.5 fF are available for routing nets 15 and 16, and 32.7 fF are available for nets 55 and 56. The layout routed by ROAD is shown in Figure 5.8. Extraction results for capacitances yield

$$C_{15} = 70.3 \text{ fF}$$
$$C_{16} = 70.4 \text{ fF}$$
$$C_{55} = 20.6 \text{ fF}$$
$$C_{56} = 18.0 \text{ fF}$$

Simulation results after extraction of the routed layout give a delay of 6.5 ns and an offset of 756 μV.

5.5 COMPACTION

SPARCS-A [79] is a mono-dimensional constraint-graph (CG) longest-path compactor, implementing algorithms to enforce symmetry and parasitic constraints. The role of compaction in the constraint-driven approach is important for two reasons:

1. Constraints enforced by the previous layout steps, such as parasitic bounds, symmetries, and shields, should not be disrupted for the sake of area minimization. The compactor must be able to respect and if necessary to enforce such constraints.

2. The compactor can recover design-rule errors and constraint violations. Hence the requirements on placement and routing in terms of constraint enforcement can be relaxed. Since compaction has generally higher computational efficiency than routing, the overall CPU cost of layout design can be substantially reduced. Thus, in addition to reducing chip area, compaction also improves the efficiency

and robustness of the entire analog synthesis process by permitting the use of more aggressive techniques during placement and routing.

The algorithm implemented in SPARCS-A takes advantage of the high speed of the CG-based technique to provide a good starting point to a Linear Programming (LP) solver. Mono-dimensional compaction is iterated alternatively in the two orthogonal directions, until no area improvement is achieved. The algorithm used in each iteration is the following:

1. With the CG technique, solve the spacing problem without symmetry constraints.

2. Use the simplex linear programming algorithm to solve the symmetry constraints, using the CG solution as initial starting point.

3. Round off the coordinates of the elements laying out of the critical path

4. Verify that all constraints are satisfied

Using the CG solution as starting point is key to a significant speed-up in the solution of the linear problem. Compared to previous approaches [233] solving compaction using an LP solver, this algorithm represents a substantial improvement. In fact in our case the LP solver starts from a feasible configuration which is already close to the final solution. The range of cases that can be managed with acceptable computational complexity is therefore significantly broadened [192]. Control over cross-coupling capacitances is enforced by modifying the constraint-graph before computing the longest path. Proper distances between parallel interconnection edges are kept to maintain cross-coupling capacitances below their bounds. This is achieved by employing a heuristic which adds extra spacing between wire segments, based on the need for decoupling and on their length, which has a direct impact on the overall area. The procedure implementing this heuristic is the following:

procedure modify-graph
/* Purpose: Add constraints to improve capacitive decoupling*/
/* Since this procedure is called only if some performance violation*/
/* has been found, we know that at least one bound has been exceeded.*/
foreach cross-coupling C_j **such that** $C_j > C_j^b$
    let δ_j = current min. distance between any two parallel segments contributing to C_j
    $\delta_j = \delta_j + mindist(C_j, C_j^b)$
    foreach pair P_i of parallel segments
        let d_i = current minimum distance between the segments of pair P_i;
        if $d_i < \delta_j$ **then**
            Add constraint to graph requiring $d_i \geq \delta_j$ between the segments

C_j indicates the j-th cross-coupling capacitance, and C_j^b is the bound on its maximum value. Function $mindist(C_j, C_j^b)$, which depends on the model used for capacitances (see Section 6.1.2), returns the minimum distance increment to add between parallel segments of the j-th pair of wires, to reduce their cross-coupling capacitance from C_j to C_j^b. For each cross-coupling C_j exceeding its bound C_j^b, δ_j is the minimum distance to be kept between parallel segments of the j-th pair of wires. The distance increment is a function of the parasitic bound violation: the bigger the violation, the wider the extra spacing added. Notice that in procedure **modify-graph**, spacing is added not only between the nearest segments, but also between all the segment pairs whose distance is less than δ_j.

The spacing step implemented by procedure **modify-graph** can introduce overconstraints making the graph unsolvable. An example where this situation might occur is illustrated in Figure 5.9. Two wire segments are connected to terminals A and B, whose relative position is fixed with respect to the instance of a sub-cell. An overconstraint, due to a positive loop in the CG, is generated if the spacing required between the segments is $\delta_j > D - W_1 - W_2$. When a positive loop is detected, a *pruning* procedure is invoked, which removes the newly-added spacing constraints contained in the positive-weight loops. In such situation, the task of decoupling the two nets is left to the remaining segment pairs. If a feasible solution involving the remaining segment pairs does not exist, an error is reported because the constraint cannot be met.

One of the main advantages of the longest-path compaction algorithm is that right after each compaction step it provides the exact value of the minimum layout pitch. This can be exploited to add geometric constraints together with electrical performance specifications. Let us consider, without loss of generality, a horizontal compaction step. The pitch W of the longest path can be checked against the maximum size W_{max} allowed to the circuit width. If it is smaller, the difference between them is the maximum amount by which the vertical parallel wire segments belonging to the

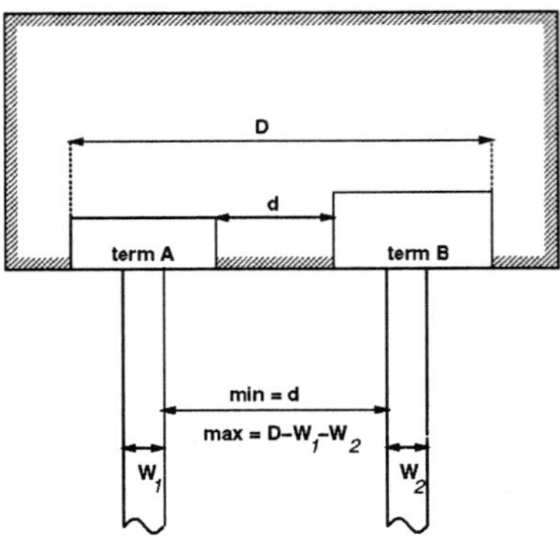

Figure 5.9   Spacing of parallel lines in SPARCS-A

longest path itself can be brought apart from each other. Otherwise, all such pairs of wire segments must be kept at their minimum distance. This corresponds to an additional constraint:

$$\sum_i \delta_i \leq \min(0,\ W_{max} - W_{lp}), \tag{5.33}$$

where the sum is extended to all the vertical segment pairs lying on the longest path. Aspect ratio constraints can be reduced to absolute-size constraints by considering, at each step, the pitch of the layout in the orthogonal direction as fixed.

Example

Consider once more the clocked comparator COMPL. As shown at the end of Section 5.4, after routing the problem instances are the following:

$$\mathbf{p}^{(0)} = \begin{bmatrix} 70.3 \text{ fF} \\ 70.4 \text{ fF} \\ 20.6 \text{ fF} \\ 18.0 \text{ fF} \\ 0.0\ \Omega \\ 0.0\ \Omega \\ 0.0\ \Omega \\ 0.0\ \Omega \\ 0.0\ \Omega \\ 0.0\ \Omega \\ 0.0\ \Omega \end{bmatrix} \qquad \mathbf{K}(\mathbf{p}^{(0)}) = \begin{bmatrix} 6.2 \text{ ns} \\ 756\ \mu\text{V} \\ -756\ \mu\text{V} \end{bmatrix}$$

$$\overline{\Delta \mathbf{K}} = \begin{bmatrix} 0.8 \text{ ns} \\ 244 \text{ } \mu\text{V} \\ -244 \text{ } \mu\text{V} \end{bmatrix}$$

The solution to the constraint-generation problem is

$$\mathbf{p}^{(\mathbf{b})} = \begin{bmatrix} 87.7 \text{ fF} \\ 87.8 \text{ fF} \\ 32.8 \text{ fF} \\ 28.4 \text{ fF} \\ 1.0 \text{ } \Omega \\ 1.0 \text{ } \Omega \\ 1.0 \text{ } \Omega \\ 1.0 \text{ } \Omega \\ 1.0 \text{ } \Omega \\ 5.2 \text{ } \Omega \\ 49.3 \text{ } \Omega \end{bmatrix}$$

Notice that now resistive mismatch has become critical because of the shrunk margin allowed to offset degradation. Of the capacitive parasitics, C_{55} and C_{56} have been recognized as more critical than the others, and their bounds have been further tightened with respect to their previous values used to drive the router. Other bounds on C_{15} and C_{16} have been relaxed as a consequence. The layout produced with this set of bounds is shown in Figure 5.10. Capacitive extraction from this layout yields the following values

$$C_{15} = 73.9 \text{ fF}$$
$$C_{16} = 75.2 \text{ fF}$$
$$C_{55} = 18.8 \text{ fF}$$
$$C_{56} = 17.2 \text{ fF}$$

Simulation showed that in the compacted layout performance specifications were met with an offset of 743 μV and a delay of 6.7 ns.

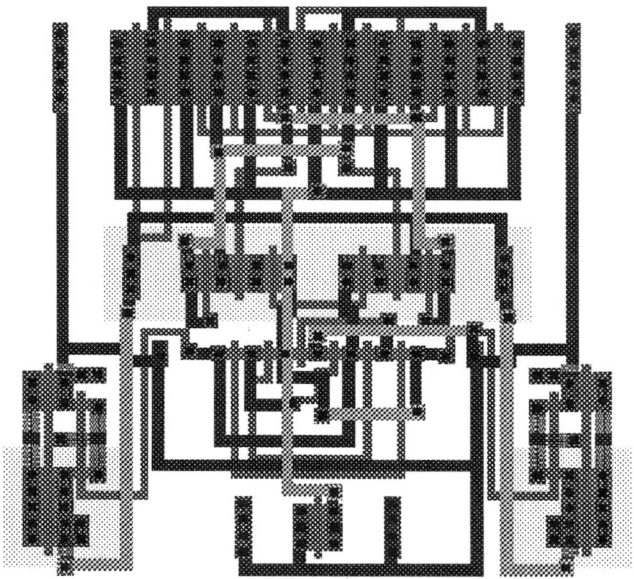

Figure 5.10   Compacted layout for COMPL

6

BOTTOM-UP VERIFICATION

Constraint-driven layout synthesis concludes the top-down process. At this point in general a detailed *bottom-up verification* phase is necessary to ensure the correctness of the design. Since behavioral model parameters have to be "estimated" at the higher levels of the hierarchy during the top-down process, bottom-up verification is essential to *fully* characterize the behavioral models together with interconnects and parasitics.

Bottom-up verification starts at the layout level with full extraction of the circuit including all *relevant* parasitics. Due to the complexity of mixed-signal ICs, layout extraction needs to be carried out in such a way that details which introduce no measurable effect on the performance measures are ignored. The resulting schematic of the complete circuit is therefore much smaller and becomes tractable by the means of a reasonably efficient and accurate analysis. These issues are addressed in Section 6.1. The need for parasitic schematic simplification has become even more acute with the emergence of substrate conduction as a major performance degradation effect. In Section 6.2 the problem of efficient substrate extraction is addressed and the techniques for handling large sets of contacts are outlined.

After layout a detailed extraction is performed and a technique known as *hierarchical behavioral/circuit simulation* is used to verify the functionality and the performance of the whole system in the presence of the extracted layout parasitics. Loading effects and interactions between components, which might have been ignored during the top-down process are accounted for during this phase. The behavioral/circuit simulation techniques used in bottom-up verification are the same ones which were used for early verification, design space exploration and architectural mapping during the top-down process. Hierarchical behavioral/circuit simulation is carried out starting with *transistor-level* verification and performance characterization of all the components of the system that can be *handled* with transistor-level simulation. A SPICE-like simulator (providing the types of analyses discussed in Section 3.1.1) is used for most verification tasks at this level.

A very critical problem in analog and mixed-signal design is related to transistor-level verification of circuit performance in the presence of *electrical noise*, i.e. small current and voltage fluctuations, such as thermal, shot and flicker noise, generated within the integrated circuit devices. For example, in the wireless communications market, it is desirable to build *low power* and fully *integrated* RF communications devices. The push for lower power solutions increases the relative effect of inherent electrical noise on the performance of the analog RF/IF signal processing sub-systems of RF communications devices, since there is a definite trade-off between low power and low noise performance. Accurate transistor-level prediction of the noise performance of the analog components (e.g. mixers, oscillators) in an RF transceiver system is crucial to ensure that the overall system will work correctly. In Section 6.3, we present a transistor-level noise simulation algorithm for nonlinear circuits for the verification of the analog circuit performance in the presence of electrical noise. It is crucial that noise simulation for final verification is performed on the circuit with fully extracted parasitic capacitors and resistors from the layout.

The whole system, or even a part of the system, is often too large and complex for complete transistor-level verification. Hence, transistor-level simulation (i.e. SPICE simulation, noise simulation described in Section 6.3) is only used to verify the performance of the small components at the bottom of the hierarchy. Then behavioral model parameters are extracted from transistor-level performance characterization results to be used in the behavioral simulations to verify the performance at a higher level of hierarchy for a larger portion of the system. This bottom-up verification process is terminated when the performance of the whole system is verified with behavioral simulations. The bottom-up verification will end with success unless we have ignored some important second-order effects in the behavioral models we have used in the top-down design process.

6.1 PARASITIC EXTRACTION AND SCHEMATIC SIMPLIFICATION

In this section the techniques used during the physical verification phase are discussed. The extraction accuracy is carefully controlled during the process so as to minimize the time used in deriving parasitic information associated with objects which are not critically degrading performance. Critical parasitics, on the contrary, are characterized

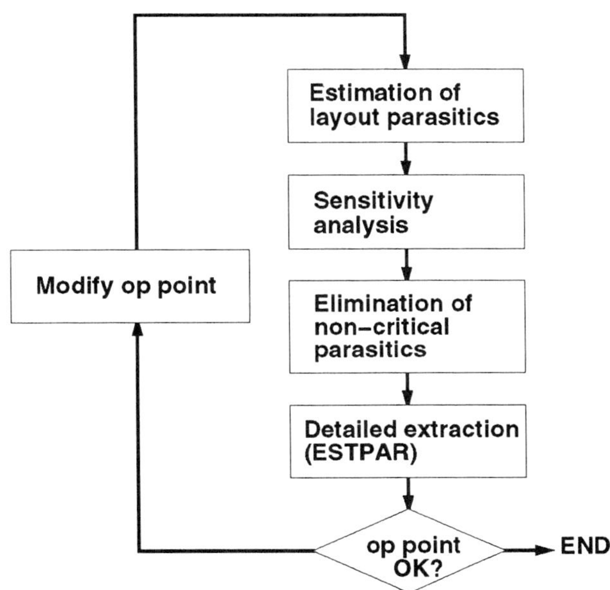

Figure 6.1 Methodology of selective extraction

in full detail using *ad hoc* analytical models generated for a number of common layout configurations.

Figure 6.1 shows the flow graph of the extraction methodology. In the first phase, a conservative estimate of all parasitics is computed. For each pair of terminals in the circuit the line resistance is computed as if the worst-case routing scheme were used under the highest possible congestion conditions. Capacitances to substrate are computed similarly, whereas analytical models [52] are used to accurately account for fringing effects. Capacitive cross-coupling is always assumed to take place at least once between any pairs of nets.

In the second phase, performance sensitivities are efficiently computed using the methods discussed in Chapter 5. From these data, performance sensitivities with respect to parasitic mismatch are also computed. Based on the previously obtained data, an integer program is constructed and solved using either standard methods or *ad hoc* heuristics developed by us. During this phase a list of critical paths, candidates for extraction, is built.

In the final phase all critical parasitics are extracted in detail. Analytical models for accurate computation of line-to-substrate and line-to-line capacitance extraction are used, thus ensuring that all important parasitic effects are accounted for. Assuming that the region of operation for the circuit is maintained in the neighborhood of the nominal operating point, this procedure ensures that in a selectively extracted layout, performance degradations are within the limits imposed by the constraints. In case of strongly nonlinear circuits a verification phase at the end of the selective extraction must be performed in order to check whether the new operating point is close enough to the previous one. If this condition is not satisfied the process must be iterated using the new operating point for the sensitivity analysis.

6.1.1   Schematic Simplification

The aim of *schematic simplification* is the reduction of complexity of the extracted schematic. This is often desirable for speeding-up circuit simulation and verification in large systems. In this section a constraint-based strategy is presented for achieving simplification and some considerations on the applicability and resulting accuracy are discussed. The essential idea underlying the method consists of identifying those parasitics which can be neglected, keeping the accuracy of the circuit simulation within pre-defined bounds. Let us consider circuit C, assume that the performance array $\mathbf{K}$ is available along with a set of specifications in form of inequalities (5.1). Moreover, assume that approximated linearized expressions exist for performance degradation $\mathbf{\Delta K}$ (5.3) and that accurate estimations of layout parasitics are also available. Using (5.3) and (5.1) one can determine if performance $\mathbf{K}$ will be within specifications. Suppose that only a certain number of parasitics can be extracted exactly, although conservative estimates can be derived for all of them. Moreover, suppose one is interested in measuring performance with a certain accuracy. To accomplish this objective one would need to determine the set of all parasitics that are essential to ensure the required accuracy. Call P^x this set. In the reminder of this section we will address the problem of minimizing the size of P^x, while ensuring accuracy.

Let us introduce some useful concepts.

Definition 6.1.1 *The set P of all the parasitics of circuit C is said to be non-critical if there exists a partition $\mathcal{P} = (P^x, P^e)$ such that: P^e is a set of all non-critical parasitics, i.e. those parasitics whose cumulative effect on performance is a fraction α of performance specification $\overline{\mathbf{\Delta K}}$. $P^x = P - P^e$ is the set of all remining parasitics.*

Definition 6.1.2 *Let us define* performance estimation accuracy A *as*

$$A = \max_{i=1\ldots N_k} \left| \frac{\Delta K_i^{(max)} - \Delta K_i^{(x\_max)}}{\overline{\Delta K_i}} \right|, \tag{6.1}$$

where $\Delta K_i^{(max)}$ *is the upper-bound on the true performance degradation* ΔK_i, $\Delta K_i^{(x\_max)}$ *is the upper-bound on the extracted value of* ΔK_i, *and* $\overline{\Delta K_i}$ *is the specification on* K_i.

Lemma 6.1.1 *If P is non-critical, then for all partitions $\mathcal{P}_m = (\mathcal{P}^x, \mathcal{P}^e)$, $A = \alpha$*

Proof : Since $P = P^x \cup P^e$ and given that upper-bounds for parasitics are known, $\Delta K_i^{(max)} = \sum_{j \in P^x} S_{i,j} p_j^{(max)} + \sum_{j \in P^e} S_{i,j} p_j^{(max)} = \Delta K_i^{(x\_max)} + \Delta K_i^{(e\_max)}$. By definition 6.1.1, $\Delta K_i^{(e\_max)} \leq \alpha \overline{\Delta K_i}$, moreover $\Delta K_i^{(max)} \geq \Delta K_i^{(x\_max)}$, hence $|\Delta K_i^{(max)} - \Delta K_i^{(x\_max)}| \leq \alpha \overline{\Delta K_i}$. Since $\overline{\Delta K_i} \geq 0 \ \forall i$, α must necessarily be the maximum over performance array **K**. ∎

Next, for each performance measure K_i we need to find a partition $\mathcal{P}_i$, that minimizes the size of P^x. The problem can be formulated in terms of the following optimization:

$$\underset{\mathcal{P}_i}{maximize} \qquad |P^e| \tag{6.2}$$

$$s.t. \quad \sum_{j \in P^e} S_{i,j} p_j^{(max)} \leq \alpha \overline{\Delta K_i}$$

Let us consider now the general problem of an array of performance measures **K**. Assume that each problem (6.2) has a unique solution $\mathcal{P}_i$. Then, the general partition $\mathcal{P}_K$, derived as

$$\mathcal{P}_K = \bigcap_{i=1}^{N_k} \mathcal{P}_i \,, \tag{6.3}$$

is also unique.

Moreover, (6.2) can be rewritten as a standard integer program. Let e be a vector of size $N_p = |P|$, whose components e_j are boolean variables, e_j being one if parasitic $p_j \in P^e$, zero otherwise. Then the problem reduces to

$$\underset{e}{maximize} \qquad \sum_{j=1}^{N_p} e_j \tag{6.4}$$

$$s.t. \quad \sum_{j \in P} e_j S_{i,j} p_j^{(max)} \leq \alpha \overline{\Delta K_i}$$

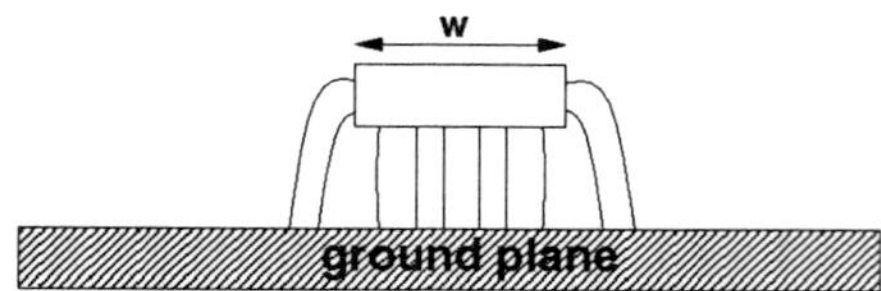

Figure 6.2   Simple interconnect line

If the components of e are permitted to vary continuously and constraint

$$0 \leq e_j \leq 1, \quad \forall j$$

is added to the integer program, problem (6.4) can be processed by a conventional linear program solver [172]. The above procedure was used in all extraction problems presented in Chapter 5. In our tests we used PARCAR's engine to obtain the $\mathcal{P}_K$. Various extraction examples are presented in Chapter 5.

6.1.2   Parasitic Extraction

Computation of interconnect parasitics is impractical for most industrial-size circuits. Numerical methods based on finite difference, finite element, integral equation or other techniques can be used to extract the exact electrical parameters of any structure [258][256][25][148][149][47]. Due to the relatively high computational cost, however, these methods cannot be efficiently used for a very large number of objects such as those present in a complex layout structure.

A better approach consists of creating models of interconnect which can be easily generated by CAD tools for layout extraction and estimation. Analytical models have been proposed for very specific capacitive structures in [10][112]. More recently automated methods have been proposed for automatic generation of models for capacitive parasitics in a wide range of structures used in semiconducting circuits [52][53]. Hereafter a short description of the models used in all our parasitic estimation tools is presented. The techniques used to derive the models are described in detail in [53] and were implemented in a tool called CAPMOD.

Substrate Capacitance

Figure 6.2 shows a simple interconnect line as it appears in most IC technologies. The substrate capacitance of the line is represented in terms of its parallel plate and fringe components. For a given technology for each wiring layer the parallel plate capacitance

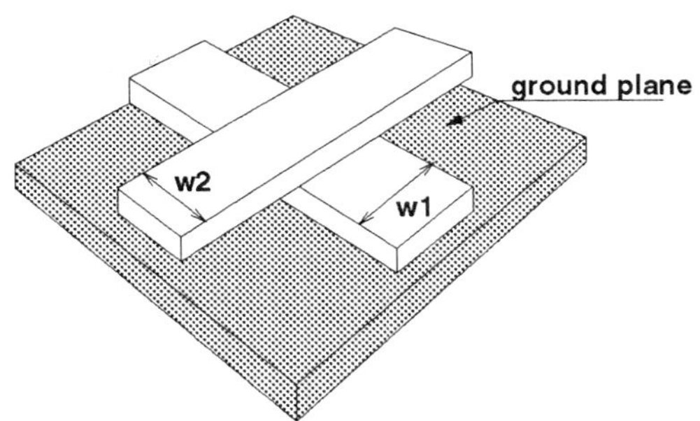

Figure 6.3   Crossover configuration

per unit length only depends on the interconnect width. The fringe capacitance per unit length on the contrary is fixed. The following equation models the substrate capacitance

$$C_0 = k_0 + k_1 w, \tag{6.5}$$

where k_0 and k_1 are real constants dependent on technology and on the type of wiring used. The coefficients are calculated only *once* for a given technology using CAPMOD.

Crossover Configuration

The crossover configuration, shown in Figure 6.3, is modeled by the cross-coupling

$$C_{12} = k_0 + k_1 w_1 + k_2 w_2 + k_3 w_1 w_2, \tag{6.6}$$

where k_0, k_1, k_2 and k_3 are real constants and w_1, w_2 the widths of the two lines. Due to the shielding effect of the lower line over the upper one and vice-versa, the substrate capacitance of both lines, calculated as in (6.5) needs be corrected by the following factors

$$
\begin{aligned}
C_{c\_1} &= k_0 + k_1 w_2 \\
C_{c\_2} &= k_0 + k_1 w_1 + k_2 w_1 w_2.
\end{aligned}
\tag{6.7}
$$

Parallel Lines on the Same Layer

Same layer parallel running lines are dominated by fringing effects, since no vertical overlapping is present and generally the ratio height/width is small[1]. Figure 6.4 shows such a configuration. The model for the cross-coupling is as following

---

[1] Sub-micron technologies are currently reversing this trend. However similar reasoning can be used for the generation of appropriate models which take into account high horizontal parallel plate fields.

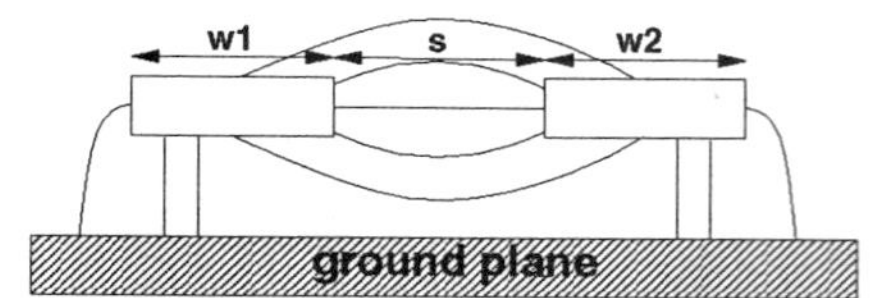

Figure 6.4   Parallel interconnect lines

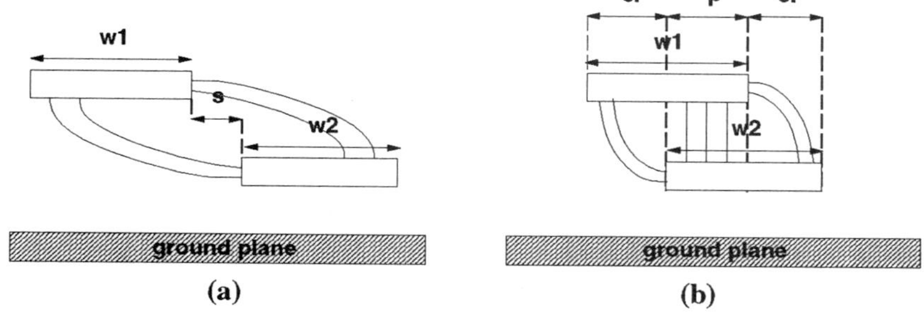

Figure 6.5   Interconnect lines on different layers: (a) non-overlapping; (b) overlapping

$$C_{12} = \mathcal{P}(1/s) + \mathcal{P}(1/(s + w_1)) + \mathcal{P}(1/(s + w_2)), \qquad (6.8)$$

where $\mathcal{P}(.)$ is a given polynomial [53]. The correction components for substrate capacitances are

$$
\begin{aligned}
C_{c\_1} &= \mathcal{P}(1/(s + s_0)) + \mathcal{P}(1/(s + s_0 + w_2)) \\
C_{c\_2} &= \mathcal{P}(1/(s + s_0)) + \mathcal{P}(1/(s + s_0 + w_1)),
\end{aligned}
\qquad (6.9)
$$

where s_0 is a technology- and layer-dependent constant.

Interconnect Lines on Different Layers

There exist two types of configurations for interconnect lines on different layers: the overlapping and the non-overlapping configuration. A non-overlapping configuration is shown in Figure 6.5a. The cross-coupling capacitance is characterized by the following model

$$C_{12} = \mathcal{P}(1/(s + s_0)) + \mathcal{P}(1/(s + s_0 + w_1)) + \mathcal{P}(1/(s + s_0 + w_2)), \qquad (6.10)$$

while correction components are

$$
\begin{aligned}
C_{c\_1} &= \mathcal{P}(1/(s + s_0)) + \mathcal{P}(1/(s + s_0 + w_2)) \\
C_{c\_2} &= \mathcal{P}(1/(s + s_0)) + \mathcal{P}(1/(s + s_0 + w_1)).
\end{aligned}
\qquad (6.11)
$$

Overlapping interconnect lines on the contrary (Figure 6.5) can be modeled as

$$C_{12} = k_0 p + \mathcal{P}(1/(e_l + e_{l0})) + \mathcal{P}(1/(e_r + e_{r0})), \tag{6.12}$$

where p, e_l and e_r are the cross-, left and right overlaps, respectively. e_{l0} and e_{r0} are constants determined using CAPMOD. The correction terms are determined as

$$\begin{aligned}
C_{c_1} &= \mathcal{P}(1/(p + p_0)) + \mathcal{P}(1/(e_l + e_{l0})) + \mathcal{P}(1/(e_r + e_{r0})) \tag{6.13} \\
C_{c_2} &= k_0 p + \mathcal{P}(1/(e_l + e_{l0})) + \mathcal{P}(1/(e_r + e_{r0})).
\end{aligned}$$

6.1.3 Technology Gradient Effects: Mismatch Modeling

A number of authors have attempted to find compact and accurate models for parasitic mismatch. In this section we focus on capacitive and resistive mismatches for passive components and on MOS transistor parameter mismatches for active devices. Let us consider a pair of parasitic components p_i and p_j, mismatch M is defined as follows

$$M = 2\,\frac{p_i - p_j}{p_i + p_j}.$$

Consider next a pair of active devices d_i, d_j, each characterized by a set of parameters $\mathbf{\Pi_i}$ and $\mathbf{\Pi_j}$ respectively. The mismatch of a pair of parameters is defined similarly as

$$M_{\ell m} = 2\,\frac{\pi_\ell - \pi_m}{\pi_\ell + \pi_m}, \ \forall \pi_\ell \in \mathbf{\Pi_i} \text{ and } \pi_m \in \mathbf{\Pi_j}.$$

In general, M and $M_{\ell m}$ are non-deterministic measures dependent of process variations, relative distance and geometry of the object pair. For this reason, mismatch is often characterized based on its mean μ_M and variance σ_M^2. Several models exist in the literature describing mismatch [240][109][168][81][111]. In the tools described in this chapter a technology-independent implementation of the models proposed by Pelgrom [240] and Lakshmi Kumar [168] is being used.

In what follows a short description of the mismatch models used in this book for both passive and active devices is presented. Consider a MOS device and its current-voltage relationship in triode and saturation region, respectively

$$I = K(V_{GS} - V_T - V_{DS}/2)V_{DS} \ ; \ \ I = \frac{K}{2}(V_{GS} - V_T)^2, \tag{6.14}$$

where I is the drain current, K the transconductance constant, V_T the threshold voltage, V_{DS} the drain-to-source voltage, and V_{GS} the gate-to-source voltage. Let us define the means of measures I, K and V_T as $\overline{I}$, $\overline{K}$ and $\overline{V}_T$. Let us assume that all devices are in saturation, then the variance of current mismatch can be derived [168] as

$$\frac{\sigma_I^2}{\overline{I}^2} = \frac{\sigma_K^2}{\overline{K}^2} + 4\frac{\sigma_{VT}^2}{(V_{GS} - \overline{V}_T)^2} - 4r\frac{\sigma_{VT}}{V_{GS} - \overline{V}_T}\frac{\sigma_K}{\overline{K}},$$

where r is the correlation coefficient between mismatches in V_T and K, while σ_K and σ_{V_T} are the standard deviations of K and V_T respectively.

Using a set of well-known analytical models, one can derive expressions for σ_{V_T} and σ_K. Consider first the threshold voltage. V_T is expressed as

$$V_T = \Phi_{MS} + 2\Phi_B + \frac{Q_B}{C_{ox}} + \frac{Q_f}{C_{ox}} + \frac{qD_I}{C_{ox}},$$

where Φ_{MS} is the gate-semiconductor function difference, Φ_B the Fermi potential in the bulk, Q_B the depletion charge density, Q_f the fixed oxide charge density, D_I the threshold adjust implant dose, q the electron charge, and C_{ox} the gate oxide capacitance per unit area. It can be shown that Q_B, Q_f, D_I and C_{ox} are all statistically independent variables, hence σ_{V_T} can be written as

$$\sigma_{V_T}^2 = \frac{1}{\overline{C}_{ox}^2}(\sigma_{Q_B}^2 + \sigma_{Q_f}^2 + q^2\sigma_{D_I}^2) + \frac{\sigma_{C_{ox}}^2}{\overline{C}_{ox}^4}\left(\overline{Q}_B^2 + \overline{Q}_f^2 + q^2\overline{D}_I^2\right) \qquad (6.15)$$

Substituting the values of the individual parametric variances into (6.15), one obtains

$$\sigma_{V_T}^2 = \frac{q}{LW\overline{C}_{ox}^2}\left(\overline{Q}_B + \overline{Q}_f + q\overline{D}_I\right) + A_{geom}^2\left(\overline{Q}_B^2 + \overline{Q}_f^2 + q^2\overline{D}_I^2\right),$$

where terms $\overline{L}$ and $\overline{W}$ are the mean channel length and width of the devices, and A_{geom}^2 is a geometry-dependent parameter. Thus, given the relative position and orientation of a device pair in terms of vector $\boldsymbol{\Delta v} = [\Delta x, \Delta y, \Delta r]^T$, a compact model of the threshold mismatch in the pair can be derived [240] as

$$\sigma_{\Delta V_T}^2 = \frac{A_p^2}{LW} + S_p^2(\boldsymbol{\Delta v}),$$

where A_p^2 models the area and $S_p^2()$ the spacing dependence for parameter V_T. Terms A_p^2 and $S_p^2()$ are process-dependent and can be obtained empirically from measurements on wafer test patterns.

The transconductance K is calculated as

$$K = \mu C_{ox}W/L, \qquad (6.16)$$

where μ is the channel mobility, C_{ox} the gate capacitance per unit area, W the channel width and L its length. Assuming independence for all the factors on the right-hand side of (6.16), $\sigma_K^2/\overline{K}^2$ can be written as

$$\frac{\sigma_K^2}{\overline{K}^2} = \frac{\sigma_L^2}{\overline{L}^2} + \frac{\sigma_W^2}{\overline{W}^2} + \frac{\sigma_\mu^2}{\overline{\mu}^2} + \frac{\sigma_{C_{ox}}^2}{\overline{C}_{ox}^2}.$$

Substituting into (6.16) the values of the individual variances for each parameter, $\sigma_K^2/\overline{K}^2$ can be re-written as

$$\frac{\sigma_K^2}{\overline{K}^2} = \frac{1}{LW}(A_\mu^2 + A_{ox}^2) + \frac{\sigma_L^2}{\overline{L}^2} + \frac{\sigma_W^2}{\overline{W}^2},$$

where A_μ^2 is a technology-dependent geometry-insensitive model parameter and A_{ox}^2 is a position-dependent parameter [168]. The transconductance mismatch can be represented compactly as

$$\frac{\sigma_{\Delta K}^2}{\overline{K}^2} = \frac{A_K^2}{LW} + S_K^2(\mathbf{\Delta v}) + \frac{A_L^2}{WL^2} + \frac{A_W^2}{W^2L}, \tag{6.17}$$

where A_K^2, A_L^2, A_W^2 and $S_K^2(\mathbf{\Delta v})$ are geometry-dependent factors. Expressions for σ_W^2 and σ_L^2 have been derived using the fact that $\sigma_L^2 \propto 1/W$ and $\sigma_W^2 \propto 1/L$ [240]. However, the latter terms are generally small with respect to the other expressions in (6.17), hence they can be neglected, yielding

$$\frac{\sigma_{\Delta K}^2}{\overline{K}^2} \approx \frac{A_K^2}{LW} + S_K^2(\mathbf{\Delta v}).$$

In the literature, functions $S_p(\mathbf{\Delta v})$ and $S_K(\mathbf{\Delta v})$ have been generally reported as being linear with $|v|$, however nonlinear expressions can be easily supported in our tools.

Using the value derived in [168] for the variance of C_{ox}, one can easily obtain formulae for the mismatch of parasitic capacitances and resistances. Consider first capacitive mismatch. Let $C = C_{ox}WL$ be the value of a parasitic capacitance of an interconnect line W wide and L long. Then, the variance of C is given by

$$\frac{\sigma_C^2}{\overline{C}^2} \propto \frac{1}{LW}.$$

Thus, the variance of the mismatch ΔC is

$$\frac{\sigma_{\Delta C}^2}{\overline{C}^2} = \frac{A_{ox}^2}{LW} S_C^2(\mathbf{\Delta v}),$$

where A_{ox}^2 is an area factor related to the oxide thickness and $S_C^2(\Delta\mathbf{v})$ is the spacing dependence for capacitance pairs.

Let us now characterize stray resistance mismatches. Let $R = R_{\square}L/W$ be the value of the interconnect stray resistance. Then, assuming spatial independence of $R_{\square}$, the variance of R is given by

$$\frac{\sigma_R^2}{\overline{R}^2} = \frac{\sigma_L^2}{\overline{L}^2} + \frac{\sigma_W^2}{\overline{W}^2}.$$

Hence, by the same reasoning as before, the variance of the mismatch ΔR becomes

$$\frac{\sigma_{\Delta R}^2}{\overline{R}^2} = \frac{A_L^2}{\overline{W}\,\overline{L}^2} + \frac{A_W^2}{\overline{W}^2\,\overline{L}}.$$

These models were implemented in both the placement and the extraction tools for mismatch-aware synthesis and analysis.

6.2   EXTRACTION OF SUBSTRATE-RELATED PARASITICS

With the continuous increase of chip complexity, device density and circuit speed, power density has become a major source of concern in the design of reliable VLSI ICs. High power density affects circuit performance directly and indirectly. Firstly, by concentrating power in a limited area of the chip, the transient and steady state temperature of large number of devices can be far from the nominal value for which the circuit was designed. This may result in catastrophic as well as parametric faults at circuit and system level.

Furthermore, due to relatively small distances between high-swing high-frequency noise sources and sensitive devices, the substrate becomes a major carrier for spurious signals within the circuit. In mixed-signal circuits in particular, coupling of analog, relatively slow signals with quickly switching digital ones is often disastrous. To alleviate the problem, heavily over-designed structures are generally adopted, thus seriously limiting the advantages of innovative technologies. For this reason substrate modeling has received attention from mixed-signal circuit designers, attempting to integrate RF analog and baseband digital circuitry on the same chip.

In purely digital systems each signal is represented by a sequence of finite number of binary digits; therefore, these signals can take on discrete values only. Due to the binary nature of signals, digital circuits are realized using gates with only two states, each state being defined in some range of the continuous signal. This makes digital circuits largely immune to various noise and parasitic sources inherent in ICs. On the contrary, due to the non-discrete nature of circulating signals, analog circuits are in general much more sensitive to electrical noise. A noise signal can be classified based on its nature (*thermal, shot, flicker and switching noise*) and propagation medium (*interconnect, substrate, power supply busses*). The first three noise types relate to intrinsically generated parasitic currents caused by the various physical phenomena, over which little control is possible during the physical assembly. The last noise type is more interesting to us in our view, since with opportune design of the medium one can control the impact it may have on performance.

This section presents an overview of the techniques used for the computation of substrate related parasitic effects. The techniques for substrate extraction outlined in this section have been implemented in a package called SUBRES [96][38], which is the basis for the Green function-based substrate analysis used by our extraction tools. In the reminder of this section we will focus on the main phases required by an effective substrate extraction methodology, namely appropriate modeling for injection, reception and transport mechanisms within substrate.

6.2.1 Modeling Injection Patterns of Complex Circuits

In typical digital circuits, a large number of gates undergo periodic transitions. When a transition occurs a spike of current is absorbed from a power bus and used to charge a load in the signal path. In general, a significant portion of this current is discharged to a ground bus through direct feed through, or it is injected directly into the substrate through two main mechanisms: capacitive coupling and impact ionization. The cumulative contribution of currents associated with switching gates in the circuit results in ripple noise, generally observable in power and ground busses. Similar spurious noise currents are injected into substrate at various locations, in direct proximity to switching circuitry. The collection of all spurious currents and voltages directly and indirectly generated by switching activity is called *switching noise*. Spatial location and rate of occurrence of transitions are generally time-variant and difficult to characterize exactly. In addition, due to the complexity of typical digital circuits, an exact waveform characterization of switching noise is impractical.

Alternatively, switching noise is often modeled based on its macroscopic appearance. For example, if the number of switching gates is large enough and the global switching activity of the digital circuit is uniformly distributed over a large section of the spectrum, this noise can be modeled as a single Gaussian white or pink noise source. Switching noise in this discrete form can be used to easily estimate its impact on relatively complex analog circuits, hence allowing a designer or an automated tool to determine "safe" floorplannings or global routings.

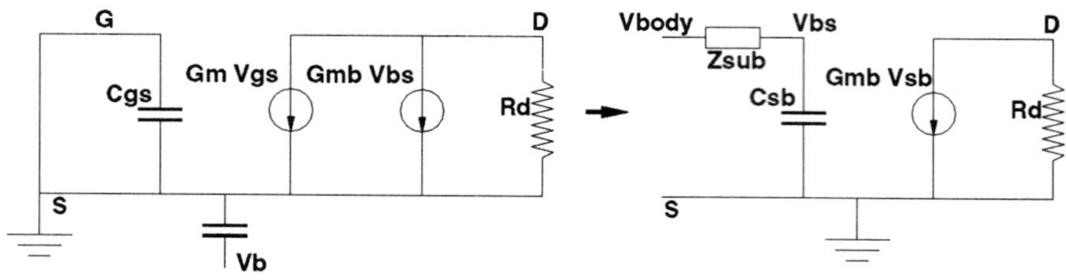

Figure 6.6 Body Effect in MOSFETs (Courtesy of Ranjit Gharpurey)

Compact but reasonably accurate injection models have been efficiently generated for digital circuits using event-driven simulation techniques in combination with a complete characterization of noise injection patterns associated with the library used to synthesize the circuit. A detailed discussion of the method can be found in [209]. The use of these models and their application to industrial strength circuits is outlined later in this section and in the examples presented in this book.

6.2.2 Modeling Substrate Reception Mechanisms

Capacitive sensing is the most common mechanism of noise reception in surface devices, as bipolar transistors, capacitors, resistors and interconnect lines. The junction with substrate in lateral pnp devices consists of the n-type base region. If the pnp device is used in a gain stage, then the base of the device must be carefully shielded, or connected to a low impedance node. Otherwise the substrate noise will be amplified by the gain of the circuit.

In addition to capacitive pickup through the source and drain depletion junctions, MOS devices also exhibit a more severe form of substrate interaction due to the *body effect*. In MOS devices threshold voltage V_t is a strong function of the substrate potential. For a uniform surface impurity concentration N_A, this dependence is given by [105]

$$V_t = V_{t0} + \frac{\sqrt{2q\epsilon N_A}}{C_{ox}} \left(\sqrt{2\Phi_f + V_{sb}} - \sqrt{2\Phi_f} \right) ,$$

where ϵ is the substrate dielectric permittivity, N_A the substrate doping, C_{ox} the per unit oxide capacitance, $2\Phi_f$ the surface inversion potential and V_{sb} the source-to-body potential. The effect can be represented by a linearized model parameter g_{mb} in the small-signal device model [105]. Shorting gate and source of transistor in Figure 6.6, a gain stage exists between substrate S and drain D. With suitable approximations

[105] it can be shown that

$$\frac{g_{mb}}{g_m} = \frac{\sqrt{2q\epsilon N_A}}{2C_{ox}\sqrt{2\Phi_f + V_{sb}}} \; ,$$

where g_m is the small-signal transconductance of the device. Parameter g_m relates the drain current to the gate-to-source voltage. In typical processes the g_{mb}/g_m ratio varies from 0.1 to 0.3. The parasitic body-to-drain gain is thus only 14-20 dB lower than the gate-to-drain gain. This fact makes MOSFET devices especially vulnerable to substrate noise reception at low to medium frequencies. On the contrary capacitive pickup, exhibited by most other devices, becomes significant only at relatively high frequencies (above 1 MHz).

6.2.3 Modeling Substrate Transport

Extensive literature has recently appeared on electrical modeling and analysis of silicon substrate. Compact macromodels for both lightly and heavily doped substrates have been proposed by several authors, e.g. [143][293]. A major drawback of this approach is the lack of accuracy in the presence of large numbers of contacts and a strong dependence on technology and the physical implementation of the circuit.

More recently, attempts to accurately model the effects of substrate on medium-sized integrated circuits have been made using a numerical finite difference method, e.g. [309]. The technique is versatile and general in nature, since it can handle lateral and vertical resistivity variations and arbitrary substrate geometries. However, to obtain accurate substrate characterization, a fine mesh is required, thus making the storage and computational effort often prohibitive. Boundary element methods have also been used for parasitic extraction. In [285][286] a Green function is described in a finite uniform medium, with a condition of zero normal electric field, using the technique of the separation of variables. Despite their efficiency however, all these techniques are still too computational intensive to be used during any interactive analysis or in an optimization loop. In addition, substrate analysis of complex structures or of thousands of contacts may be impractical or extremely time consuming.

In what follows the techniques used in our extraction tools are described and their suitability in optimization problems is discussed. In general, silicon substrates in ICs are composed by one or more lightly doped epitaxial layers and a highly doped "core." Hence, differently conductive areas are present in the vertical section of the chip, while lateral resistivity variations are due to device and well implants as well as other integrated components. The latter are junctions with the substrate and may therefore be considered equipotential. Calculating resistances between any contact locations on

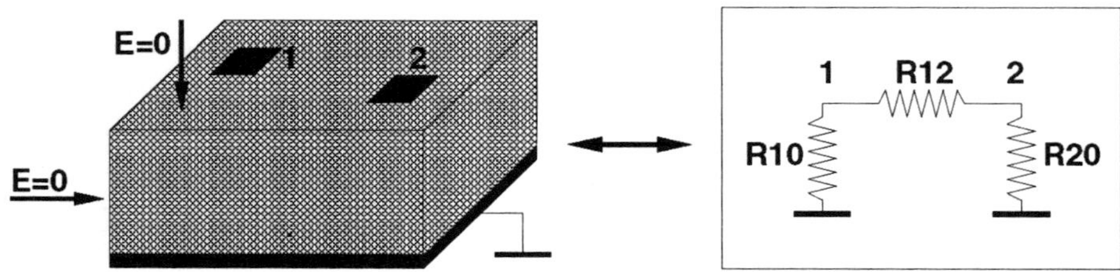

Figure 6.7   Substrate boundaries and contact resistance modeling

the substrate requires the computation of electric potential $\Phi(x, y, z, t)$ at any location (x, y, z) in the bulk. From Maxwell's equations one can show that

$$\frac{1}{\rho}\nabla \bullet \nabla\Phi(x, y, z, t) + \epsilon\frac{\partial}{\partial t}(\nabla \bullet \nabla\Phi(x, y, z, t)) = 0, \tag{6.18}$$

holds, where ϵ and ρ are respectively the local dielectric permittivity and resistivity of the substrate. In the electrostatic case (6.18) reduces to the Laplace equation

$$\nabla^2\Phi = 0 \,, \tag{6.19}$$

with either Dirichlet or Neumann boundary conditions[2] or a combination of those on the surfaces. Equation (6.19) is often solved numerically using techniques based on some discretization of the workspace into atomic volumes or *nodes*. To find the impedance matrix **R** associated with each contact pair, it is sufficient to set all the nodes associated with a given contact to 1 volt and to measure the current flowing out of each other contact. The resulting system of simultaneous equations is diagonally dominant and sparse, since only seven elements in each row are non-zero. Hence, standard techniques for the solution of sparse linear systems can be applied [264].

Alternatively, (6.19) can be solved only at the points **r** located on the surface of the silicon substrate. A solution of this problem can be obtained by simple two-dimensional integration of the Green function. The Green function $G(r, r')$ is the potential at r due to a point charge placed at a point r'. The main problem that remains to be solved is the computation of the Green function. This problem can be significantly simplified by recognizing the essentially resistive nature of substrate at frequencies up to 4-5 GHz.

Consider the structure in Figure 6.7. Suppose one would like to compute the complex impedances between contacts **1** and **2** and the impedance toward ground. In the electrostatic case, the problem of computing the resistance between a substrate contact and all the others can be translated into that of computing the charge at the contact

---

[2] Dirichlet conditions impose a given potential, Neumann conditions a given electrical field.

when set at a potential of 1 volt, while the other contacts and the backplane are grounded. Hence, by solving the capacitance problem (after appropriate scaling of medium susceptances), it is possible to easily obtain all substrate conductances of the resistance problem. The next step is therefore the computation of the potential everywhere due to charges placed in the substrate. The potential of a contact is computed as the result of averaging all internal contact partitions. If this calculation is extended to all contact partitions on the surface of the substrate, one obtains a matrix relating the potentials of every contact pair. Such matrix, called *coefficient of potential* **P**, once inverted, is the basis for computing the impedance matrix **R**.

Given the above assumptions, the Green function corresponds to an infinite series of sinusoidal functions [96]. From the relation between coefficients of potential equation and charges, adapted for surface contacts[3], one can derive an expression for the average potential at contact i due to the charge on contact j as

$$\bar{\Phi}_i = \frac{Q_j}{S_j S_i} \int_{S_i} \int_{S_j} G(s_j, s_i) ds_j ds_i \,, \tag{6.20}$$

where $\bar{\Phi}_i$ is the average potential of contact i and Q_j is the total charge of contact j, while S_j and S_i are the surfaces of the respective contacts. The components p_{ij} of matrix **P** are computed explicitly as the ratio of $\bar{\Phi}_i$ and Q_j.

The doubly infinite series of component p_{ij} tends to converge slowly. The problem can be eliminated by rewriting p_{ij}, after proper scaling, as a cosine series [96]. By replacing the ratios of contact coordinates and the substrate dimensions with ratios of integers and summing over finites limits, p_{ij} can be represented in terms of a two-dimensional discrete cosine transform (DCT) of the Green function after only minor manipulations. Several efficient techniques exist for efficient computation of the DCT, e.g. FFT-based techniques only require a computation complexity $O(n \, log \, n)$, n being the number of nodes in the backplane. Note that the value of the Green function is solely dependent on the properties of the substrate in z-direction. Hence, for a given substrate structure the DCT needs be derived *only once*. Any modification in the relative position of one or more nodes is captured completely by the Fourier transform, thus only matrix **P** needs be calculated and inverted. However, due to the relatively small size of **P**, this process does not require a significant CPU time. Non-abrupt doping profiles can be analyzed at low CPU cost by simply discretizing in z-direction with a gradually changing value of permittivity.

The inversion of **P**, the last step before the computation of the impedance matrix **R**, is performed using LU decomposition. The DCT-based substrate extraction implemented in SUBRES also includes built-in sensitivity calculation and technology trend

---

[3] A version for 3-D contacts can found in [95].

analysis [38]. Schematic simplification is also available for complex problems (more than 100 contacts). The scheme, which uses provably accurate methods, is based on iterative inversion techniques and has demonstrated its effectiveness in containing the computational complexity of problems of the size of 2,500 contacts [95][38].

6.3   TRANSISTOR-LEVEL NOISE SIMULATION

6.3.1   Introduction

A time-domain, non-Monte Carlo method for computer simulation of electrical noise in nonlinear dynamic circuits with arbitrary excitations and arbitrary large-signal wave-forms is presented [64][65]. This time-domain noise simulation method is based on results from the theory of stochastic differential equations. The noise phenomena considered in this work are caused by the small current and voltage fluctuations, such as thermal, shot and flicker noise, that are generated within the integrated-circuit devices themselves. The existence of electrical noise is basically due to the fact that electrical charge is not continuous but is carried in discrete amounts equal to the electron charge. Electrical noise is associated with *fundamental* processes in integrated-circuit devices and represents a *fundamental* limit on the performance of electronic circuits [106]. Accurate prediction of the noise performances of especially analog sub-blocks in a mixed-signal system design is crucial to ensure that the overall system will work correctly.

Noise sets the lower limit to the size of an electrical signal that can be amplified by a circuit without significant deterioration in signal quality. It also results in an upper limit to the useful gain of an amplifier, because if the gain is increased without limit the output stages of the circuit will eventually begin to cut off or saturate on the amplified noise from the input stages [106]. The influence of noise on the performance is not limited to amplifier circuits. For instance, active integrated mixer circuits, which are widely used for down conversion in RF receivers, add noise to their output. It is desirable to be able to predict the noise performance of a given mixer design [135][136]. Most of the time, amplifier circuits operate in small-signal conditions, that is, the operating-point of the circuit does not change. For analysis and simulation, the amplifier circuit with a fixed operating-point can be modeled as a linear time-invariant network by making use of the small-signal models of the integrated-circuit devices. On the other hand, for a mixer circuit, the presence of a large local-oscillator signal causes substantial change in the active devices' operating points over time. So, a linear time-invariant network model is not accurate for a mixer circuit. There are many other kinds of circuits which do not operate in small-signal conditions, such as voltage-controlled oscillator

(VCO) circuits. Noise simulation for these circuits requires a method which can handle nonlinear dynamic circuits with arbitrary large-signal excitations and arbitrary large-signal waveforms.

In Section 6.3.2 below, previous work on computer simulation of noise in electronic circuits is reviewed with comparisons to our method. Formulation of the circuit equations with noise is described in Section 6.3.3. In Section 6.3.4, noise models for integrated-circuit devices are discussed. Section 6.3.5 describes our time-domain, non-Monte Carlo method for computer simulation of electrical noise in nonlinear dynamic circuits. In Section 6.3.6, the implementation of the noise simulation method, in a nodal-analysis circuit simulation program (SPICE), is described. Two examples of noise simulation, using the implementation described in Section 6.3.6, are presented in Section 6.3.7. Conclusions and future work are stated in Section 6.3.8.

6.3.2 Previous Work

Frequency Domain Methods

Noise Simulation with Linear Time-Invariant Transformations

The electrical noise sources in passive elements and integrated-circuit devices have been investigated extensively. Small-signal equivalent circuits, including noise, for many integrated-circuit components have been constructed [106]. The noise performance of a circuit can be analyzed in terms of these small-signal equivalent circuits by considering each of the uncorrelated noise sources in turn and separately computing its contribution at the output. In this method, a nonlinear circuit is assumed to have *time-invariant* (DC) large-signal excitations and also *time-invariant* steady-state large-signal waveforms. Then, the nonlinear circuit is *linearized* around the fixed operating point to obtain an LTI network for noise analysis. It is also assumed that both the noise sources and the *noise at the output* are *wide-sense stationary (WSS)* stochastic processes. The spectral density for the WSS output noise arising from a specific noise source is then given by

$$S_o(f) = |H(f)|^2 S_i(f) \tag{6.21}$$

where $H(f)$ is the transfer function for the LTI transformation from the noise source to the output, and $S_i(f)$ and $S_o(f)$ are the spectral densities of the noise source and the output noise respectively. Noise simulation with this method is, basically, the numerical calculation of the transfer functions between the noise sources and the output for a range of frequencies.

Implementation of this method based on the interreciprocal *adjoint network* concept [253][206] results in a very efficient computational technique for noise analysis, which is available in almost every circuit simulator and widely used by analog circuit designers.

Unfortunately, this method is only applicable to LTI circuits (e.g. the small-signal equivalent circuits corresponding to circuits with fixed operating points). It is not appropriate for noise simulation of circuits with changing bias conditions (e.g. a mixer circuit as discussed in Section 6.3.1), or circuits which are not meant to operate in small-signal conditions (e.g. VCOs).

Noise Simulation with Linear Periodically Time-Varying Transformations

As a generalization of the method using LTI transformations (described in the previous section), a noise simulation method that uses *linear periodically time-varying* (LPTV) transformations was proposed [135][136][234]. In this method, a nonlinear circuit is assumed to have *periodic large-signal* excitations and also *periodic steady-state* large-signal waveforms. Then, the nonlinear circuit is *linearized* around the periodic steady-state "operating-point" to obtain an LPTV network for noise analysis. It is also assumed that both the noise sources and the *noise at the output* are *cyclostationary* [92] stochastic processes. The "average" spectral density for the cyclostationary output noise arising from a specific noise source is then given by

$$\langle S_o(f) \rangle = \sum_{n=-\infty}^{\infty} |H_n(f)|^2 \, S_i(f + n \frac{1}{T}) \tag{6.22}$$

where $H_n(f)$ is the transfer function for the LPTV transformation from the noise source to the output, $S_i(f)$ is the spectral density of the WSS noise source, $\langle S_o(f) \rangle$ is the "average" spectral density for the output noise and T is the period [92]. Noise simulation with this method is, basically, the numerical calculation of the transfer functions between the noise sources and the output for a range of frequencies centered at a number of harmonics. Noise sources may be assumed to be WSS by capturing the cyclostationarity of the noise sources in the transfer functions [136].

This noise analysis technique is applicable to only a limited class of nonlinear circuits with periodic excitations (e.g. mixer circuits, switched capacitor circuits). Moreover, with this method, it is assumed that the noise at the output is cyclostationary, which justifies the calculation of an "average" spectral density for the output noise. An open-loop (free running) oscillator circuit has time-invariant excitations and periodic large-signal steady-state waveforms, but noise in an oscillator circuit is *not* cyclostationary [67], hence this method as presented in [136][234] is not applicable to the characterization of phase noise in open-loop oscillators.

Time Domain Methods

Monte Carlo Noise Simulation for Nonlinear Dynamic Circuits

Noise simulation in time-domain has traditionally been based on the Monte Carlo technique [17][202]. In these methods, noise sources in a circuit are "simulated" with either a sum of sinusoids (with random phases) [17] or a random amplitude pulse waveform [202] representation using a random number generator. The circuit with the noise sources is simulated using standard transient analysis. "Many" transient analyses of the circuit are performed with different sample paths of the noise sources. Then the probabilistic characteristics of noise are calculated using the data obtained in these simulations.

This method has several drawbacks. Shot and thermal noise sources in electronic circuits are modeled as "white" noise sources. To simulate "white" noise sources accurately, one must either include very high frequency components in the sum of sinusoids representation, or set the pulse width to a very small value (to the correlation width of the noise sources, which is approximately 0.17 ps for thermal noise at room temperature [92]) in the random amplitude pulse waveform. This limits the maximum time-step in transient analysis to a very small value making the simulation highly inefficient, which has to be repeated "many" times in a Monte Carlo fashion.

The noise content in a waveform obtained with a single run of the transient analysis will be much smaller when compared with the magnitude of the desired signal in the waveform. As a result, the waveforms obtained for different sample paths of the noise sources will be very close to each other. It is known that, in a simulator, these waveforms are only numerical approximations to the actual waveforms, therefore they contain numerical noise. The rms value of noise is calculated by taking a difference of these waveforms. That is, two large numbers, which have uncertainty in them, are being subtracted from each other. Consequently, the rms noise calculated with this method, in fact, includes the noise generated by the numerical algorithms. This degrades the accuracy of the results obtained by this method.

Pseudo-random number generators often do not generate a large sequence of independent numbers, but reuse old random numbers instead. This can also become a problem if a circuit with many noise sources is simulated. This is usually the case, because every device has several noise sources associated with its model.

Proposed Method

Our method, unlike the frequency domain methods, is not restricted to circuits with time-invariant or periodic large-signal steady-state waveforms with WSS or cyclostationary noise at the output. Our time-domain noise simulation method is general in the following sense: Any nonlinear dynamic circuit with any kind of excitation and large-signal waveforms, which can be simulated by the transient analysis routine in a circuit simulator, can be simulated by our noise simulator in time-domain to produce the noise variances and covariances of circuit variables as a function of time, provided that noise models for the devices in the circuit are available. Our time-domain noise simulation method is based on results from the theory of stochastic differential equations. There are no pseudo-random number generators involved in the simulation. The simulation of the noiseless waveforms and the simulation of noise are separated, even though they are done concurrently. Thus, the numerical noise problem that arises in Monte Carlo methods is avoided.

For WSS stochastic processes, the autocorrelation

$$Z(\tau) = \mathcal{E}[y(t)\, y(t + \tau)] \tag{6.23}$$

(function of a single variable which is the time difference [92]) completely characterizes the *second-order* probabilistic characteristics in time-domain, and the spectral density

$$S(f) = \mathcal{F}\{Z(\tau)\}. \tag{6.24}$$

(the Fourier transform of the autocorrelation) does it in the frequency domain. The method using LTI transformations (described in Section 6.3.2) calculates the spectral density for the WSS output noise. Noise in a nonlinear dynamic circuit with arbitrary excitations and arbitrary time-varying large-signal waveforms is, in general, *nonstationary*. The autocorrelation

$$Z(t + \tau, t) = \mathcal{E}[y(t)\, y(t + \tau)] \tag{6.25}$$

for a nonstationary stochastic process is a function of two variables. Our method is capable of calculating the variances and the covariances (that is, the covariance matrix) for the noise content in the node voltages and other circuit variables as a function of time. Furthermore, noise correlations between circuit variables at different time points can also be calculated. The combination of the time-varying covariance matrix and correlations at different time points completely characterize the *second-order* probabilistic characteristics of any vector of *nonstationary* stochastic processes. An "instantaneous" spectral density

$$S(t, f) = \mathcal{F}\{Z(t + \tau, t)\} = \int_{-\infty}^{\infty} Z(t + \tau, t)\, e^{-j2\pi f\tau}\, d\tau \tag{6.26}$$

(function of two variables, one of which is time and the other is frequency) can be defined for a nonstationary process [92]. The noise simulation method which uses LPTV transformations (described in Section 6.3.2) assumes that the noise at the output is cyclostationary and characterizes it in the frequency domain by calculating an "average" spectral density. The instantaneous spectral density for a cyclostationary process is a periodic function of the "time" variable [92]. On the other hand, an average spectral density for a general nonstationary process does not characterize the second-order probabilistic characteristics adequately. For instance, noise in an open-loop oscillator is nonstationary, and not cyclostationary. Hence, the noise analysis methods described in [136][234] can not be used for open-loop oscillators. Our method can calculate the complete autocorrelation for the nonstationary noise and can be used for phase noise characterization of open-loop oscillators [67].

Finally, the implementation of our method fits naturally into a circuit simulator (such as SPICE) which is capable of doing time-domain transient simulations. Noise simulation is performed along with the transient simulation over the time interval specified by the user.

6.3.3 Formulation of the Circuit Equations with Noise

Without loss of generality, the noise simulation problem will be formulated assuming that modified nodal analysis (MNA) [265] is used for the formulation of circuit equations[4].

The MNA equations for any circuit, *without the noise sources*, can be written compactly as

$$F(\dot{x}, x, t) = 0 \qquad x(0) = x_0 \qquad (6.27)$$

where x is the vector of the circuit variables with dimension n, $\dot{x}$ is the time derivative of x, t is time and F is mapping x, $\dot{x}$ and t into a vector of real numbers of dimension n. The time dependence of x and $\dot{x}$ will not be written explicitly for notational simplicity. In MNA, the circuit variables consist of node voltages and some branch currents, e.g. currents through inductors and voltage sources. The circuit equations consist of the node equations (KCL) and branch equations of the elements for which branch currents are included in the circuit variables vector. Under some rather mild conditions (which are satisfied by well modeled circuits) on the continuity and differentiability of F, it can be proven that there exists a unique solution to (6.27) assuming that a fixed initial value $x(0) = x_0$ is given [265]. Let x_s be the solution to (6.27). Transient analysis in circuit simulators solves for x_s using numerical methods for ordinary

---

[4]MNA is used by most of the existing circuit simulators including SPICE.

differential equations (ODEs) [265]. The initial value vector is obtained by a DC analysis of the circuit before the transient simulation is started. For a circuit, there may be several different DC operating-points.

The first-order Taylor's expansion of F around x_s, to be used later, is expressed as

$$F(\dot{x}, x, t) \cong F(\dot{x}_s, x_s, t) + \frac{\partial}{\partial x} F(\dot{x}, x, t) \Big|_{\substack{x = x_s \\ \dot{x} = \dot{x}_s}} (x - x_s)$$

$$+ \frac{\partial}{\partial \dot{x}} F(\dot{x}, x, t) \Big|_{\substack{x = x_s \\ \dot{x} = \dot{x}_s}} (\dot{x} - \dot{x}_s). \tag{6.28}$$

If the noise sources are included in the circuit, the MNA formulation of the circuit equations can be written as

$$F(\dot{x}, x, t) + B(x, t)\, v = 0 \qquad x(0) = x_0 + x_{noise,0} \tag{6.29}$$

where $B(x, t)$ is an $n \times p$ matrix, the entries of which are a function of x, and v is a vector of p standard white Gaussian stochastic processes. A one-dimensional standard Gaussian white noise is a *stationary* Gaussian process $\xi(t)$, for $-\infty < t < \infty$, with mean $\mathcal{E}[\xi(t)] = 0$ and a constant spectral density on the entire real axis. The autocorrelation function of $\xi(t)$ is given by $\mathcal{E}[\xi(t + \tau)\xi(t)] = \delta(\tau)$, where δ is Dirac's delta function [7]. Such a process does not exist as a physically realizable process, since it has infinite variance. Nonetheless, the white Gaussian noise $\xi(t)$ is a very useful mathematical idealization for describing random processes that fluctuate rapidly and hence are virtually uncorrelated for different instants of time. Thermal and shot noise in electronic circuits can be modeled in terms of the standard white Gaussian noise $\xi(t)$. Flicker noise sources are modeled by synthesizing them using white noise sources. (See Section 6.3.4 for a discussion of noise models). v in (6.29) is simply a combination of p independent one-dimensional standard white Gaussian noise processes as defined above. These noise processes actually correspond to the current noise sources which are included in the models of the integrated-circuit devices. Since the noise models are to be employed here in the context of an MNA circuit simulator (SPICE), noise sources in the devices are all modeled as *uncorrelated* current sources.

$B(x, t)$, in (6.29), contains the *intensities* (see Section 6.3.4 for the definition) for the noise sources in v. Every column in $B(x, t)$ corresponds to a current noise source in v, and has either one or two nonzero entries. The rows of $B(x, t)$ correspond to either a node equation (KCL) or a branch equation. There are no nonzero entries in the rows which correspond to the branch equations.

6.3.4 Noise Models

The most important types of electrical noise sources (thermal, shot and flicker noise) in passive elements and integrated-circuit devices have been investigated extensively, and appropriate models have been derived [106][144][311]. In noise analysis with LTI transformations, these noise models are inserted into the small-signal equivalent (at a fixed operating point) circuits of the devices as stationary noise sources [106]. In this section, we describe the adaptation of these noise models for use in our time-domain noise simulation method. In our method, the noise sources are inserted into the large-signal models of the integrated-circuit devices and they are, in general, nonstationary.

Thermal and Shot Noise

Thermal noise is due to the random thermal motion of electrons. It exists in resistive materials and is unaffected by the presence or absence of a direct current [106]. It can be shown (using a thermo-dynamical argument) that the bandwidth of thermal noise at room temperature is approximately 6000 GHz, and hence, time samples of thermal noise that are separated by more than 0.17 ps are uncorrelated [92]. For practical purposes, thermal noise associated with a time-invariant resistor is modeled as a shunt current noise source $g_{th}(t)$, which is a stationary white Gaussian stochastic process with the autocorrelation function

$$\mathcal{E}[g_{th}(t)\, g_{th}(t + \tau)] = 2\, k\, T\, G\, \delta(\tau) \tag{6.30}$$

where k is the Boltzmann's constant, T is the temperature in degrees Kelvin and G is the conductance in mhos [92]. In (6.30) G is a constant, independent of time. The channel material in MOSFETs is resistive and exhibits thermal noise, which is the major source of noise in FETs. The resistance of the channel could be time-varying during circuit operation. The autocorrelation function of the current noise source $h_{th}(t)$, modeling the thermal noise for a time-varying resistor, is then given by

$$Z_{thermal}(t + \tau, t) = \mathcal{E}[h_{th}(t + \tau)\, h_{th}(t)] = 2kT\, G(t)\, \delta(\tau) =$$
$$IN_{thermal}(t)^2\, \delta(\tau) \tag{6.31}$$

where $IN_{thermal}(t) = \sqrt{2kT\, G(t)}$ is the *intensity* and $G(t)$ is the time-varying conductance. This noise process is no longer stationary, but, as a "delta-correlated" process, it still has independent values at every time point, which is the reason we were able to generalize (6.30) to obtain (6.31) for the time-varying case. We can obtain $h_{th}(t)$ from the standard white Gaussian noise process $\xi(t)$ (defined in Section 6.3.3) by

$$h_{th}(t) = IN_{thermal}(t)\, \xi(t). \tag{6.32}$$

Intensities for the noise sources are placed in $B(x,t)$ in the equation formulation (6.29).

Shot noise is due to the random emission of charge carriers across a pn-junction. It is always associated with a direct current flow and it is present in diodes and bipolar transistors. It can be shown that (by modeling the emission times with a Poisson point process) bandwidth of shot noise is inversely proportional to the transit time of carriers through the junction. Hence, bandwidth for shot noise is well into the high gigahertz region [92]. For practical purposes, shot noise associated with a time-invariant current through a pn-junction is modeled as a current noise source $g_{sh}(t)$, which is a stationary white Gaussian stochastic process with the autocorrelation function

$$\mathcal{E}[g_{sh}(t)\, g_{sh}(t+\tau)] = q\, I_D\, \delta(\tau) \tag{6.33}$$

where q is the electron charge and I_D is the current through the junction [92]. In (6.33) I_D is a constant, independent of time. When the current is time-varying, the shot noise becomes nonstationary and the autocorrelation function for this *modulated* shot noise [225] $h_{sh}(t)$ is then given by

$$Z_{shot}(t+\tau,t) = \mathcal{E}[h_{sh}(t+\tau)\, h_{sh}(t)] = q\, I_D(t)\, \delta(\tau) =$$
$$IN_{shot}(t)^2\, \delta(\tau) \tag{6.34}$$

where $IN_{shot}(t) = \sqrt{q\, I_D(t)}$ is the *intensity* and $I_D(t)$ is the time-varying current. We can obtain $h_{sh}(t)$ from the standard white Gaussian noise process $\xi(t)$ as in (6.32)

$$h_{sh}(t) = IN_{shot}(t)\, \xi(t). \tag{6.35}$$

If we calculate the instantaneous spectral density (defined by (6.26)) for the nonstationary shot and thermal noise associated with a time-varying resistance or current, we obtain

$$S(t,f) = (IN(t))^2 \tag{6.36}$$

which is a constant function of frequency since the noise sources are "delta-correlated."

Flicker Noise

Flicker noise is found in all active devices, as well as some discrete passive elements such as carbon resistors. One cause for this type of noise is the random capture and release of charge carriers at traps due to contamination and crystal defects. Actually, flicker noise has been observed as fluctuations in the parameters of many systems, many of which are completely unrelated to semiconductors [152]. For practical purposes, flicker noise associated with a direct time-invariant current is modeled as a "stationary" stochastic process with the spectral density

$$S_{flicker}(f) = K \frac{I^a}{f^b} \qquad (6.37)$$

where K is a constant for a particular device, a is a constant in the range 0.5 to 2, b is a constant approximately equal to 1, and f is the frequency. If $b = 1$, the spectral density has a $1/f$ frequency dependence. Flicker noise is not a "delta-correlated" process, and hence does not have independent values at every time point (unlike thermal and shot noise). Actually, its present behavior is *equally* correlated with both the recent and distant past [152]. We can not express a flicker noise source in terms of the standard white Gaussian noise process $\xi(t)$ as we did it for thermal and shot noise in (6.32) and (6.35). In Section 6.3.3, the circuit equations with noise were formulated using white noise processes only (v in (6.29)). This restriction is required by the stochastic differential equation theory. In order to include flicker noise sources in the formulation, we need to synthesize them using white noise sources along with a filtering network. A promising approach for $1/f$ (flicker) noise generation is to use the summation of Lorentzian spectra [241]. (See [65] for details).

The spectral density equation in (6.37) was given for flicker noise associated with a *time-invariant* current. As for thermal and shot noise, we are interested in generalizing the flicker noise model to the time-varying case. This was straightforward for thermal and shot noise, since they are "delta-correlated" noise processes having independent values at every time point, and they were modeled as "modulated" white noise processes. In [106], it is pointed out that no flicker noise is present in carbon resistors until a direct current is passed through the resistor. This suggests that the model given by (6.37) is still valid for the flicker noise associated with a time-varying current when we use the "average" value of the time-varying current (DC part) for I in (6.37) [163]. But, still, either a theoretical or experimental derivation of a model for flicker noise associated with a time-varying current is needed.

6.3.5 Development of the Noise Simulation Method

Background on Stochastic Differential Equations[5]

Equations such as (6.29), which involve rapidly and irregularly fluctuating random processes (i.e. white noise), were first treated in 1908 by Langevin in the study of the

---

[5] Most of the material in this section is summarized from the "Introduction" of [7].

Brownian motion of a particle in a fluid. The differential equation

$$\dot{Y}_t = f(t, Y_t) + G(t, Y_t)\,\xi_t \qquad Y_{t_0} = c \tag{6.38}$$

can not be treated as an ordinary differential equation using classical differential calculus, because it involves the rapidly and irregularly fluctuating white noise process ξ_t as defined in Section 6.3.3. (6.38) involves random variables and stochastic processes as opposed to real numbers and deterministic functions. The notions of convergence and limit for sequences of random variables are quite different than the ones for sequences of real numbers [110]. Actually, there are several different notions of convergence for sequences of random variables.

For a mathematically rigorous treatment of equations of type (6.38), a new theory was necessary. It turns out that, whereas "white noise" is only a "generalized" stochastic process, the indefinite integral

$$W_t = \int_0^t \xi_s\,ds \tag{6.39}$$

can nonetheless be identified with the Wiener process (i.e. Brownian motion). This is a Gaussian stochastic process with continuous (but nowhere differentiable) sample functions, with mean $\mathcal{E}[W_t] = 0$ and with covariance $\mathcal{E}[W_t\,W_s] = \min\,(t, s)$.

If we write (6.39) symbolically as

$$dW_t = \xi_t\,dt \tag{6.40}$$

(6.38) can be put in the differential form

$$dY_t = f(t, Y_t)\,dt + G(t, Y_t)\,dW_t \qquad Y_{t_0} = c. \tag{6.41}$$

This is a stochastic differential equation (Ito's) for the process Y_t. It should be understood as an abbreviation for the integral equation

$$Y_t = c + \int_{t_0}^t f(s, Y_s)\,ds + \int_{t_0}^t G(s, Y_s)\,dW_s. \tag{6.42}$$

Since the sample functions of W_t are with probability 1 continuous though not of bounded variation in any interval, the second integral in (6.42) cannot be regarded in general, even for smooth G, as an ordinary Riemann-Stieltjes integral with respect to the sample functions of W_t, because the value depends on the intermediate points in the approximating sums [89]. In 1951, Ito defined integrals of the form

$$Y_t = \int_{t_0}^t G(s)\,dW_s \tag{6.43}$$

for a broad class of so-called non-anticipating functionals G of the Wiener process W_t and in doing so put the theory of stochastic differential equations on a solid foundation. This theory has its peculiarities. For example, the solution of the equation

$$dY_t = Y_t \, dW_t \quad Y_0 = 1 \tag{6.44}$$

is not $\exp(W_t)$, but

$$Y_t = \exp(W_t - t/2) \tag{6.45}$$

which does not derive by purely formal calculation according to the classical rules.

The calculus of stochastic differential equations deals with the random variable Y_t and its variation, and hence belongs to the probabilistic or direct methods. It deals directly with the time-wise development of the state Y_t. An equation of the form (6.41) or (6.42) represents a rule used to construct the trajectories of Y_t from the trajectories of a Wiener process and an initial value c.

In the development of the noise simulation method, we will treat (6.29) as a system of stochastic differential equations and make use of the stochastic calculus developed by Ito [7].

Derivation of the System of Stochastic Differential Equations for Noise from MNA Formulation of the Nonlinear Circuit Equations

(6.29) is a system of nonlinear stochastic differential equations (SDEs) describing the dynamics of the circuit with the noise sources. Our goal is to develop an algorithm to calculate the various "probabilistic characteristics" of the noise in the circuit so that we can characterize its noise performance. Let x_{sn} be the solution of (6.29). x_{sn} is a vector of stochastic processes, since it is the solution of the circuit equations with the noise sources included, and satisfies

$$F(\dot{x}_{sn}, x_{sn}, t) + B(x_{sn}, t) \, v = 0 \quad x_{sn}(0) = x_0 + x_{noise,0} \tag{6.46}$$

where x_0 is deterministic, and $x_{noise,0}$ is a vector of n zero-mean random variables. Actually, *second-order* probabilistic characteristics of $x_{sn}(t)$ are sufficient in almost all noise performance prediction problems. In fact, the noise in the circuit (noise at the output, not the noise sources which were already modeled as Gaussian stochastic processes) is "approximately" Gaussian, and hence can be completely characterized by the second-order probabilistic characteristics. In particular, we would like to be able to calculate the time evolution of the mean and correlation function of $x_{sn}(t)$:

$$\mathcal{E}[x_{sn}(t)] = ? \quad \mathcal{E}[x_{sn}(t_1) \, x_{sn}(t_2)^T] = ? \tag{6.47}$$

Unfortunately, for the system of nonlinear stochastic differential equations given in (6.46), $\mathcal{E}[x_{sn}(t)]$ and $\mathcal{E}[x_{sn}(t_1)\,x_{sn}(t_2)^T]$ do not satisfy any simple equation [7]. In particular, closed expressions for $\mathcal{E}[x_{sn}(t_1)\,x_{sn}(t_2)^T]$ can not be obtained. The equations that are derived for $\mathcal{E}[x_{sn}(t_1)x_{sn}(t_2)^T]$ involve higher-order moments [247].

Up to this point we have not made any assumptions on the "magnitude" of the noise in the circuit, but, in any useful circuit, the "magnitude" of the noise content in a signal is much smaller when compared with the "magnitude" of the signal itself. With this motivation, we use (6.28) in (6.46) to approximate $F(\dot{x}_{sn}, x_{sn}, t)$, and obtain

$$F(\dot{x}_s, x_s, t) + \left.\frac{\partial}{\partial x}F(\dot{x}, x, t)\right|_{\substack{x = x_s \\ \dot{x} = \dot{x}_s}} (x_{sn} - x_s) +$$

$$\left.\frac{\partial}{\partial \dot{x}}F(\dot{x}, x, t)\right|_{\substack{x = x_s \\ \dot{x} = \dot{x}_s}} (\dot{x}_{sn} - \dot{x}_s) + B(x_{sn}, t)\, v \cong 0 \qquad (6.48)$$

$$x_{sn}(0) = x_0 + x_{noise,0}.$$

Define

$$x_{noise} = x_{sn} - x_s. \qquad (6.49)$$

x_{noise} is the difference between the solutions of the circuit equations, with and without the noise sources. In other words, x_{noise} is the noise content in x_{sn}. x_{noise} is much smaller when compared with x_s, which validates the above approximation.

For notational simplicity, define

$$A(t) = \left.\frac{\partial}{\partial x}F(\dot{x}, x, t)\right|_{\substack{x = x_s \\ \dot{x} = \dot{x}_s}} \qquad C(t) = \left.\frac{\partial}{\partial \dot{x}}F(\dot{x}, x, t)\right|_{\substack{x = x_s \\ \dot{x} = \dot{x}_s}} \qquad (6.50)$$

where $A(t)$ and $C(t)$ are $n \times n$ matrices with time-dependent entries. Furthermore, we approximate

$$B(x_{sn}, t) \cong B(x_s, t) \qquad (6.51)$$

and define

$$B(t) = B(x_s, t). \qquad (6.52)$$

If (6.49), (6.50), (6.51) and (6.52) are substituted in (6.48) we obtain

$$F(\dot{x}_s, x_s, t) + A(t)\, x_{noise} + C(t)\, \dot{x}_{noise} + B(t)\, v \cong 0$$

$$\qquad (6.53)$$

$$x_{noise}(0) = x_0 + x_{noise,0} - x_s(0).$$

Since x_s is the solution of (6.27) we have

$$F(\dot{x}_s, x_s, t) = 0 \qquad x_s(0) = x_0 \tag{6.54}$$

and if we substitute (6.54) in (6.53), we obtain

$$A(t)\, x_{noise} + C(t)\, \dot{x}_{noise} + B(t)\, v \cong 0$$

$$x_{noise}(0) = x_{noise,0}. \tag{6.55}$$

(6.55) is a linear SDE [7] in x_{noise} with time-varying coefficient matrices. $A(t)$, $B(t)$ and $C(t)$ are functions of x_s, and they do not depend on x_{noise}.

$C(t)$ in (6.55) is, in general, not full-rank. Hence, some of the nodes in the circuit may have infinite noise power [65]. This is a result of the "white" noise models used for thermal and shot noise. Of course, in a practical circuit no node can have infinite noise power, since there are always parasitic capacitors from any node to ground, and shot and thermal noise are, in fact, "band-limited." If we eliminate the variables in x_{noise} with infinite power, we arrive at

$$\dot{x}^1_{noise} = E(t)\, x^1_{noise} + F(t)\, v \qquad x^1_{noise}(0) = x^1_{noise,0} \tag{6.56}$$

where x^1_{noise} (m-dimensional) is a reduced version of x_{noise} (n-dimensional), $E(t)$ is $m \times m$ and $F(t)$ is $m \times p$. $E(t)$ and $F(t)$ are obtained from $A(t)$, $B(t)$ and $C(t)$ by performing some matrix operations [65].

System of Ordinary Differential Equations for the Noise Correlation Function

(6.56) is a system of stochastic differential equations which is *linear in the narrow sense* (right-hand-side is linear in x^1_{noise}, and the coefficient matrix for the vector of noise sources is independent of x^1_{noise}). Just as with *linear* ordinary differential equations, a much more complete theory can be developed for *linear* stochastic differential equations [7].

Our goal is to determine the *mean* and the correlation function of x^1_{noise} as a function of time in the time interval desired. If x^1_{noise} is a Gaussian stochastic process, then it is completely characterized by its mean and correlation function.

Since $\mathcal{E}[v(t)] = 0$ and $\mathcal{E}[x^1_{noise,0}] = 0$, we obtain

$$m^1(t) = \mathcal{E}[x^1_{noise}(t)] = 0. \tag{6.57}$$

Next, we would like to determine the correlation matrix of the components of x^1_{noise} as a function of t, which is given by

$$K^1(t) = \mathcal{E}[x^1_{noise}(t)\, x^1_{noise}(t)^T]. \tag{6.58}$$

Using stochastic (Ito) calculus, one can obtain [65]

$$\dot{K}^1(t) = E(t)\, K^1(t) + K^1(t)\, E(t)^T + F(t)\, F(t)^T \tag{6.59}$$

where $K^1(t)$ is the unique symmetric nonnegative-definite solution of the matrix equation (6.59). The differential equation for $K^1(t) = K^1(t)^T$, (6.59), satisfies the Lipschitz and boundedness conditions in the time interval of interest, so that a unique solution exists [7]. (6.59) represents (in view of symmetry of $K^1(t)$) a system of $m(m + 1)/2$ linear ordinary differential equations. (6.59) can be solved for $K^1(t)$ using a numerical method for the solution of ODEs.

$K^1(t)$ represents the noise correlation matrix of circuit variables as a function of time. So, the information about the noise variances of circuit variables, or the noise correlations between circuit variables at a given time point are contained in $K^1(t)$. In some problems, one might be interested in the noise correlations of circuit variables at different time points [67], which can be expressed as

$$K^1(t_1, t_2) = \mathcal{E}[x^1_{noise}(t_1)\, x^1_{noise}(t_2)^T]. \tag{6.60}$$

One can also derive

$$\frac{\partial}{\partial t_2}\, K^1(t_1, t_2) = K^1(t_1, t_2)\, E(t_2)^T \tag{6.61}$$

with the initial condition $K^1(t_1, t_1) = K^1(t_1)$ [247]. Integrating (6.61) at various values of t_1, one can obtain a number of sections of the correlation function $K^1(t_1, t_2)$ at $t_2 > t_1$. Then, $K^1(t_1, t_2)$ at $t_2 < t_1$ is determined by

$$K^1(t_1, t_2) = K^1(t_2, t_1)^T. \tag{6.62}$$

We were able to derive closed form expressions ((6.59) and (6.61)) for the second-order moments of x^1_{noise}. This was possible, because (6.55) is a system of *linear* stochastic differential equations, which was obtained from a system of *nonlinear* stochastic differential equations, (6.46), making use of the small noise perturbation approximation (6.28).

6.3.6   Implementation in SPICE

The noise simulation method, along with the noise models described, was implemented inside the circuit simulator SPICE3 [248]. Time-domain noise simulation is performed along with the transient simulation in the time interval specified by the user. The transient simulation in SPICE3 solves for x_s, which is the solution of (6.27). The initial value vector $x(0) = x_0$ in (6.27) is obtained by a DC analysis before the transient simulation is started. The noise simulation (numerical solution of (6.59) and (6.61)) is performed concurrently with the transient simulation. (6.59) represents a system of $m(m+1)/2$ linear differential equations in a special form. We use the trapezoidal scheme or the backward differentiation formula to discretize these equations in time. The transient simulation and noise simulation have separate automatic time-step control mechanisms, but the two are synchronized in the sense that the transient analysis is forced to take the time steps required by the noise analysis but not vice versa. The coefficient matrices $E(t)$ and $F(t)$ in (6.59) and (6.61 are available only numerically, and are calculated using the transient analysis solution at the current time point. (See [65] for a detailed description of the numerical methods).

The current implementation of the noise simulation algorithm requires $O(m^3)$ flops at every time point compared with the roughly $O(m^{1.5})$ flops required by the transient analysis algorithm [65]. Hence, the CPU time usage of the noise simulation will be largely dominated by the noise analysis routines. The computational cost of noise simulation would be high for "large" circuits, but this noise simulation method is intended for evaluating the noise performances of small sub-blocks (e.g. analog blocks such as mixers and oscillators) of a mixed-signal system design. Increasing the efficiency of the numerical methods is part of the future work.

6.3.7 Noise Simulation Examples

In this section, we present two examples of noise simulation. In particular, noise simulations for a CMOS inverter circuit and a BJT active mixer circuit will be presented. For both of these circuits, we have included only the shot and thermal noise sources in the simulation. One reason for this is that flicker noise has little effect on the noise performance of these circuits. Secondly, including the flicker noise sources increases the simulation time because of the extra nodes created and large time constants introduced by the networks for flicker noise source synthesis.

Application of the noise simulation algorithm to the characterization of phase noise in open-loop oscillator circuits is presented in [67].

CMOS Inverter

A CMOS inverter loaded with a 1 pF capacitor was driven with a periodic waveform at the input, and a noise simulation was performed. In Figure 6.8, the large-signal

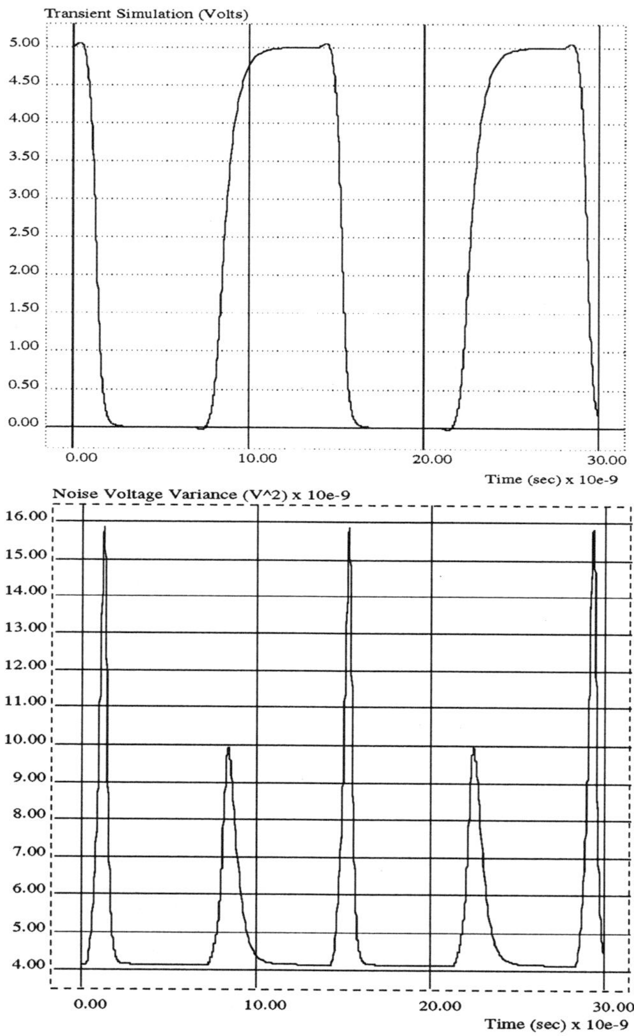

Figure 6.8   Noise Simulation for the CMOS Inverter

and the noise variance waveforms at the output of this inverter can be seen. As seen in Figure 6.8, the noise at the output is nonstationary, that is, the noise variance is not a constant as a function of time. The noise variance (mean-squared noise power) is highest during low-to-high and high-to-low transitions of the large-signal output waveform.

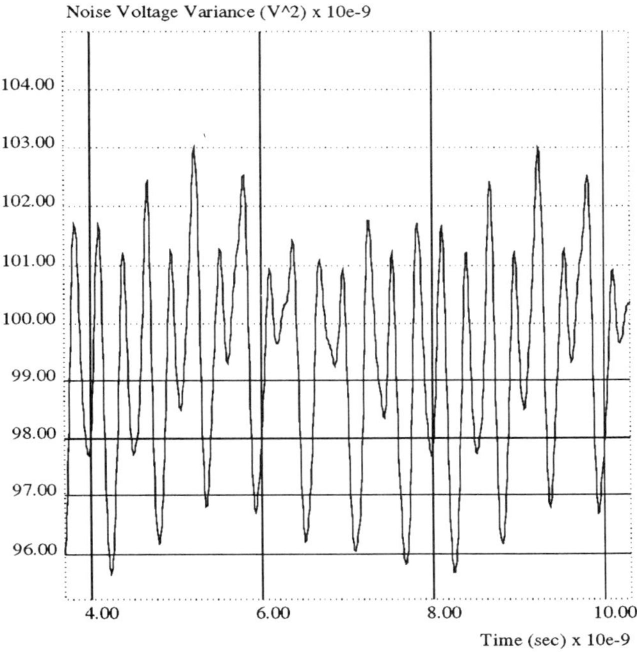

Figure 6.9   Bipolar Mixer-Noise Variance at the Output

BJT Active Mixer

This bipolar mixer circuit contains 14 BJTs, 21 resistors, 5 capacitors, and 18 parasitic capacitors connected between some of the nodes and ground. The LO input is a sine-wave at 1.75 GHz with an amplitude of 178 mV. The RF input is a sine-wave at 2 GHz with an amplitude of 31.6 mV. Thus, the intermediate frequency (IF) frequency is 250 MHz. $1/f$ noise sources are not included in the simulation, because $1/f$ noise is rarely a factor at RF and microwave frequencies [135].

This circuit was simulated to calculate the noise variance at the output as a function of time. (Figure 6.9: This waveform is periodic with a period of 4 ns; the IF frequency is 250 MHz.) The noise at the output of this circuit is not stationary, because the signals applied to the circuit are large enough to change the operating point. The noise analysis of this circuit with small-signal equivalent circuits around a fixed operating point does not give correct results. Such an analysis would predict the noise at the output as stationary, i.e. a constant noise variance as a function of time.

The noise performance of a mixer circuit is usually characterized by its *noise figure*, which can be defined by [106]

$$NF = \frac{\text{total output noise power}}{\text{output noise power due to the source resistance}}. \tag{6.63}$$

This definition is intended for circuits in small-signal operation. For such circuits, noise power is a constant function of time. In our case, the noise variance at the output of the mixer circuit changes as a function of time over one period. We can use the "average" noise power (calculated over the periodic noise variance waveform) in the above definition of noise figure. To calculate the noise figure as defined, we simulate the mixer circuit again to calculate the noise variance waveform at the output with all the noise sources turned off except the noise source for the source resistance $RS_{RF} = 50\ \Omega$ at the RF port. Then we can calculate the noise figure as below, and the result is 19.7 dB.

$$NF = 10\log\left(\frac{\text{average of total noise variance}}{\text{average of noise variance due to source resistance only}}\right). \tag{6.64}$$

This BJT mixer circuit has 65 nodes (including the internal nodes for BJTs) which are connected to capacitors. There are a total of 91 noise sources associated with the bipolar transistors and the resistors in the circuit. The noise simulation (with 400 time points) took 279 CPU seconds on a DEC Alpha Server 2100 4/200 with our current implementation (with the Bartels-Stewart algorithm used to solve the Lyapunov matrix equation). In this simulation, 2145 noise covariance matrix entries for the 65 nodes are calculated at 400 time points.

6.3.8   Conclusions and Future Work

We have presented a novel, general time-domain method for the simulation of electrical noise in nonlinear electronic circuits Based on the noise simulation algorithm presented, we have developed a "phase noise" characterization methodology for open-loop (free running) oscillator circuits [67]. In [67], the definition of *phase noise* for general oscillation waveforms is discussed. The second-order probabilistic characteristics (autocorrelation function) of phase noise for oscillator circuits are *extracted* from noise simulations of the circuit using the algorithm proposed in this paper. The phase noise characterization methodology presented in [67] has been applied to three different kinds of oscillators (ring-oscillators, harmonic oscillators and relaxation oscillators). Phase noise performance prediction results agree with analysis and experimental results in literature.

Future work includes exploring numerical methods for the efficient solution of the Lyapunov matrix equation for larger circuits. An iterative method tuned for the structure of the problem and which can exploit the sparsity of the coefficient matrix seems to be the most promising approach [131]. The proposed noise simulation algorithm will be applied to other practical noise performance prediction problems in analog and RF circuit design.

7

TESTING

7.1 BACKGROUND

7.1.1 Testing as Part of the Design Process

The presence of analog components in today's complex mixed-signal systems complicates their testing significantly. Analog circuits, in general, require much longer testing times than digital circuits because second-order effects must be considered and because few CAD tools are available to aid in the design of the test vectors. Analog testing is currently performed on a relatively ad-hoc basis; a design or test engineer relies primarily upon intuition about a circuit's internal functionality to derive the circuit's test suite. This test suite frequently defaults to the complete set of circuit specifications. This approach is becoming increasingly expensive in both test development and test execution times. The specifications of mixed analog-digital circuits are usually very large (e.g. see [18]), which not only results in long manual test development, but also in prohibitive testing times on very expensive automated test equipment (ATE) with mixed-signal capabilities; it is estimated that testing currently accounts for 30% of total manufacturing cost [272]. Furthermore, the use of sophisticated CAD tools has reduced the design cycle so that the influence of testing on time-to-market and final cost of the circuit is increasingly significant.

7.1.2 Previous Work in Analog ATPG

Most of the previous work in automatic test pattern generation (ATPG) for electrical systems has been directed at digital circuits, and efficient techniques have been developed for both combinational and sequential digital systems [268][97]. These test systems are based on the common single stuck-at-0/stuck-at-1 fault model and the controllability and observability of each fault.

Unfortunately most of these ideas cannot be directly applied to analog systems. One of the major problems is that analog systems are, in general, much more difficult to model because second- and higher-order effects must be considered to accurately predict system performance. Another major difference is that analog circuits are less susceptible to catastrophic (e.g. stuck-at) faults, because of the typically larger device and wire sizes, but much more susceptible to parametric faults, which are small variations in component values that cause the system performance to violate its specifications.

In the area of analog testing, work has been done in system modeling and test ordering, both of which are useful toward the goal of reducing testing time. In the area of system modeling, Souders and Stenbakken proposed using linear models, which can be derived either from simulation [291] or from manufacturing data [288] using QR decomposition. In the area of test ordering, Milor has described an algorithm for minimizing average test time by ordering tests in such a way that the tests which are most likely to detect faults are performed first [211].

7.2 METHODOLOGY

7.2.1 Behavioral Models

We have developed an algorithm for deriving a minimal set of test vectors for fully testing the performance specifications of a general class of analog systems. The class of systems to which the algorithm can be applied are those systems which can be modeled in a linear function space, i.e. the system output must be an *linear* combination of user-specified basis functions. Note that the basis functions themselves do not have to be linear. Mathematically the model is formulated as

$$Y = \beta_0 g_0 + \beta_1 g_1 + \beta_2 g_2 + \beta_3 g_3 + \ldots + \beta_n g_n \qquad (7.1)$$

where Y is the system output, $\{g_i\}$ is a set of arbitrary basis vectors (user-specified), and β_i is the coefficient of the ith basis vector. Many analog systems can be accurately modeled in this fashion.

There are several simple methods which can be used to choose the $\{g_i\}$ basis vectors. For extremely simple systems the basis functions may be obvious from a simple description of the expected output. For more complicated systems, the Taylor expansion can be used to derive a very useful linear model

$$f(a+x) = f(a) + x f'(a) + \frac{x^2 f''(a)}{2!} + \ldots + \frac{x^{n-1} f^{(n-1)}(a)}{(n-1)!} \qquad (7.2)$$

where a represents the nominal value of a model parameter, $f(a)$ represents the value of the output when that model parameter is at its nominal value, and x represents the amount by which that model parameter deviates from its nominal value because of manufacturing non-idealities. We wish to estimate $f(a+x)$.

A first-order Taylor series approximation is a reasonably accurate model for many common analog systems with parameters that do not deviate significantly from their nominal values. This is the model used by Stenbakken and Souders [292], and our discussion of it here will be brief. Dropping the higher-order terms and generalizing to multiple dimensions, the expansion becomes

$$\begin{aligned} f(a+x) &= f(a) + \triangledown f(a)x & (7.3) \\ &= f(a) + \frac{\partial f}{\partial a_1} x_1 + \frac{\partial f}{\partial a_2} x_2 + \ldots & (7.4) \end{aligned}$$

where a_i is the nominal value of the ith model parameter and x_i is the deviation in that parameter. The basis functions for this system are thus $\{f(a), \frac{\partial f}{\partial a_1}, \frac{\partial f}{\partial a_2}, \ldots\}$. $f(a)$ is the nominal system performance, and each of the partial derivatives represents an error signature for a particular type of manufacturing defect which can occur. The error signatures are computed by finding the sensitivity of the output to the parameters of interest at each point on the response surface. Note that these error signatures could represent either catastrophic faults, such as shorts and disconnections, or parametric faults, such as small deviations in capacitance values or process parameters.

Our testing algorithm is based upon the statistical theory of optimal experimental design, in which test vectors are chosen to be maximally independent so that the system performance Y will be characterized as accurately as possible in the presence of measurement noise and model inaccuracies. More specifically, we wish to choose the test vectors to minimize the average standard error of the predicted output, thereby maximizing the likelihood that we will be able to conclusively verify that the performance specifications have or have not been met after a minimum number of test vectors. If the minimum number of test vectors is not sufficient to conclusively verify the performance specifications, then additional test vectors are selected and applied, one at a time, until the standard error of the predicted output is low enough to verify the performance specifications. Linear regression is used to analyze the results of the tests and compute the required standard errors.

7.2.2 Testing Algorithm

Given an arbitrary basis consisting of n functions which span an n-dimensional space, at least n test vectors must be applied to the system in order to fully characterize the output function. If fewer than n input vectors are applied, then at least one dimension of the space remains unexplored and hence the output function is unconstrained in that dimension. Furthermore, because of inevitable measurement noise, n test vectors may not be sufficient to verify that the output function falls within the desired bounds. Using additional test vectors will lower the standard error of the estimates, thereby making it more likely that the system output can be verified to fall within the desired limits.

With these factors in mind, the testing algorithm that we propose is

1. Apply n "maximally orthogonal" vectors to the system, where n is the dimensionality of the space to be characterized.

2. Generate the estimated response function and confidence intervals for that response function.

3. If the confidence intervals for the response function fall definitely within the system performance specifications at all points, then accept the chip.

4. If the confidence intervals for the response function fall definitely outside the system performance specifications at any point, then reject the chip.

5. Otherwise apply additional test vectors which are "maximally orthogonal" to those already applied, one at a time, until the confidence intervals are small enough to determine whether or not the chip meets its specifications.

The selection of the test vectors is fully described in Section 7.3.

7.3 AUTOMATIC TEST PATTERN GENERATION

7.3.1 Algorithms

The choice of test vectors is a difficult optimization problem. The objective is to minimize the standard error of the estimated response function, which is a function of the choice of test vectors. Intuitively, the orthogonality of the test vectors is measured by the degree to which each test vector maximizes the contribution of one basis function while minimizing the contribution of the others.

The algorithm used to derive the maximally orthogonal test vectors is

1. Eliminate any redundant basis vectors.

2. Run the I-optimality algorithm to select best n tests, where n is the dimensionality of the function space after eliminating redundant basis vectors.

3. Run the I-optimality algorithm to select best additional vectors, one at a time, for use if the prior tests are not conclusive.

Redundant basis vectors are eliminated by computing the null space of $\{g_1, g_2, \ldots, g_n\}$. The parameters associated with error signatures that are linearly dependent are said to belong to the same *ambiguity group*, since variations in those parameters are indistinguishable at the system output. Ambiguity groups reduce the number of basis vectors needed to model the response surface and hence the number of test vectors which must be applied to fully characterize a system.

Let U be the matrix formed from these basis vectors, where g_i is the i_{th} column of U. Suppose U has dependent columns, then its null space is non-empty such that

$$UN = 0 \tag{7.5}$$

where $N \in R^{m \times r}$ is a matrix with r independent column vectors that spans the null space of U. Non-zero entries in N indicate that the corresponding components are in ambiguity groups. A component i belongs to an ambiguity group if and only if row i of N has a non-zero entry. Furthermore, components i and j are in the same ambiguity group if rows i and j of N are non-zero and not orthogonal to each other [179]. It follows that the components fall into the same group if their corresponding row vectors of N are non-zero and not orthogonal.

The null space of U can be computed using singular value decomposition (SVD) or Gaussian elimination. In the case of SVD, we first compute $U^T U$, followed by SVD

$$U^T U = X_1 X_2 N^T \tag{7.6}$$

where N spans the null space of $U^T U$. Since $U^T U N = 0$, $U N = 0$, so N is the null space of U also. The reason for computing $U^T U$ in (7.6) is that U often has many more rows than columns, so computing N for a smaller matrix $U^T U$ is more efficient. Furthermore, note that computing N and checking the rows of N for pairwise orthogonality can be performed in polynomial time.

In summary, the approach for finding ambiguity groups is

1. Given a sensitivity matrix U.

2. Find null space of U, N, using singular value decomposition (7.6) or Gaussian elimination. Let group number $g = 1$.

3. Remove the first non-zero row of N and assign to group number g.

4. Check if any remaining rows are orthogonal. If not, assign them to group g and remove.

5. Increment g and repeat 3 until all rows are removed.

Once this set is formed, the I-optimality routines, as described in Section 7.3.3, are executed to find a good set of test vectors.

Once the test vectors have been applied, the measured responses are used to estimate $\hat{\beta}$, the vector of coefficients for each of the basis vectors. For the special case when the number of test points is equal to the number of basis functions, $\hat{\beta}$ is found by solving

$$X\hat{\beta} = Y \tag{7.7}$$

for $\hat{\beta}$, where X is the design matrix as output by the I-optimality routine and Y is the vector of measured responses. When the number of test points is greater than the number of basis functions, $\hat{\beta}$ is found by solving

$$X^T X\hat{\beta} = X^T Y \tag{7.8}$$

for $\hat{\beta}$, which is a linear regression.

A 99% confidence interval for the $\hat{\beta}$ estimates over the entire response surface is computed by applying Equations 7.13 and 7.14 from Section 7.3.2, followed by

$$CI(\hat{\beta}) = \left(\hat{\beta} - 2.576\sqrt{var(\hat{\beta})}, \hat{\beta} + 2.576\sqrt{var(\hat{\beta})}\right), \tag{7.9}$$

where the factor 2.576 is used because it is the 99.5% quantile of a normal distribution, which leaves only a 1% probability that the actual value of the response at that function falls outside of the confidence interval.

Note that we have assumed that the variance σ^2 of the error term ϵ is known by the designer. If this is not the case, then S^2, an estimator of σ^2, can be derived from the observations Y as

$$S^2 = \frac{\sum_{i=1}^{n}[Y_i - \hat{\mu}(x_i)]^2}{(n - d)} \tag{7.10}$$

where Y_i is the ith observation, $\hat{\mu}(x_i)$ is the predicted value of Y_i, based on the model, n is the number of measurements, and d is the dimensionality of the model (the number of independent basis vectors). When S^2 is used instead of σ^2 then the confidence intervals are constructed by using the t-distribution quantiles with $n - d$ degrees of freedom instead of the normal quantiles. Note that the t-distribution approaches the normal distribution for large n (e.g. $n > 30 - 40$), so the normal distribution can often be used as an approximation to the appropriate t-distribution when generating the confidence intervals from S^2 instead of from σ^2.

7.3.2 Optimality Criteria

There are several different optimality criteria (A-, D-, E-, G-, and I-), the relative merits of which have been debated extensively in the relevant literature [153][23]. D-optimality, which is generally considered to be the simplest type of optimality, minimizes the average prediction variance of the model coefficients. This type of optimality is very suitable for fault diagnosis, in which we wish to estimate the actual values of each circuit component as accurately as possible. D-optimality claims nothing about the the average prediction variance of the system output, however, so it is not the best choice for verifying that the system output meets its specifications.

The two types of optimality which do consider the prediction variance of the system output are G- and I-optimality. G-optimality minimizes the *maximum* prediction variance over the response surface of interest. It would probably be the most suitable for verifying that a circuit meets its specifications, since specifications are frequently stated as worst-case bounds. G-optimality is difficult to optimize upon, however, because it is not continuously differentiable. Hence I-optimality, which is continuously differentiable, was chosen for this research. I-optimality minimizes the *average* prediction variance over the response surface of interest. To formulate these ideas mathematically, let

$$y = f(g_1, g_2, \ldots, g_p) + \epsilon \tag{7.11}$$

where y is the response variable, g_i are the independent basis vectors, and ϵ represents the measurement and modeling errors, which are assumed to be independent with mean 0 and variance σ^2.

Let X be the design matrix, which contains one row for each of the n test vectors.

$$X = \begin{bmatrix} g_1(x_1) & g_2(x_1) & g_3(x_1) & \cdots & g_p(x_1) \\ g_1(x_2) & g_2(x_2) & g_3(x_2) & \cdots & g_p(x_2) \\ & & \vdots & & \\ g_1(x_n) & g_2(x_n) & g_3(x_n) & \cdots & g_p(x_n) \end{bmatrix} \tag{7.12}$$

The design moment matrix M_X can be calculated as

$$M_X = \frac{1}{n} X^T X \tag{7.13}$$

and the prediction variance at an arbitrary point x on the response surface is

$$var \ \hat{y}(x) = \frac{\sigma^2}{n} f(x) M_X^{-1} f(x)^T. \tag{7.14}$$

An I-optimal design is one which minimizes the average of this variance over the response surface R

$$I = \frac{n}{\sigma^2} \int_R var \ \hat{y}(x) d\mu(x). \tag{7.15}$$

This integral simplifies [23] to give

$$I = trace\{M M_X^{-1}\} \tag{7.16}$$

where M is the moment matrix of the region of interest R.

$$M = \int_R f(x)^T f(x) d\mu(x) \tag{7.17}$$

7.3.3   Optimal Design of Experiments

Finding an exactly I-optimal design is believed to be NP-complete [59] and hence only feasible for very small problems. For larger problems, several heuristic algorithms have been successfully used to find "good" solutions to this and other related problems in the area of optimal experimental design. These heuristic algorithms include simulated annealing [59], greedy swap techniques [224], and gradient descent techniques. For this research we used the gradient descent techniques implemented in the software package GOSSET, which was recently developed by Hardin and Sloane at AT&T Bell Laboratories [122]. The primary focus of GOSSET is low-order polynomial models, which are of only limited use in characterizing typical analog circuits. For our research in automatic test pattern generation, GOSSET was extended to utilize arbitrary Lipschitz continuous functions, such as the piecewise linear output of common behavioral simulators [182] and SPICE [141].

GOSSET uses an optimization algorithm known as *Hooke and Jeeves pattern search* [133], which is based on steepest descent. The optimization begins by selecting a random point on the response surface, calculating the gradient at that point, and proposing a set of small perturbations in the direction of the gradient. If this set of perturbations causes the objective function to improve, then this "move" is accepted and the step size is increased by a constant factor. Otherwise the set of perturbations is rejected and a smaller move is attempted.

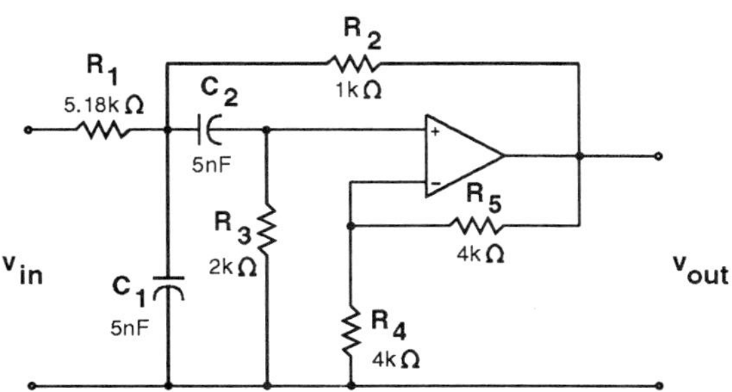

Figure 7.1   Bandpass filter with center frequency at 24.5 kHz.

7.3.4 Bandpass Filter

As an example of automatic test pattern generation, consider the bandpass filter shown in Figure 7.1. As basis functions we select the constant function, the nominal frequency response, and the sensitivities of the nominal response with respect to R_1, C_1, R_2, C_2, R_3, R_4, and R_5.

Running the ambiguity algorithm reveals that many of the component sensitivities are linearly dependent; the constant function, the nominal response, and the sensitivities with respect to C_2, C_1, and R_1 are sufficient to fully characterize the filter.

Since there are five basis functions in the model, at least five test frequencies will be needed to estimate the system response. We impose a constraint that the test frequencies lie between 15 kHz and 40 kHz, since that is the region of the response in which we are interested, and run the I-optimality algorithm. The five test frequencies which the algorithm selects are shown in Table 7.1. Note that the fifth test point is pushed to the user-imposed limit of 40kHz, while the remaining test points sample the response at intervals of approximately 3 kHz near the nominal center frequency. Furthermore, if the constant function is eliminated from the model, then the test point at 40.00 kHz is no longer needed and disappears from the test set.

Applying the five selected test frequencies to a simulated circuit produces the estimated output and 99% confidence intervals shown in Figure 7.2. According to the testing algorithm outlined in Section 7.2.2, these confidence intervals would be compared against the filter specifications to determine whether the component should be accepted or rejected, or whether additional test vectors should be applied to tighten the confidence intervals.

Frequency	| Output |
19.32 kHz	0.914
22.57 kHz	1.65
24.89 kHz	1.99
28.42 kHz	1.30
40.00 kHz	0.482

Table 7.1   Test frequencies chosen for bandpass filter.

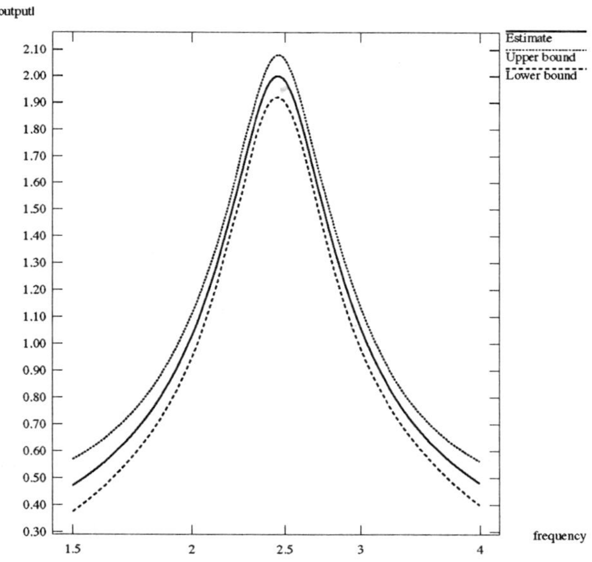

Figure 7.2   Estimated output and 99% confidence intervals for bandpass filter.

V_{GS}	V_{DS}	I_{DS}
5.0	1.0	19.04910×10^{-3}
3.2	0.9	10.00540×10^{-3}
5.0	0.1	2.66950×10^{-3}
5.0	10.0	33.62560×10^{-3}
2.0	10.0	6.24432×10^{-3}
5.0	2.2	27.45550×10^{-3}

Table 7.2   Test points chosen for MOS transistor.

7.3.5   MOS Transistor

Suppose we wish to test an MOS transistor to verify that its drain current I_{DS} falls within certain bounds over all values of V_{GS} and V_{DS}. The manufacturer has provided a level 3 SPICE model for the device[1] As basis functions, we select the constant function, the nominal performance, and the sensitivities with respect to V_{T0}, k', γ, t_{ox}, and θ. The normalized basis functions are shown graphically for three values of V_{GS} in Figure 7.3. Although only three values of V_{GS} are shown, both V_{DS} and V_{GS} are treated as continuous variables, so the response surface is 2-dimensional and continuous.

Running the ambiguity group algorithm, we find that k' and the nominal performance are linearly dependent on the entire response surface, which can be seen in Figure 7.3(a) and 7.3(c). So the k' vector is dropped from the model, and there are 6 remaining independent basis functions. Therefore we will need at least 6 test points to characterize the device.

To prevent the I-optimality algorithm from selecting unreasonable test points, we impose constraints on the inputs V_{GS} and V_{DS} such that $0.1 \text{ V} \leq V_{DS} \leq 10.0 \text{ V}$ and $2.0 \text{ V} \leq V_{GS} \leq 5.0 \text{ V}$. We then run the I-optimality algorithm; it selects the test points shown in Table 7.2.

Figure 7.4 shows the estimated response curves for three values of V_{GS} after applying the indicated 6 test vectors to a device, along with the 99% confidence intervals for those estimates. The confidence intervals are based upon a measurement accuracy of 0.1%. The expected value of the model error at $V_{GS} = 3.5V$ is shown in Figure 7.5,

---

[1] This example is based on the HP CMOS26B 0.8 μm model.

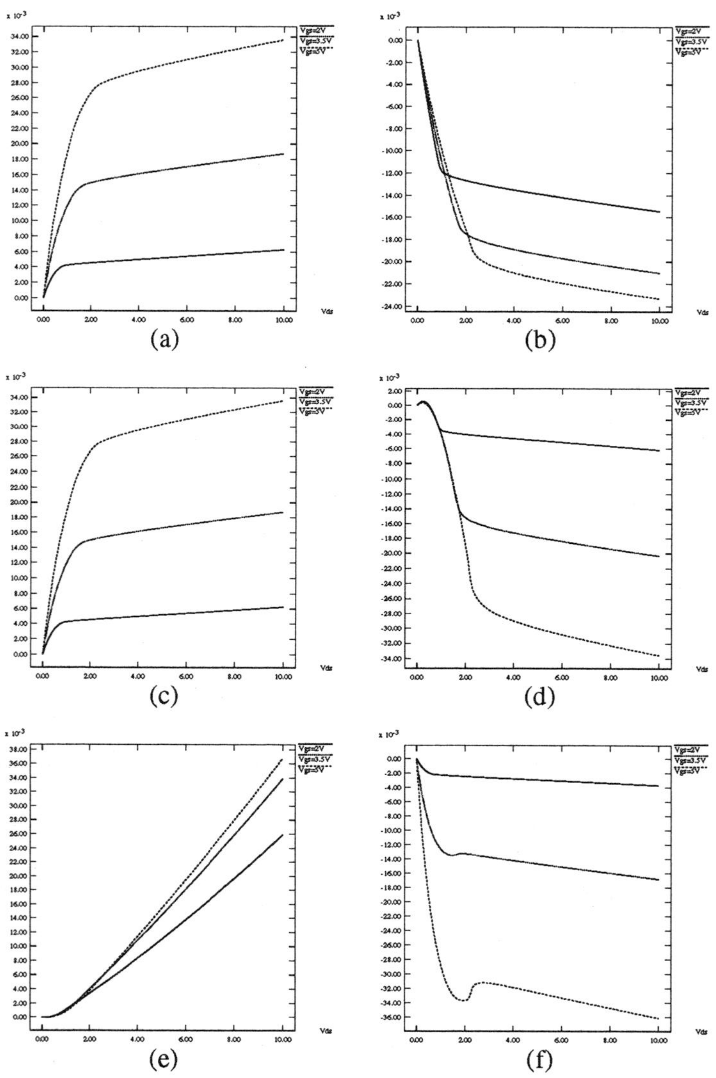

Figure 7.3 (a) Nominal I_{DS} vs. V_{DS} transistor curves. (b) Sensitivity with respect to V_{T0}. (c) Sensitivity with respect to k'. (d) Sensitivity with respect to γ. (e) Sensitivity with respect to t_{ox}. (f) Sensitivity with respect to θ.

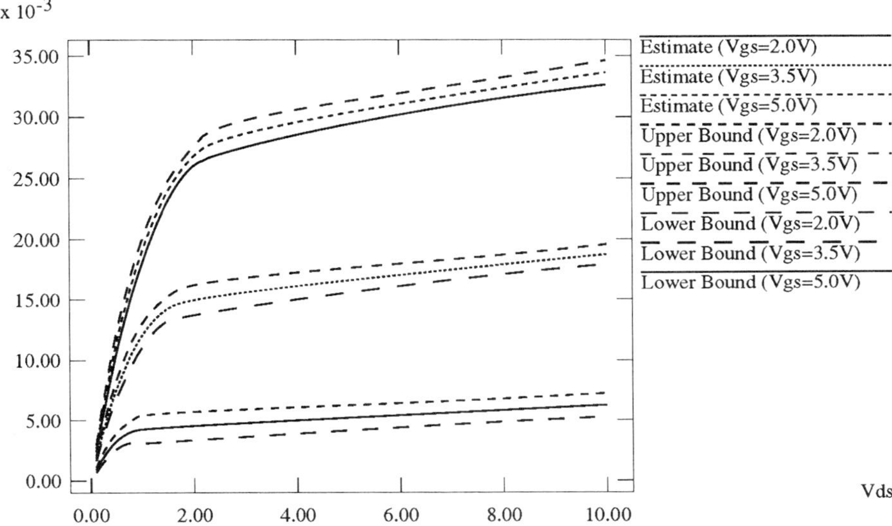

Figure 7.4 Estimated response and confidence intervals for MOS transistor from seven test points.

from which we conclude that our estimates are least accurate near $V_{GS} = V_T$. This fact is not surprising, since that region of transistor operation is difficult to model.

7.3.6 Nyquist-Rate D/A Converter

A 6-bit Nyquist-rate D/A converter based on binary-weighted current sources is shown in Figure 7.6. The basis vectors for the system are chosen to be $\{1, x_5, x_4, x_3, x_2, x_1, x_0\}$, where the constant function 1 is used to model the converter offset. Since there are seven independent basis functions in the model, at least seven tests must be performed to fully characterize the system. The I-optimal design is shown in Table 7.3, along with the next seven extra points which would be chosen, in succession, to tighten the confidence intervals on the estimated performance.

Application of the seven initial test vectors to a simulated D/A produces the results shown in Figure 7.7. The upper and lower confidence intervals illustrate how the entire performance of the D/A can be modeled quite accurately after the application of only seven well-chosen test vectors. Furthermore, we may be able to draw some conclusions

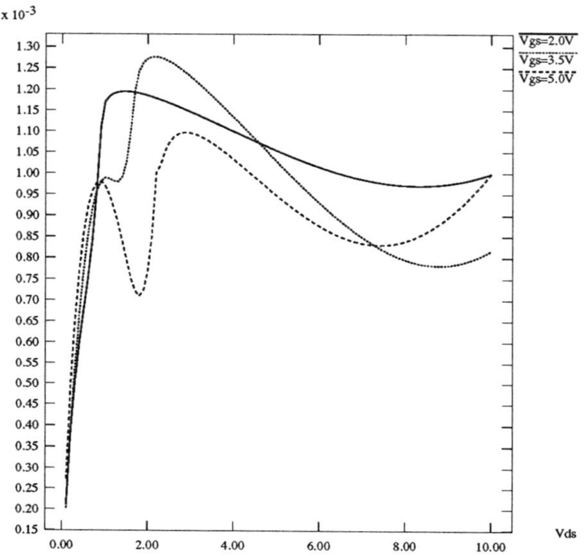

Figure 7.5   Standard error of estimated response for MOS transistor from seven test points.

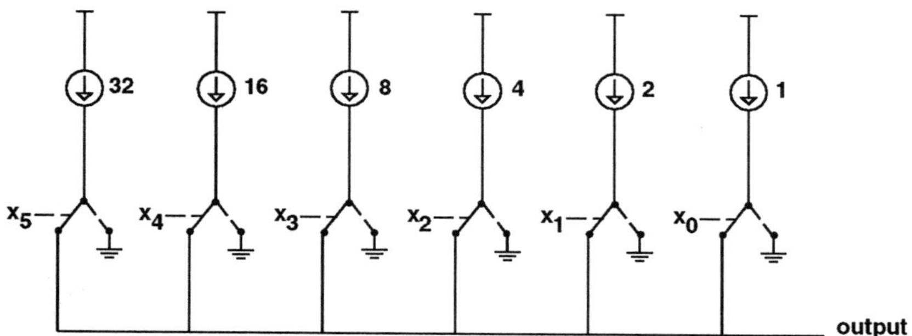

Figure 7.6   6-Bit binary-weighted current source D/A converter.

Code	Inputs						I-Value
	x_5	x_4	x_3	x_2	x_1	x_0	
8	0	0	1	0	0	0	
15	0	0	1	1	1	1	
21	0	1	0	1	0	1	
22	0	1	0	1	1	0	
35	1	0	0	0	1	1	
44	1	0	1	1	0	0	
59	1	1	1	0	1	1	1.27778
61	1	1	1	1	0	1	1.12500
38	1	0	0	1	1	0	0.97619
1	0	0	0	0	0	1	0.83333
48	1	1	0	0	0	0	0.70000
48	1	1	0	0	0	0	0.58333
26	0	1	1	0	1	0	0.55263
10	0	0	1	0	1	0	0.52222
All 64 codes							0.10938

Table 7.3   Test vectors chosen for D/A converter.

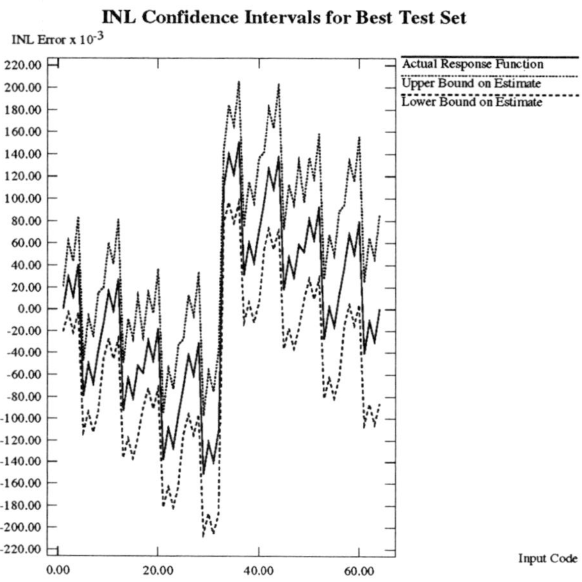

Figure 7.7   Upper and lower bounds on INL error from seven test vectors.

regarding the acceptability of this D/A, depending upon the INL specification. If the INL specification is greater than 0.2 LSB, then the D/A should be accepted with no further tests. If the INL specification is less than 0.1 LSB, then the D/A should be rejected with no further tests. If the INL specification falls between these bounds, then additional test vectors must be applied to tighten the confidence intervals.

7.4   ATPG FOR A/D CONVERTERS

In [287][288] a *linear model* for data converters along with a test selection strategy was presented. The model for an N-bit A/D converter represents the $2^N - 1$ transition points of the converter as a linear function of the component errors,

$$t = S_v v \qquad (7.18)$$

where t is an $2^N - 1$-dimensional vector that represents the transition points, v is an m-dimensional vector that represents the component errors, and S_v is the $2^N - 1$ by m sensitivity matrix with full column rank.

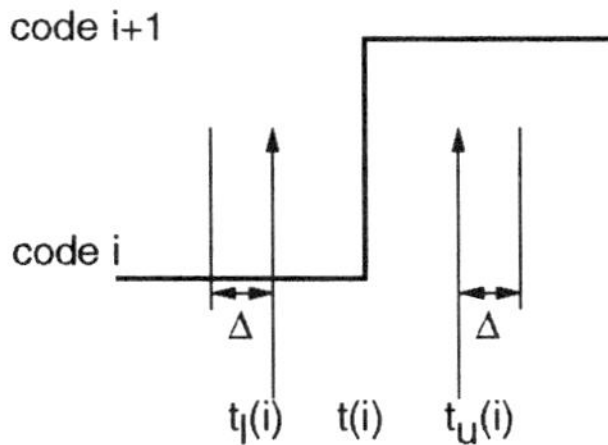

Figure 7.8 Detection thresholds

Assuming no measurement noise, only m linearly independent test points are required to estimate the m model coefficients. The objective is to find the optimal subset of the full set of $2^N - 1$ test points that minimizes the prediction variance of m. Stenbakken [291] introduced an algorithm for a near optimal solution based on QR factorization with pivoting. The algorithm first chooses the row of S_v with the largest norm, orthogonalizing all remaining rows to it using a modified Gram-Schmidt orthogonalization procedure, then choosing the row of largest norm of those remaining, and repeating the process. The algorithm records the row indices during pivoting, and chooses the m pivot rows of the initial S_v for S_v'.

7.4.1 Measurement and Detection Threshold

Previous testing strategies [288][129] have difficulty in accurate measurements of circuit performance such as A/D converter transition points in the presence of measurement noise. To circumvent the problem, we propose a simpler measurement that is robust against noise. In contrast to the traditional approach where the exact value of a performance parameter such as a transition point is measured, we verify that *a circuit performance parameter falls within certain detection thresholds in the presence of measurement noise.* For example, the test for A/D converter transition point, $t(i)$, consists of only *two* A/D conversions for input thresholds $t_l(i)$ and $t_u(i)$. A successful test corresponds to outputs of the converter for $t_l(i)$ and $t_u(i)$ being $o_l \leq i$ and $o_u \geq i+1$, respectively. In this case, the transition point plus uncertainty due to noise, $t(i) + \delta$, will lie within the range given by $(t_l(i), t_u(i))$, where δ is the uncertainty due to noise. Assuming noise δ falls in the range of $(-\Delta, \Delta)$, $t(i)$ satisfies the following inequality in the presence of measurement noise (Figure 7.8),

$$t_l(i) - \Delta \leq t(i) \leq t_u(i) + \Delta. \tag{7.19}$$

7.4.2 Testing Nonlinearity Errors

Successful tests for integral and differential nonlinearity establish bounds on these parameters for *all* inputs. For a 10-bit converter, there are a total of 1024 possible inputs. Fortunately, it suffices to test a subset, T, of the set of all inputs, S, due to correlations between outputs of data converters.

We can bound integral and differential nonlinearities for individual inputs using the proposed measurement scheme. The user specifies the *INL test threshold*, s_b, from which we compute the needed test input thresholds.

Each successful test i establishes upper and lower bounds on $s(i)$. Due to correlations between the transition points, INL at other points may also be bounded. In test selection, we choose the best subset, T, of test points, S, such that all untested INLs are bounded by a *maximum INL performance*, s_m. Let U be the set of unbounded INL. *Test coverage* is defined as

$$1 - \frac{|U|}{|S|} \tag{7.20}$$

where $|\cdot|$ denotes the cardinality of a set. *Full coverage* refers to the case where U is empty and hence test coverage is one.

We formulate the test selection problem as follows. Let Δ be the magnitude of the measurement noise in (7.19), s_m be the specified maximum INL performance, s_b' be the INL threshold in the absence of noise, $s_b = s_b' + 2\Delta$ be the INL threshold in the presence of noise, T be the test set, choose the minimum T such that

$$-s_m \leq -s_b \leq s(i) \leq s_b \leq s_m, i \in T \tag{7.21}$$

$$-s_m \leq s(i) \leq s_m, i \notin T \tag{7.22}$$

where s is given by a linear transformation of some independent variables from the behavioral model:

$$s = U_s c_s + \mu_s \tag{7.23}$$

where $c_s \sim normal(0, \Sigma_{cs})$ is a zero mean multivariate normal distribution with r_s statistically *independent* variates.

Due to linear dependence, we use *linear programming* [246] to check the bounds on untested INL, $s(i)$, in (7.22). We formulate the problem as follows. We check that

$$\max_{c_s} s(i) \leq s_m, i \notin T \tag{7.24}$$

$$\min_{c_s} \; s(i) \geq -s_m, i \notin T \qquad (7.25)$$

such that

$$-s_b \leq s(j) \leq s_b, j \in T \qquad (7.26)$$

$$-k\sqrt{\Sigma_{cs}(i,i)} \leq c_s(i) \leq k\sqrt{\Sigma_{cs}(i,i)}, i = 1 \ldots r_s, \qquad (7.27)$$

where k is a user specified parameter to limit the distribution, c_s. The limit prevents signatures with small magnitudes from being scaled unrealistically with unbounded c_s in the linear program. Typically, we choose $k = 5$ for a 10σ statistical limit. If c_s is unbounded, then the statistical information provided by the distribution of c_s is lost.

The optimum test selection problem is likely intractable. Thus, we propose a heuristic solution. First, we rank the tests according to the algorithm proposed in [291] based on QR factorization with pivoting.

1. Choose the row of U_s with the largest norm,

2. Orthogonalize all remaining rows to it using a modified Gram-Schmidt orthogonalization procedure,

3. Choose the row of largest L_2 norm of those remaining,

4. Repeat (2) and (3) until all rows are chosen,

5. Record the row indices of all chosen rows, and list them in the order they were chosen.

As a result, we produce a list of INLs with decreasing order of importance where the jth item on the list is $s(i)$.

Next, we use the following algorithm to find T.

1. Let $j = 1$, and T be empty.

2. If j is larger than the number of test points, done.

3. Pick the jth item on the list, $s(i)$. Check $s(i)$ for bounds specified in (7.22) using linear programming.

4. If the linear programming is infeasible, quit due to performance specifications too tight.

5. If the bounds are not satisfied, add i to T.

6. $j = j + 1$ and go to (2).

Unless the specified performance is infeasible, the algorithm guarantees full coverage since every INL is either bounded or constrained by a test. To tradeoff test size against coverage, the algorithm should stop after a certain coverage is achieved.

7.5 MOS MISMATCH MODELING

Mismatch in a certain component can be defined as the variation in the value of identically designed components. Mismatch can be divided into two categories: *random* and *systematic*. Systematic mismatch is that part of the total mismatch where a deterministic trend can be observed in the mismatch values of the various transistors. The remainder of the mismatch, in which no apparent trend is observed, falls under the category of random mismatch.

As a result of testing three architectures of fabricated D/A converters, we have developed some new mismatch models. We propose modeling the systematic component as a linear gradient across the die, with the direction of the gradient being a function of the location of the die on the wafer. Our preliminary study shows that the systematic component accounts for nearly half of the total mismatch. We have observed that the direction of the gradient across the die is consistent with a radial mismatch pattern across the wafer. With respect to the random component of the mismatch, we have proposed a model which is 60% more accurate, on average, than the previous models.

7.5.1　Extraction Methodology

Our methodology for the measurement of mismatch is based on the idea that extracting the mismatch information from the observed behavior of a functional circuit is often more efficient and more accurate than individually probing the devices. In our research, for example, transistor mismatch information was extracted from measurements on a D/A converter. The converter contains a regular array of transistors, and mismatch between the transistors in this regular array directly manifests itself as INL errors in the output of the D/A converter. Measurements of these INL errors, therefore, can be used to compute the mismatch of each transistor in the array.

This new methodology is more efficient because no probing of the wafer is required; all measurements can be made by applying various test vectors to the already-packaged chip. Furthermore, the new methodology is more accurate than measuring one device at a time because each component is, in effect, sampled multiple times. The D/A functions by turning on various combinations of transistors, so the output of the D/A for a given input code is the sum of the transistors which are ON for that code. Since each transistor is ON for more than one sample, the current contribution from that transistor is measured multiple times and hence the effective measurement noise is reduced.

7.5.2   Systematic Mismatch

The systematic component of the mismatch represents that portion of the mismatch which can be precisely predicted, given the process gradients. The random mismatch, on the contrary, represents that portion of the mismatch which is stochastic and hence cannot be predicted. The objective of a mismatch model, therefore, is to maximize the percentage of mismatch which can be systematically predicted and, with regard to the random mismatch, to characterize the σ^2 of the random mismatch as accurately as possible.

To determine what portion of the mismatch is systematic we generate 3-D plots of the actual mismatch values for the transistor arrays. A very strong linear gradient has been observed in all the arrays. We model this gradient by a plane which minimizes the mean square error from the actual values. Figure 7.9(b) shows the plot of the best fit planes corresponding to the plots of actual mismatch in Figure 7.9(a). We observe that the best fit planes have a large inclination, indicating the importance of the systematic component. Notice that for a completely random mismatch, the best fit plane would be flat. Furthermore, note that these planes are oriented in different directions, so modeling the mismatch as a one-dimensional function of distance would not produce a very accurate model. To make any reasonably accurate predictions about the mismatch we need to know the relative placement of the transistors of interest, the orientation of the best fit plane, and the inclination of the best fit plane.

The next problem in the modeling of systematic mismatch is to find the dependence of systematic mismatch on the position of the die on the wafer, since the direction of the best fit plane is different for different dies. Under the assumption that the origin of systematic mismatch is *deterministic* in nature, it should be possible to predict the direction of the gradient based on the position of the die on the wafer.

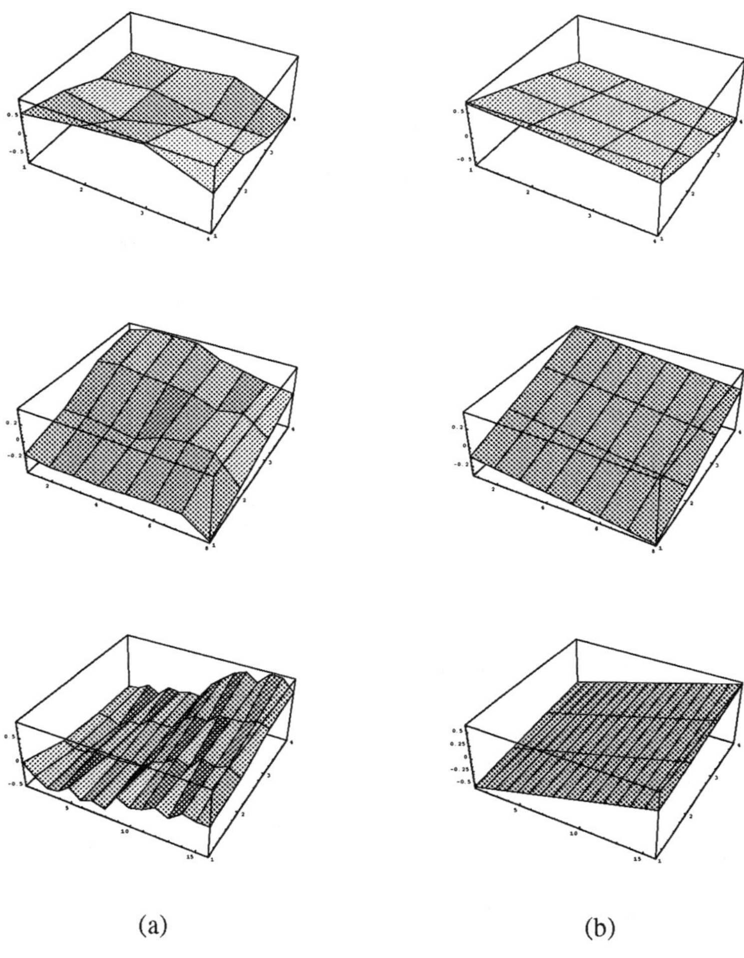

(a)   (b)

Figure 7.9   (a) 3-D plots of actual intra-die mismatch, for three linear arrays. (b) Systematic mismatch approximated by a linear gradient.

7.5.3 Random Mismatch

Random mismatch is characterized by its variance σ^2, and we propose to develop a model for accurately predicting this variance as a function of $V_{GS} - V_T$ and device area.

For the binary D/A converters, we estimate the variance of random mismatch by using the maximum likelihood estimation criterion. For the linear array we take the difference of the mismatch value from the best fit plane to get the random mismatch. Taking the mean of the square of these values, we obtain the variance of random mismatch. We have measured the mismatch values at two different reference currents, 1 mA and 0.6 mA.

The drain current in the saturation region is given by

$$I = \frac{K}{2}(V_{GS} - V_T)^2 \tag{7.28}$$

and the variance in the drain current, ignoring the correlation between V_T and K, can be written as

$$\frac{\sigma_I^2}{\bar{I}^2} = \frac{\sigma_K^2}{\bar{K}^2} + 4\frac{\sigma_{V_T}^2}{(V_{GS} - \bar{V}_T)^2} \tag{7.29}$$

where σ_K and σ_{V_T} is given by

$$\frac{\sigma_K^2}{K^2} = \frac{k_1}{Area} \tag{7.30}$$

and

$$\frac{\sigma_{V_T}^2}{(V_{GS} - V_T)^2} = \frac{k_2}{Area(V_{GS} - V_T)^2} + \frac{k_3}{(V_{GS} - V_T)^2} \tag{7.31}$$

The above model differs from the model proposed in [238] in two ways. First, we do not have any distance-dependent term because we have already accounted for the systematic mismatch. Second, we have added the k_3 term, which does not appear in the Pelgrom model, because in fitting Equation (7.31) to our data we observed that this k_3 term, which does not contain an area dependence, was much more statistically significant than the corresponding area-dependent term (corresponding to k_2). The addition of this k_3 term leads to an average improvement of nearly 60% as compared to the previous models.

The values for the various variables in Equations (7.29), (7.30), and (7.31) are shown in Table (7.4). Equation (7.28) is used to calculate the nominal value of $V_{GS} - V_T$.

Description	Transistor Count	W (μm)	L (μm)	$Area$ (μm^2)	$V_{GS} - V_T$ (V)	Measured σ_I/I
DAC46_lin_1mA	192	121	24	2904	0.679	0.00374
DAC46_bin_1mA	768	48	110	5280	0.288	0.01153
DAC55_lin_1mA	384	21	24	504	1.163	0.00374
DAC55_bin_1mA	384	15	48	720	0.344	0.01005
DAC64_lin_1mA	768	28	21	588	0.660	0.00424
DAC64_bin_1mA	192	22	47	1034	0.278	0.01029
DAC46_lin_0.6mA	192	121	24	2904	0.526	0.00458
DAC46_bin_0.6mA	768	48	110	5280	0.223	0.01374
DAC64_lin_0.6mA	768	28	21	588	0.511	0.00500
DAC64_bin_0.6mA	192	22	47	1034	0.216	0.01300

Table 7.4   Measured mismatch for 10 sets of measurements on 6 "optimal" test structures.

Model Used	Mean percentage error, relative to actual values
Our Model	11.78%
Pelgrom's Model [238]	30.04%
Lakshmikumar's Model [169]	30.44%

Table 7.5   Comparison of various models for Random Mismatch.

Values of the parameters k_1, k_2, and k_3 were obtained using nonlinear programming. In the nonlinear programming we minimize the square of the difference between the actual and predicted values, subject to the constraint that each of the coefficients must be greater than or equal to 0. The value of k_2 comes out to be zero in our model, validating our claim that the contribution of k_3 is much more significant than the k_2 term.

In Table 7.4 we present the actual values of σ_I/I vs. the values predicted by our model, along with the percentage errors. The average error from the actual values is approximately 12%. A comparison of our model with that in [238] and [169] is presented in Table 7.5. In this comparison, only the random part of the total mismatch was used. The mean percentage error is reduced by 60%, from approximately 30% to 12%, by using our model over the previous models.

7.6 CONCLUSION

We have presented a new CAD algorithm which automatically generates a minimal set of test vectors for characterizing a general class of analog circuits, namely those circuits which can be efficiently modeled as an additive summation of user-defined basis functions. The algorithm chooses the set of test vectors so as to minimize the average prediction variance of the model. Applying the minimal set of test vectors to a circuit produces an estimate of the circuit's performance for all possible input vectors and, more importantly, confidence intervals on those estimates which can be used to determine whether the component should be passed or failed, or whether additional test vectors should be applied to tighten the confidence intervals.

Because these techniques generate the tightest possible confidence intervals after a minimum number of test vectors, they represent the most efficient way of fully characterizing system performance. Tight confidence intervals will lead to reduced testing time for analog systems because more components will be fully verifiable, to a desired confidence level, with the minimum number of test vectors. We have applied the algorithm to several analog systems and have shown it to be efficient and effective.

8

MODULE GENERATION

8.1 DISCUSSION

Several performance-driven module generators have been developed to implement specific analog structures. Module generators can be implemented using an expert system approach or a parameterized schematic of known architectures approach. Expert system implementation for simple analog blocks such as operational amplifiers and comparators have been reported in earlier years [124]. Unfortunately, switched-capacitors and A/D converters are much more complex. No clear way exists to set the rules for their implementation.

A parameterized schematic of known architectures implementation is a better approach, because it can generate highly-effective circuits based on known architectures for complex analog functions in a flexible environment by optimizing the device sizes with respect to various performance measures [159][62][146]. This approach fits nicely into our methodology.

In the remainder of this section, discussed briefly will be our existing module generators, OPASYN (operational amplifier synthesis) [159], ADORE (switched-capacitor synthesis) [320][321], and CADICS (A/D converter synthesis) [146].

8.1.1 OPASYN

OPASYN [159] is an automatic synthesis tool for the generation of operational amplifiers. It is based on analytic circuit models. Given a set of specifications, OPASYN performs the necessary optimizations for transistor sizing. Layout is generated, and a performance summary is presented. Originally, this program only considered the generation of operational amplifiers. Recently, it has been expanded to include comparators and output buffers.

The design synthesis phase requires two steps: (1) a circuit topology is selected based on the specifications, and (2) the circuit is optimized to meet the required specifications. To simplify the optimization process, for each given topology, abstracted design parameters are selected to reduce the number of variables in the circuit optimization problem. The range of possible solutions is *not* affected by this step. Such a parameter is bias current. Bias current defines the sizes of the biasing devices. Rather than having two variables (width and length) for each bias device, the problem is greatly simplified by lumping these variables into one variable representing bias current. Sets of equations representing the function and performance for each amplifier are defined using these parameters. OPASYN then uses these equations to optimize the parameters solving a nonlinear optimization problem, meeting all required specifications, and minimizing a cost function.

Physical assembly consists of three steps: (1) leaf cell generation for the layout blocks, (2) floorplanning using slicing trees, and (3) custom routing and layout spacing [159].

8.1.2 ADORE

ADORE is a switched-capacitor filter module generator [320]. It places special emphasis on the physical design aspect. ADORE separates the switched-capacitor filter design problem into a filter synthesis problem and a layout problem. The FILSYN [299] program automates the design of switched-capacitor filters. It mimics the decision making process of a filter designer. For layout, ADORE uses a sophisticated algorithm to generate the capacitor arrays required, and then uses standard operational amplifier and switches to complete the layout. The operational amplifiers, switches, and capacitors are ordered to optimize the wiring required.

This program is currently being modified to better fit into our design methodology and to expand its capabilities such as implementing fully differential switched-capacitor filter architectures.

8.1.3 CADICS

A performance-driven CMOS A/D module generator, CADICS [146], has been developed. Given a set of specifications and data on the technology, the design-rule, device-matching, and the floorplan, CADICS will not only generate a complete netlist

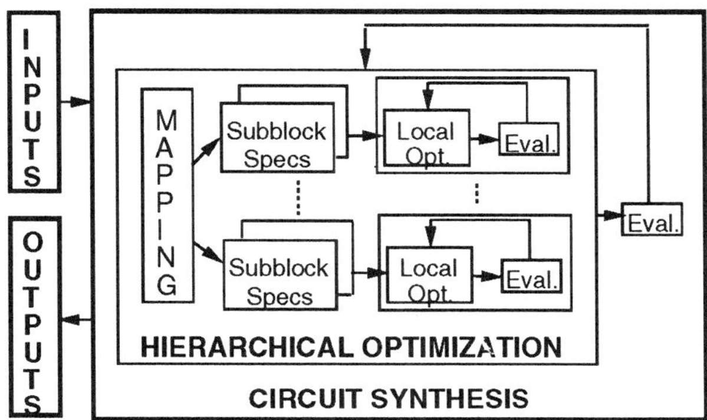

Figure 8.1   Hierarchical Optimization Procedure

but also generates the layout of the A/D converter. The implementation of this A/D module generator is divided into two parts: circuit synthesis and layout synthesis.

Circuit synthesis takes as its inputs a set of specifications for an A/D converter such as resolution, conversion-speed, maximum input signal frequency, maximum DNL, maximum INL, total power dissipation, supply voltage, reference voltage, area, layout aspect-ratio range, and technology. In addition, statistical information such as capacitor and transistor matching can be provided.

Recognizing that A/D circuits are realized using different functional blocks such as sample/hold blocks, gain-of-two circuits, comparators, digital circuitry, and others, CADICS is implemented by using a *hierarchical* optimization approach. The optimization procedure is shown in Figure 8.1. First, the A/D specifications and its respective relative priorities are mapped into sub-block specifications and their respective priorities. This mapping function can be realized by a set of rules given by experienced analog designers, by running a behavioral simulation of the A/D circuit, or by a combination of both. Then the specifications and priorities of each sub-block are fed into the local optimizer to generate the device sizes. This local optimizer can be a low-level synthesis block such as OPASYN or LAGER [28] (to generate digital circuitry), or library cells can be used. When all of the sub-blocks have been optimized locally, a behavioral simulator is used to obtain the A/D performance. If the performance is unsatisfactory, the entire operation is repeated by relaxing the given specifications until the desired performance is obtained, or the iteration limit of optimization cycles is exceeded.

In addition to the input files for circuit synthesis, layout synthesis requires: (1) a structured A/D converter netlist and (2) device structure information from the circuit synthesis phase. A mixed top-down/bottom-up approach [146] is used. This method consists of: netlist rearrangement, circuit partitioning, floorplan optimization, leaf-cells generation, sub-block place and route, and global place and route. In all steps, emphasis is placed on the *fine details of layout* which influence analog circuit performance. To isolate capacitors from noisy signals, capacitors are placed on top of wells which are connected to an analog ground line. Depending on the floorplan and netlist, well and substrate contacts are created during placement. Fully differential architectures and symmetry in layout are maintained. Sub-block routing is done with ROAD [191].

8.2   EXAMPLE

Sample inputs and outputs for ADORE are presented in this section. Given a SPICE-like netlist for a switched-capacitor filter circuit and a file containing technology information, the leaf cells are generated and placed. Routing is left as a post processing step. The circuit is composed of three primary components—switches, capacitors, and operational amplifiers. The switches and capacitors are generated by ADORE. The operational amplifiers can be generated by hand, obtained from a library, or generated using OPASYN.

A sample schematic input is shown in Figure 8.2. The technology file, the other input file, contains information on capacitances, on which layers to build the capacitor, and on the sizing of nets. The layout output from ADORE is shown in Figure 8.3. At the top are the switches. The capacitor arrays are in the center. And at the bottom are the operational amplifiers. Input/output vias are located along the sides of the cell.

```
*** Low-pass Filter with 2 poles / 2 zeroes
*** Integrators and their integrating capacitors
e1 2 0 0 1 new_twost:symbolic
ca   1 2 8.0p
e2 4 0 0 3 new_twost:symbolic
cb   4 3 8.0p
*** Switched capacitors
c3 c3_a c3_b 7.0p
c1 c1_a c1_b 1.0p
c1s in   3     4.0p
c2 c1_b c2_b 7.0p
C4 c2_b c3_b 1.0p
*** Switches
sc3_1 2 phi2 c3_a phi1 0
sc3_2 3 phi2 c3_b phi1 0
sc1_1 in phi2 c1_a phi1 0
sc1_2 1 phi1 c1_b phi2 0
sout 4 phi2 ou phi1   10
sc4 0 phi1 c2_b phi2 4
*** input
vin in 0
***output
out ou 0
```

Figure 8.2 Example Input for ADORE

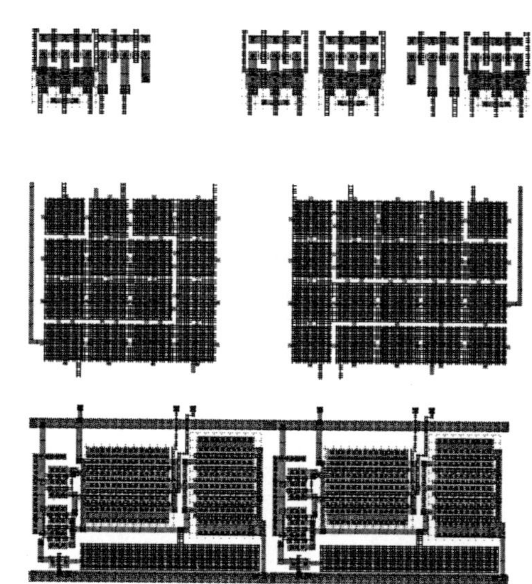

Figure 8.3   ADORE Generated Output

9

CURRENT SOURCE DIGITAL-TO-ANALOG CONVERTER DESIGN EXAMPLE

9.1 INTRODUCTION

Presented in this chapter is the first of a series of industrial strength design examples of increasing levels of design complexity to illustrate the methodology and to show its effectiveness. *Design complexity* is difficult to quantify, because there are many factors to consider, e.g. designer level of expertise, availability of tools, and aggressiveness of performance specifications desired. However, to show where our examples fit into the "big picture," an attempt is made to rank complexity levels in Figure 9.1. Our first example, current source D/As, falls somewhere in the mid-range for analog circuits.

The complete design flow for this interpolative switched current source D/A converter is given. This chapter presents the following: (1) the design and physical synthesis phases; (2) data on the three fabricated 10-bit D/A converters; and (3) a testing methodology for the fabricated parts.

Using the design paradigm, the design time for this type of D/A has been greatly reduced, and a significant amount of data has been gathered for fault diagnosis. To further speed up the design process, a new module generator for this class of D/A converter has also been developed. As an example of the design speed-up, the second of the two chips fabricated required only *five days* from the time of receiving the design specifications to when it was sent out for fabrication. All three D/As required only one fabrication iteration.

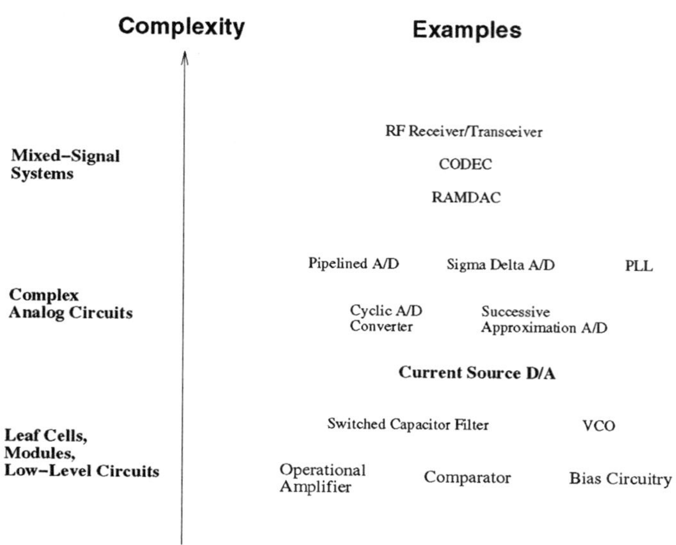

Figure 9.1   Complexity Levels

9.2 DESIGN SPECIFICATIONS

The architecture for the converters that were built is an interpolative switched current source D/A [267] (Figure 9.2). Given an N-bit digital word (D_{in}) and a reference current (I_{ref}), it returns an output current (I_{out}) which is proportional to D_{in}.

The converter has two stages. The first stage consists of unity or linearly weighted current sources each having a nominal value of $\frac{1}{2^{N_1}} I_{ref}$. This stage decodes the first N_1 bits of D_{in}. Depending on these bits, the currents are either channeled to the output (I_{out}), the current mirror, or the current sink (I_{sink}). If it is determined that n current sources need to be switched to I_{out} then the first n current sources are switched to I_{out}; the $(n+1)^{th}$ current source is switched to the current mirror, and the remaining unused current sources are switched to I_{sink}. The second stage is made of binary weighted current sources with the smallest current source having a nominal value of $\frac{1}{2^N} I_{ref}$. It decodes the remaining $(N - N_1)$ bits (N_2) of D_{in}. Depending on the remaining bits, these current sources are either switched to I_{sink} or to I_{out}. The currents are summed at the I_{out} node producing the analog output current.

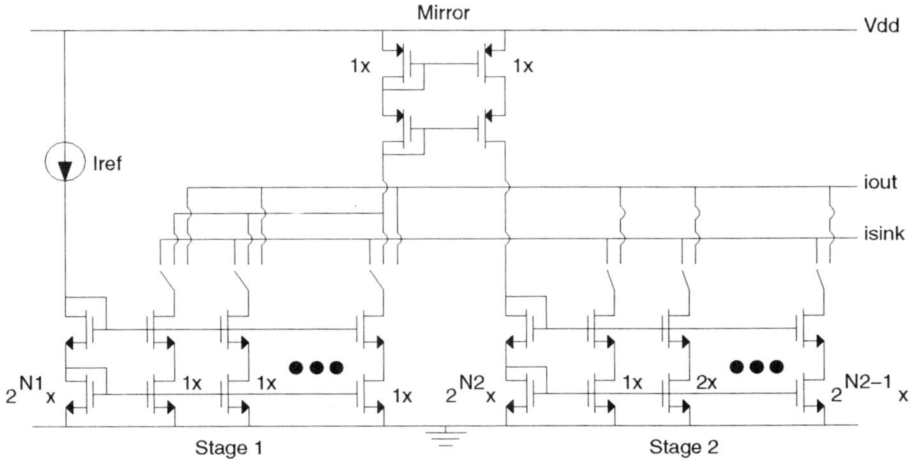

Figure 9.2 D/A Architecture

Type	Specifications
Architecture	N_1, N_2
Performance	Max 3σ bound on INL, max 3σ bound DNL
Operation	Supply voltages, output voltage (V_{out}) range, I_{ref}
Technology	Design rules, SPICE parameters, process variation (σ_W, σ_L), threshold voltage variation ($\sqrt{a_{VFB}}$) [240]
Layout	Input/output terminal locations, non-default cell placement, minimum wire widths for critical signals, use of bonding pads for a stand-alone part, etc.
Optimization	Non-default cost function(s) Objective for project: Minimize area (default)

Table 9.1 Design Specifications

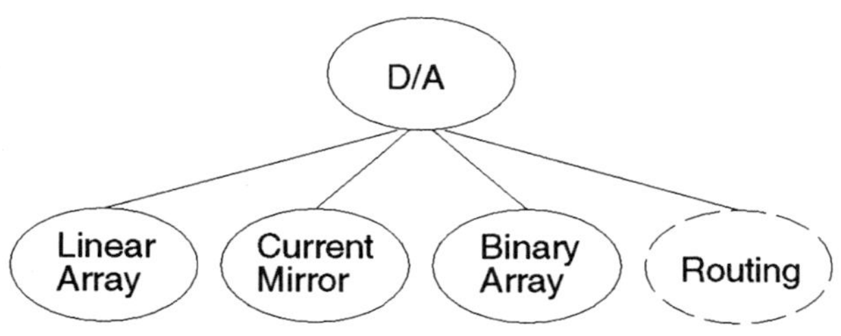

Figure 9.3 D/A Hierarchy

Table 9.1 lists the set of design specifications for the D/A. They are broken down into different categories; each of which is handled differently as will be described later. The user enters these parameters to the D/A module generator.

We chose to design and fabricate three 10-bit D/A converters. Three pairs of N_1/N_2 values were selected: 4/6, 5/5, 6/4. In all cases, we specified that the maximum 3σ bound on INL and DNL to be 2.0 LSB and 0.5 LSB respectively. The output voltage was set at half the supply voltage, and the reference current was set at 1 mA. The MOSIS scalable CMOS (SCMOS) design rules were used. The target technologies were the MOSIS 2.0 μm N-well process and the MOSIS 2.0 μm P-well process. Threshold variation data were obtained from [208]. Since the 5/5 D/A converter was a stand-alone part, bonding pad locations were specified. However, the 4/6 and 6/4 D/As were designed to be part of a larger system (a chip with both D/As), so I/O terminal locations were specified instead. The overall objective for all of the designs was to minimize layout area.

9.3   SYNTHESIS PATH

Figure 9.3 shows the two-level hierarchy used in the synthesis process. The five partitions or blocks into which the D/A has been divided each contain a set of constraints and a set of parameters to be determined by the generator for that block. The synthesis begins at the top-most block and descends as values for the parameters of that block are chosen. These chosen values then become constraints for the next block. This continues until all of the final low-level parameters (e.g. transistor sizes) are chosen.

Parameter	Description
$\sigma_{iL}, \sigma_{iM}, \sigma_{iB}$	Standard deviations of the current sources in the linear, mirror, and binary array respectively.
R_L, R_M, R_B	Output resistances of the current sources in the linear, mirror, and binary array respectively.
ΔINL, ΔDNL	Errors introduced from block level routing.

Table 9.2 Parameters for D/A Block

9.3.1 High-Level Synthesis

The top block in the design as shown in Figure 9.3 is the D/A. The constraints at this level are the overall D/A performance specifications, INL and DNL. The parameters for this block are shown in Table 9.2. Two tools are applied to select values for these parameters.

The first is the behavioral simulator [177]. It uses the *intermediate level parameters* (listed in Table 9.2) as inputs as well as some of the architectural, operational, and technological information found in Table 9.1. The simulator returns the statistical INL and DNL characteristics for the D/A converter. A sample output from the behavioral simulator is shown in Figure 9.4 for a 10-bit D/A converter. INL_{max} is approximately 1.5 LSB and DNL_{max} is approximately 0.3 LSB for this particular design. Recall that behavioral simulation is *implementation independent* [98]. These simulations are valid for any current source circuit that can be characterized by the parameters in Table 9.2.

A nonlinear optimizer [216] is the other tool that is invoked. This tool requires the performance and optimization information found in Table 9.1. The performance specifications are used as constraints for the optimization problem. The objective is to *maximize design flexibility*. From Equation 4.1, the parametric form of the flexibility function for σ_i is

$$flex_\sigma(\sigma) = -a\sigma^2 - \frac{b}{\sigma} + c$$

where $a = 0$. Similarly, the parametric form of the flexibility function for R is:

$$flex_R(R) = -aR^2 - \frac{b}{R} + c$$

where $b = 0$. An alternative approach to the one discussed in Section 4.1 was used to calculate the coefficients for the flexibility functions. One "moderate" difficulty point, assigned a value of 0, was still used. The second point, however, was a desired slope rather than another point. For σ_i, a 1% change from the "moderate" value in mismatch

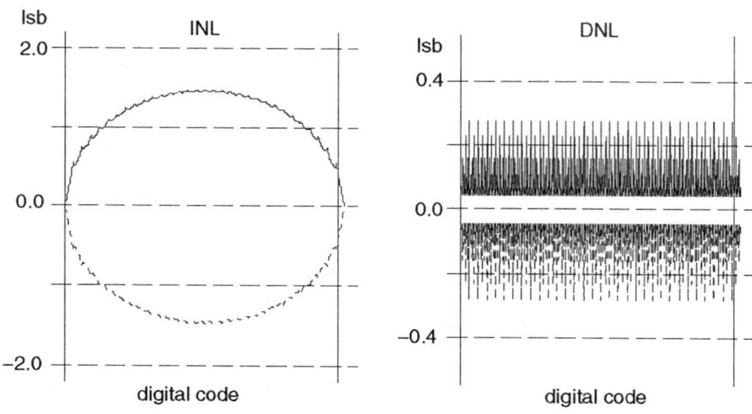

Figure 9.4   INL and DNL behavioral simulation output for a 10-bit ($N_1 = 5$, $N_2 = 5$) D/A

was set to have a slope of $+1$. While for R, a pre-defined resistance change with respect to r_o, the small-signal output resistance for a transistor in saturation, was set to have a slope of -1. This method was felt to be more appropriate. Ultimately, however, this design turned out to be fairly insensitive to the flexibility function. The coefficients are shown in Table 9.3 for $N_1 = 5$, $N_2 = 5$, and $I_{ref} = 1$ mA. Since mismatch is dependent on N_1, N_2, and I_{ref}, parametric flexibility functions were built with these parameters as variables. These are contained in the D/A module generator. Figure 9.5 shows example flexibility functions for σ_i and R. The optimization problem can be stated as:

$$
\begin{aligned}
maximize \quad & flex_{\sigma_{iL}}(\sigma_{iL}) + flex_{\sigma_{iM}}(\sigma_{iM}) + flex_{\sigma_{iB}}(\sigma_{iB}) + \\
& flex_{R_L}(R_L) + flex_{R_M}(R_M) + flex_{R_B}(R_B) \\
s.t. \quad & INL(\sigma_{iL}, \sigma_{iM}, \sigma_{iB}, R_L, R_M, R_B) \leq 2.0 \text{ LSB} \\
& DNL(\sigma_{iL}, \sigma_{iM}, \sigma_{iB}, R_L, R_M, R_B) \leq 0.5 \text{ LSB}
\end{aligned}
$$

Upper and lower bounds on the six design variables are also present but not shown in the problem statement. The behavioral simulator and the nonlinear optimizer, working together, chooses numerical values for the parameters in the D/A, solving this problem. The performance of the D/A is evaluated using the behavioral simulator, and more values are chosen. This process continues until all of the performance criteria have been met and the flexibility function has been maximized. The results of this high-level optimization are shown in Table 9.4 for the 5/5 D/A converter. An asterisk (*) denotes that they are the selected values– constraints for the next stage.

Parameter	Coefficients		
	a	b	c
σ_{iL}	0	1.53×10^{-4}	0.7
σ_{iM}	0	3.72×10^{-3}	0.61
σ_{iB}	0	2.99×10^{-3}	2.59
R_L	8.74×10^{-18}	0	0.45
R_M	1.36×10^{-14}	0	0.45
R_B	1.52×10^{-21}	0	2.52

Table 9.3 Flexibility Coefficients for the 5/5 D/A Parameters

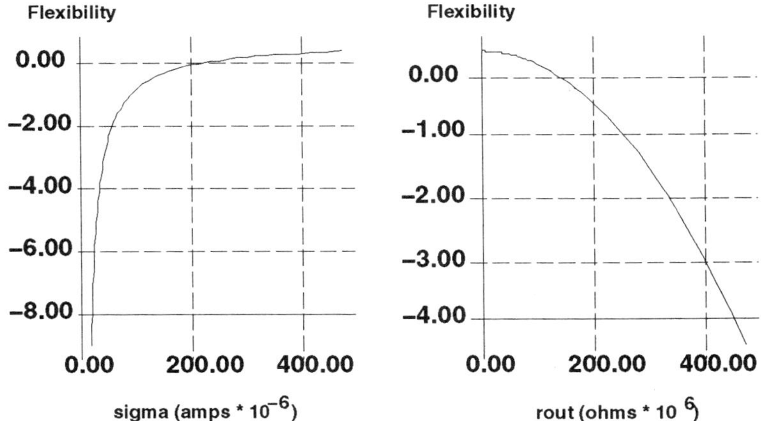

Figure 9.5 Example flexibility functions for σ_i and R.

Parameter	Value
σ_{iL}^{*}	16×10^{-5}
σ_{iM}^{*}	11×10^{-4}
σ_{iB}^{*}	28×10^{-5}
R_L^{*}	8.8×10^{7}
R_M^{*}	1.0×10^{7}
R_B^{*}	2.8×10^{9}

Table 9.4 Parameter Selection for the 5/5 D/A Block

N_1/N_2	σ_{iL}^* (10^{-5})	σ_{iM}^* (10^{-4})	σ_{iB}^* (10^{-5})	R_L^* (10^7)	R_M^* (10^7)	R_B^* (10^9)
3/7	33	2.7	2.9	4.2	1.0	5.3
4/6	23	5.4	9.1	6.0	1.0	3.8
5/5	16	11	28	8.8	1.0	2.8
6/4	11	23	85	13	1.0	2.0
7/3	7.5	50	260	19	1.0	1.4

Table 9.5 Parameter Selection for Various D/A Blocks

Block	Parameters
Linear Array	$W_{main}, L_{main}, W_{cascode}, L_{cascode}$
Current Mirror	$W_{main}, L_{main}, W_{cascode}, L_{cascode}$
Binary Array	$W_{main}, L_{main}, W_{cascode}, L_{cascode}$

Table 9.6 Parameters for Low-level Blocks

Extending this result, several optimizations on the different N_1/N_2 values were performed to compare the results. This is shown in Table 9.5. These results were used to synthesize a 4/6 and a 6/4 D/A as well as the 5/5 D/A.

9.3.2 Low-Level Synthesis

The selected values for the intermediate level parameters now become constraints for the four blocks in the next level of hierarchy. For example, σ_{iL}^* and R_L^* are the constraints for the linear array generator. As long as the linear array generator produces a current source array with $\sigma_i \leq \sigma_{iL}^*$, and an $R \geq R_L^*$, then based on the behavioral level simulation results, it can be asserted that if all of the low-level blocks meet the constraints on σ_i and R, then the D/A performance constraints (i.e. INL and DNL) will be satisfied. This assertion is checked during the final extraction and verification phase (Section 9.3.4) at the end of the design cycle.

During the low-level synthesis phase, the linear, mirror, and binary array generators are invoked to select the parameters in Table 9.6, essentially the widths and the lengths of the main and cascode MOS devices for each of the arrays. As before, a simulator and an optimizer are used. SPICE enhanced with a statistical package receives as inputs a set of Ws and Ls corresponding to the main and cascode MOS devices along with architectural, operational, and technology information from Table 9.1, and returns σ_i

	Linear		Mirror		Binary	
	Main	Cascode	Main	Cascode	Main	Cascode
	W/L	W/L	W/L	W/L	W/L	W/L
N_1/N_2	μm/μm	μm/μm	μm/μm	μm/μm	μm/μm	μm/μm
4/6	121/24	118/4	402/146	285/5	48/110	22/4
5/5	21/24	12/9	179/62	42/24	15/48	6/4
6/4	28/21	26/4	48/60	30/4	22/47	16/4

Table 9.7 Low-level Optimization Solutions

and R. A cutting plane algorithm [185] is used for the optimization.[1] The optimization problem is shown below:

$$\begin{aligned}
minimize \quad & area(W_{main}, L_{main}, W_{cascode}, L_{cascode}) \\
s.t. \quad & \sigma_i(W_{main}, L_{main}, W_{cascode}, L_{cascode}) \leq \sigma_i^* \\
& R(W_{main}, L_{main}, W_{cascode}, L_{cascode}) \geq R^* \\
& W_{main}, W_{cascode} \geq 4\ \mu m \\
& L_{main}, L_{cascode} \geq 2\ \mu m
\end{aligned}$$

The last two constraints are from the design rules. The optimizer minimizes the layout area subject to the constraints on σ_i, R, and the operating condition constraints from Table 9.1. As at the high-level, the optimizer chooses values for the parameters, in this case the Ws and Ls; invokes the circuit simulator to determine whether or not the constraints are satisfied; and continues until the objective function has been minimized and the constraints have been met. The selected values, W^* and L^* are returned. These values are shown in Table 9.7 for the 4/6, 6/4, and 5/5 D/A converters. In total, the low-level optimizer was run nine times, three times for each D/A converter. The 5/5 D/As were fabricated on a different technology and with different operating conditions than the 4/6 and 6/4 D/As. This is why no discernible pattern exists in the table, as might be expected.

The selected values, W^*s and L^*s, will satisfy the σ_i, R, and operating condition constraints. Therefore, with the previous assertion on the relationship between σ_is and Rs to INL and DNL, the D/A generated should meet the performance constraints on INL and DNL.

---

[1] The cutting plane technique was more numerically stable than the algorithm used by MINOS.

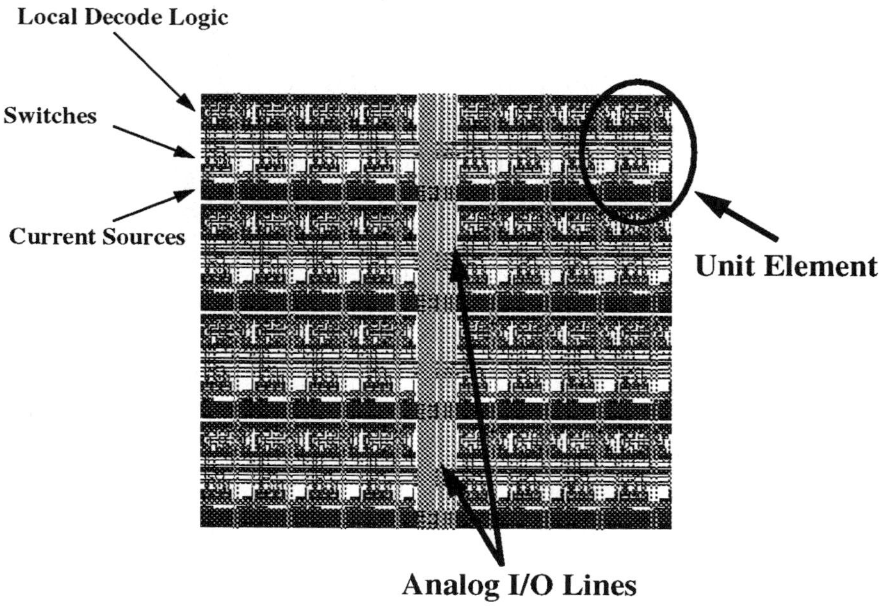

Figure 9.6   Linear Array Layout in the 5/5 D/A Converter

9.3.3 Physical Layout

The layout generation phase consists of two steps. First, the three arrays based on the results from the low-level synthesis phase are generated. The linear array, the binary array, and the current mirror generators use templates to create subcells, which are tiled to generate the desired blocks. A standard cell generator was used to synthesize the logic and latches needed for the D/A, and a module generator was used to generate the pads. In the second step, these sub-blocks are place, routed, and compacted.

The linear array uses a standard two-dimensional structure [219], with the addition of new decoding circuitry to select the interpolated output. An example array for $N_1 = 5$ is shown in Figure 9.6. It consists of 32 unit elements. A sample element is circled in the upper right hand corner. There are eight unit elements in each row with four rows in the array. To minimize sensitivity to process gradients, the linear array devices are switched using a symmetrical switching pattern [219]. As illustrated in the upper left hand corner, the unit elements are broken down into three parts. The top part consists of local decode logic. Performing part of the decode logic locally allows a tremendous savings in routing area. The three current steering switches are located in the center.

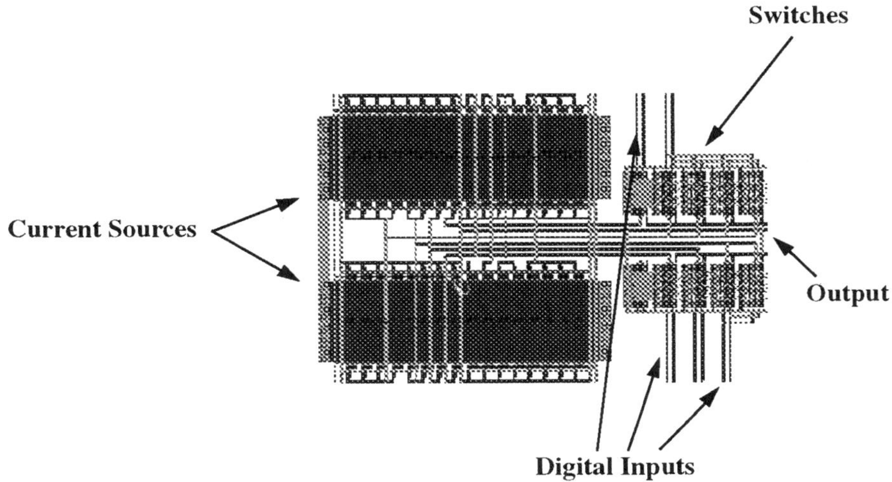

Figure 9.7 Binary Array Layout in the 5/5 D/A Converter

The current source transistors are located at the bottom with the output transistors on one side and the diode connected transistors on the other side. The analog I/O lines run vertically in the center of the array while the digital control and power lines (not shown) are connected from the left and right sides of the array.

The binary array for $N_2 = 5$ is shown in Figure 9.7. Since the binary signals control the switches directly, no decode logic is necessary. As a result, this cell is much simpler. The current sources—both the output transistors and the diode connected transistors—are shown on the left. To minimize sensitivity to process gradients, they are connected in a common centroid pattern. The current steering switches are on the right. The signal path is from the current sources to the switches to the output on the right. The digital inputs enter from both the top and the bottom on the right side of the cell. The regular nature of this array as well as the linear array makes these circuits especially well suited for tiling type layout tools (*ad hoc* cell generators) rather than general purpose placement, routing, and compaction tools.

As a part of this project, a bonding pad generator was written for the OCTTOOLS suite. Given a set of design rules, the pad generator creates a family of ten pads. These pads include analog Vdd/Vss pads, digital Vdd/Vss pads, pad ring Vdd/Vss pads, digital I/O pads, and analog I/O pads. These pads were designed to address analog noise concerns. Two digital input pads are shown in Figure 9.8. The pad on the left is a pad designed for a P-well process while the one on the right is designed for an N-well

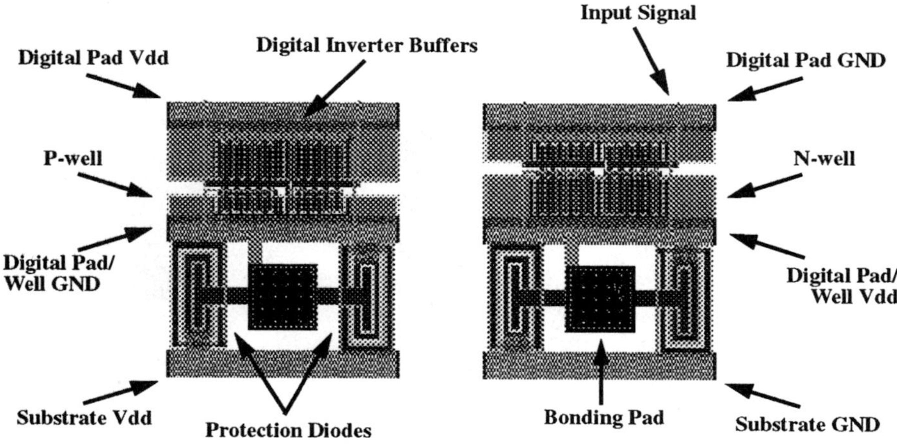

Figure 9.8 Digital Input Pads

process. An effort was made to prevent the inverting buffers from coupling signal noise into the substrate. In the P-well pad (N-substrate), the outer ring, substrate Vdd for the PMOS transistors, does not connect to the power rail for the inverter but to a separate pad (not shown). To further reduce noise coupling, the power rails for the inverters use separate nets, not the internal digital Vdd/Vss lines. Only three power rings are used. The well is shorted to the power line. A further reduction in noise could have been achieved with the addition of a fourth power line for the well, but this was felt to be unnecessary. The second D/A test chip fabricated uses these pads.

With complete schematics from the low-level blocks, ΔINL* and ΔDNL* constraints, and layout information from the design specifications, the three analog blocks, along with the digital block, are routed. [177] describes how ΔINL and ΔDNL are calculated.

This complete layout is shown in Figure 9.9. The four main sub-blocks, the linear array, the binary array, the current mirror, and the control logic, are shown in the interior of the chip. This example illustrates the point made in Section 2.4 that binary arrays consume significantly less area than linear arrays even when decoding the same number of bits. Bonding pads and bypass capacitors are shown in the exterior areas of the chip. Not necessary for the operation, but useful in debugging, test structures and probe pads are located in the unused areas of the chip.

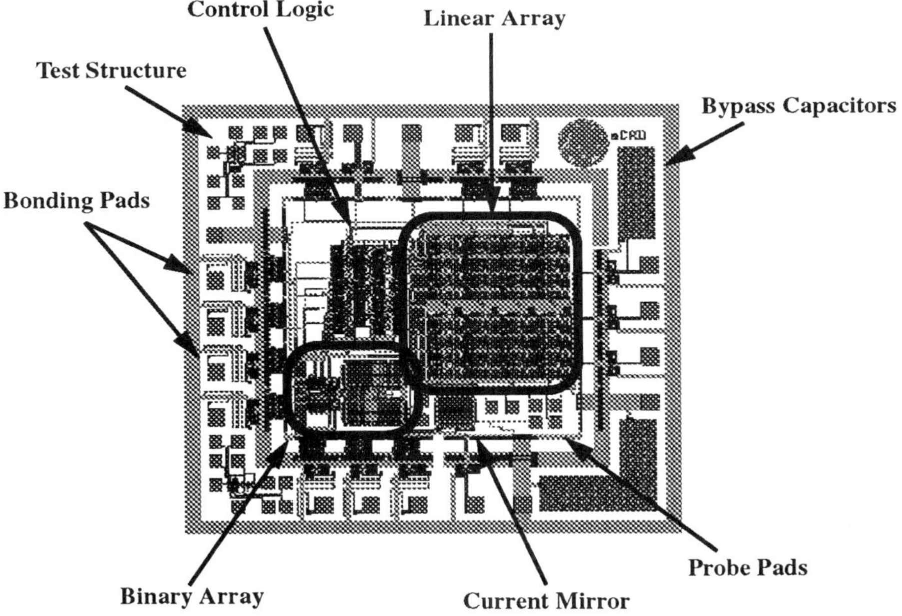

Figure 9.9 Complete Layout of the 5/5 D/A Converter

9.3.4   Extraction and Verification

During this final phase, a full layout extraction including routing parasitics is performed. The results for each block are fed into the circuit simulator which returns σ_{iL}, σ_{iM}, σ_{iB}, R_L, R_M, and R_B. Then this plus ΔINL and ΔDNL are fed into the behavioral simulator which returns the INL and DNL statistics for the D/A generated.[2]

Sensitivity analysis at the behavioral level [177] is used to insure that extra parasitics introduced in the routing has not degraded the D/As performance beyond the ΔINL and ΔDNL amount allocated. [177] shows the parasitic insensitive nature of this design with respect to the routing between the blocks.

9.4   EXPERIMENTAL RESULTS

We have fabricated three 10-bit D/As. The first chip (die photo shown in Figure 9.10) fabricated on the MOSIS 2.0 μm N-well process, contains a D/A with N_1=5, N_2=5 (5/5) partition. The active area is 1.92 mm^2. The second chip (die photo shown in Figure 9.11), fabricated on the MOSIS 2.0 μm P-well process, contains two D/As, one with N_1=6, N_2=4 (6/4) and the other with N_1=4, N_2=6 (4/6). The active areas are 2.86 mm^2 and 3.81 mm^2, respectively. The areas compare reasonably with previously published D/As [239][219]. Figures 9.12 and 9.13 depict the measured INL and DNL respectively. The top set of graphs show the data gathered from all of the parts[3] while the bottom set depict only the components which met the performance specifications.

Although not in the design specifications, transient testing on the D/A shows that for a full-scale switch at the output, the D/A can settle to 1 LSB for 10 bits from the start-of-conversion in 20 ns. The delay incurred from the falling edge of the clock to the start-of-conversion is 10 ns. A plot of this is shown in Figure 9.14. The first falling edge is the clock assertion, and the second edge is the current output fed into a current-to-voltage converter implemented using a 100 MHz operational amplifier with a feedback shunt resistor.

---

[2] All of the simulations required for this step consume on the order of 5 minutes of CPU time on a DECstation 5000/125.

[3] Twelve parts for each D/A were fabricated. Three out of the 36 total parts were not graphed due to either profound INL or DNL nonlinearities.

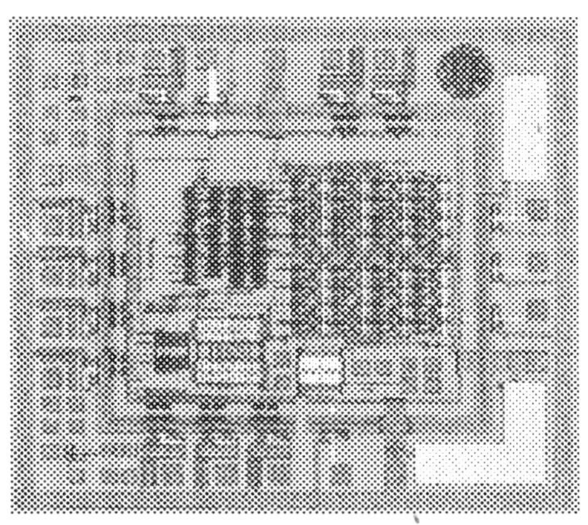

Figure 9.10 One 10-bit D/A (MOSIS 2.0μm, N-well)

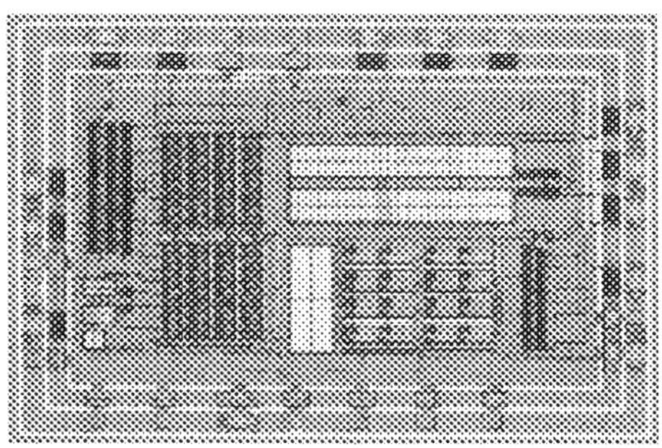

Figure 9.11 Two 10-bit D/As (MOSIS 2.0μm, P-well)

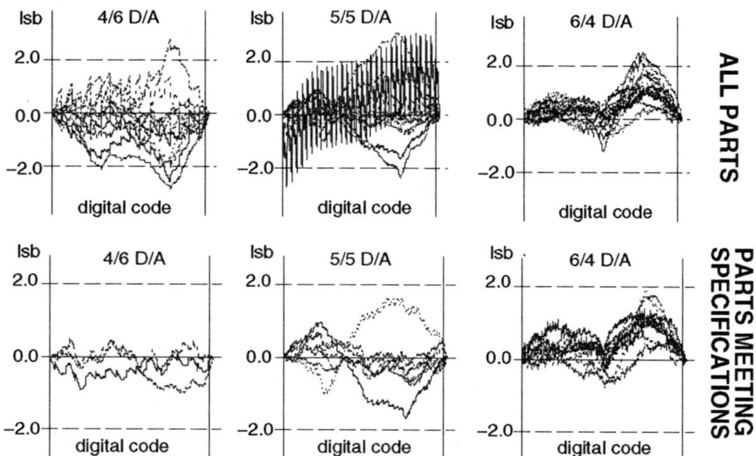

Figure 9.12   Measured INL for the three D/As

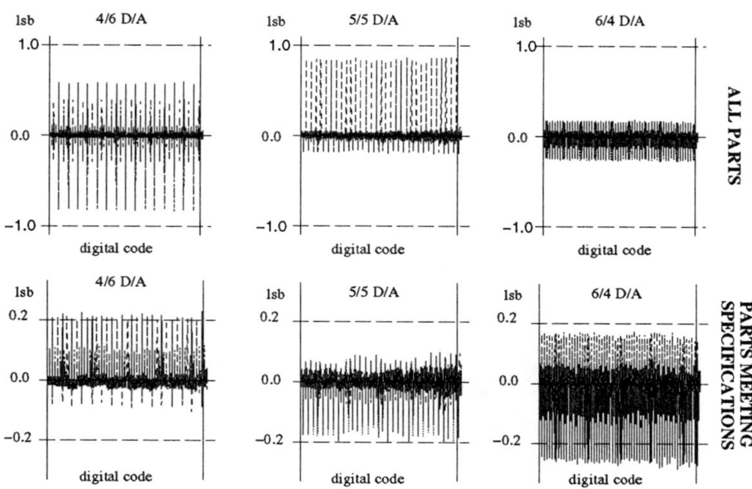

Figure 9.13   Measured DNL for the three D/As

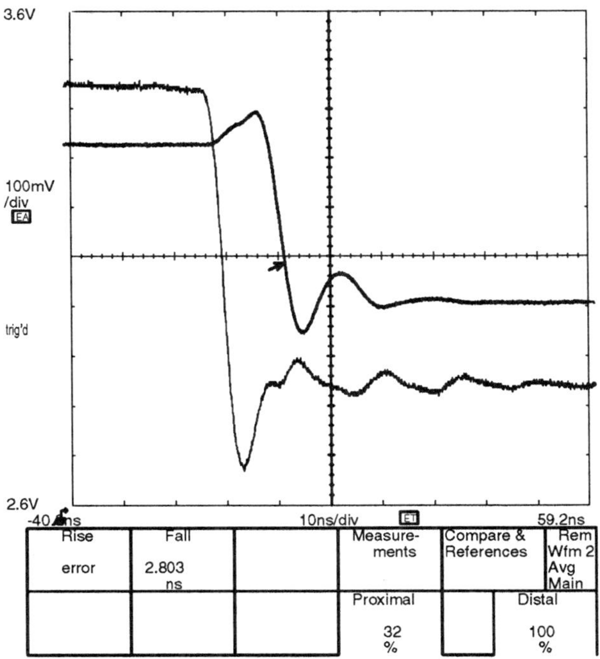

Figure 9.14   Falling Edge for the 10-bit 6/4 D/A

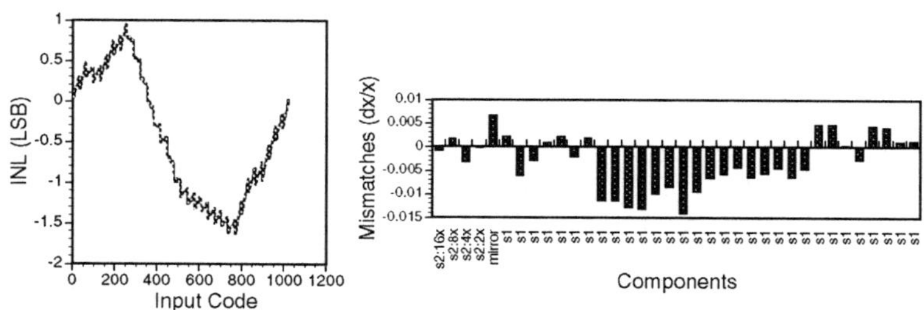

Figure 9.15 Example 1: Measured INL and Component Mismatches

9.5 MISMATCH EXTRACTION

In addition to its use during the synthesis phase, the behavioral simulator also provides a vehicle for fault diagnosis once the design has been fabricated, since the behavioral model used captures all of the essential mismatch effects. From the measured INL curves, the mismatch for each component (i.e. current sources) can be derived for all of the experimental chips. According to our model [177], the curves were assumed to be a linear function of the mismatches in each component. Using linear regression, the *best* estimate can be computed for each mismatch parameter assuming that the mismatch distributions are Gaussian.

For example, shown on the left in Figures 9.15 and 9.16 are the measured INL results for two of the 5/5 D/As. On the right, the estimated component mismatches are shown. The validity of these estimates is checked by re-computing the INL using the estimated mismatch parameters and putting these values into the behavioral simulator. The re-computed INL curve is then compared to the measured one. The graphs on the left show not only the measured INL values, but also the recomputed ones, superimposed. The maximum difference at any point on the two INL curves between the measured and re-computed values is 0.15 LSB, validating the mismatch estimates.

Having decomposed the INL into component mismatches as described above, causes for the INL characteristic can be pinpointed. For example, in Figure 9.16, the mismatch graph on the right shows that most of the nonlinearity for this chip is due to a -7% mismatch in the current mirror and a -9% mismatch in the 8x current source in the binary array. Future designs can try to improve upon this by avoiding this problem.

Component mismatches, in general, are caused by underlying basic statistical process variations. In this design, underlying process variations [240][208][168] such as the flat band voltage variations, a_{VFB}, and the standard deviation of the MOS transistor

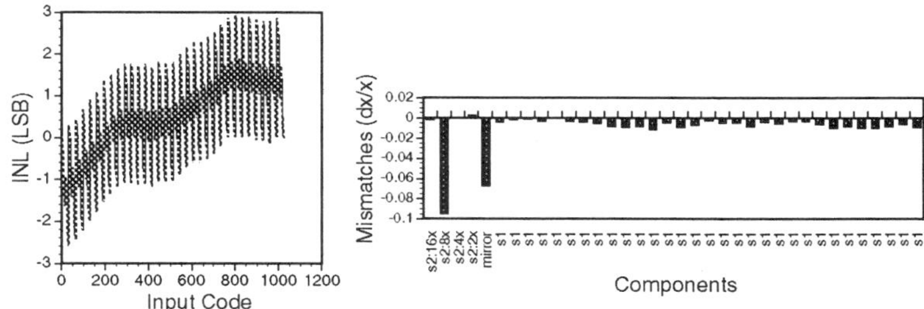

Figure 9.16 Example 2: Measured INL and Component Mismatches

Constants	units	Assumed	5/5 D/A
$\sqrt{a_{VFB}}$	V μm	0.0623	0.0555
σ_W	μm	0.05	0.0437

Table 9.8 Basic Statistical Process Variations

widths and lengths were assumed. Equation 9.1 was the expression used to model the flat band voltage.

$$\sigma_{vt}^2 = \frac{\sqrt{a_{VFB}}^2}{WL} \tag{9.1}$$

Using the component mismatch data measured at two reference currents and using a basic mismatch formula [105], the process mismatches were estimated. Table 9.8 shows the results. The close agreement between the assumption and the extracted results validates the assumptions made during design.

9.6 TESTING

A testing methodology was developed for testing all DC performance of these converters including offset error, full scale gain error, integral nonlinearity, and differential nonlinearity. In contrast to previous testing strategies based on linear models that require accurate measurements of circuit performance, this new strategy uses a simpler measurement to verify that a circuit performance parameter falls within certain detection thresholds in the presence of measurement noise. Using this new strat-

D/A Converter	# Fabricated	# Which Met Specs
6/4	12	8
5/5	12	7
4/6	12	4

Table 9.9   Yield for three fabricated D/A converters

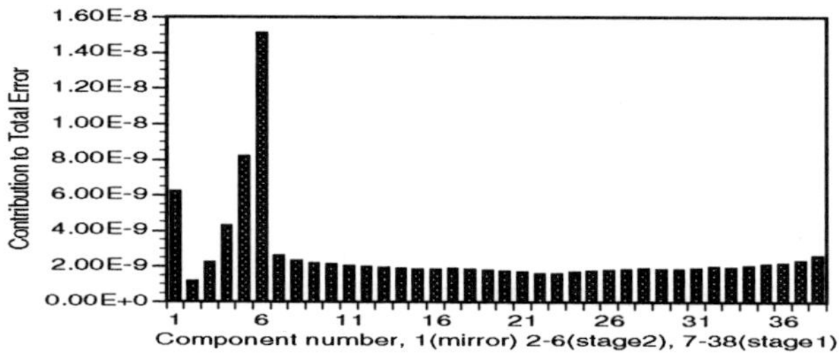

Figure 9.17   Relative Importance of Component Errors

egy, we evaluated tradeoffs between test set size, test coverage, detection thresholds, measurement noise, chip performance, and estimated yield [183].

The three D/A converters were fully tested against their specifications; the yields are reported in Table 9.9. Since some of the converters did not meet all of the specifications, a fault diagnosis methodology was developed to determine the cause of failure for those converters which failed one or more tests. A key contribution from this methodology is a linear-time algorithm for determining all of the untestable groups of components in a linear system. These groups are called ambiguity groups. They are untestable because the faults that occur on components within the group cannot be distinguished at the circuit output. The ambiguity groups are found in linear time by computing the null space of the circuit's sensitivity matrix [179]. A ranking of the most critical components is shown in Figure 9.17. Once the ambiguity groups were found and removed from the linear system of equations describing the D/A converter behavior, linear regression was used to find the best estimate for each behavioral model parameter. This information pinpointed the parameters which caused each of the failed chips to exceed the specifications. Using the behavioral model, automatic test patterns were generated for the D/A [82]. These results showed that the D/As could be characterized in under 40 tests.

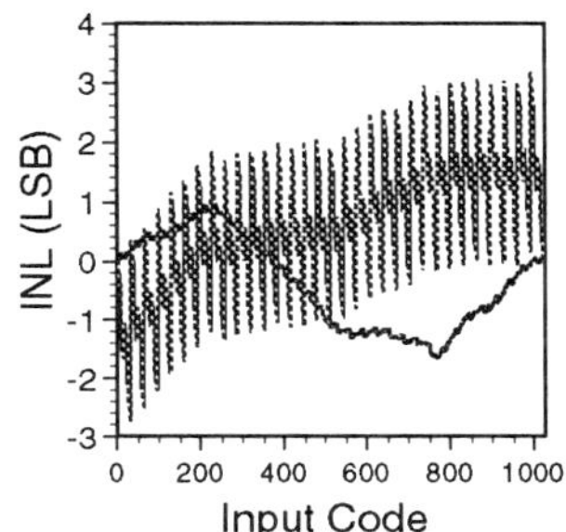

Figure 9.18 Measured INL of two sample silicon chips

9.6.1 Automatic Test Pattern Generation

Figure 9.18 shows the measured INL for all 1024 input codes for two of the chips. For an INL performance specification of 3 LSB, an INL detection threshold of 2 LSB, an assumed measurement noise of 0.1 LSB, and expected process variation parameters, the number of needed tests was computed to be 76 using a total of 446 seconds of DEC 5000 CPU seconds. The first 5 test inputs, in descending order of importance, are 479, 736, 352, 255, and 928. The first failed chip was detected at test number 1 using input code 479. Another was detected at test number 37 using input code 762. The last two failed chips were detected at test number 2 using input code 736. The INL for all input codes for the last failed chip is shown in Figure 9.18 which shows clearly that the INL exceeds 2 LSB for input code 736.

9.7 DESIGN TIMES

One measure of the success or failure of this methodology is the amount of *actual* design time required for the design and physical synthesis phases. This includes not only decision making times and CPU time but also the time required *to setup the tools*. This is a true measure of time spent on the project. Table 9.10 categorizes these design times for the 5/5 D/A. This shows that the first time through the design cycle required approximately 8 person months (6 months real time).

This was our first cut through *any* serious design problem where behavioral simulation and optimization were involved. Much time was spent on the development of the methodology for the various phases of design. For example, flexibility functions were

Design Phase	Time Required
Behavioral model development	1 month
High-level optimization	2 months
Low-level synthesis	2 months
Incorporation of new unit cell to layout tool	1 month
Layout generation	2 months
Extraction/Verification	1 week

Table 9.10   Design Times for 5/5 D/A

Design Phase	Time Required
Behavioral model development	4 hours
High-level optimization	10 hours
Low-level synthesis	5 hours
Incorporation of new unit cell to layout tool	2 days
Layout generation	2 hours
Extraction/Verification	5 hours

Table 9.11   Second Round Design Times

formulated. So, though this does not seem to give any advantage to hand design, this first design was expected to require more time, and six months is not unreasonable.

The design times for the second design (shown in Table 9.11), however, are much shorter. This reflects the total amount of time required for the generation of two D/As— the 6/4 and 4/6 converters. Since a module generator had been built during the first design cycle, these times are more of a reflection of the time to set up the tools and of the CPU time consumed by the tools. They provide good estimates on how long it would take to re-do each of these phases given a completely *new* set of design specifications (those found in Table 9.1). The total person time to go from specifications to fabrication was only *five* days.

9.8   CONCLUSION

Following the top-down, constraint-driven paradigm, a complete design cycle for a class of D/As has been performed. The cycle was begun with a set of high-level specifications and low-level assumptions on the technology, and was completed with the verification of both the design specifications and the gathered technology information from the finished products. Two critical measures on the quality of the methodology are shown. The fabricated parts worked the first time, and the design times are shown in Table 9.10 and 9.11.

Future work on the D/A design example include adding transient constraints into the design specifications and the introduction of new circuits such as ones for higher speed or ones with self-calibration based on test results.

10

Σ-Δ ANALOG-TO-DIGITAL CONVERTER DESIGN EXAMPLE

10.1 INTRODUCTION

This Σ-Δ A/D design example is the second in a series of design examples that we take through fabrication. We show that even though the circuits and architecture are different from the ones in the first example, the methodology is applied in the same way. The two examples significantly differ for the following reasons: (1) the subcomponents are different, (2) the overall performance specifications and the reasons for their degradation are different, and (3) the sub-blocks are not in regular arrays. Its relative level of complexity is shown in Figure 9.1.

10.2 DESIGN SPECIFICATIONS

The Σ-Δ A/D converter that has been implemented is a second-order system with a 1-bit quantizer similar to the one found in [20]. The basic architecture of the Σ-Δ is shown in Figure 10.1. Figure 10.3 shows the components of the complete system, i.e. the Σ-Δ converter itself, the clock generator, the bias circuitry, etc. The input anti-aliasing filter and the decimation filter to convert the bit-stream output to data words is to be done off chip. Figure 10.2 defines the symbol used for transmission gates.

The Σ-Δ converter consists of two switched capacitor integrators, a comparator, and a 1-bit D/A. The differential signal enters the system where it is noised shaped by the two integrators. The D/A in the feedback loop drives the output bit stream to be equal to the input signal thus giving the proper A/D conversion operation.

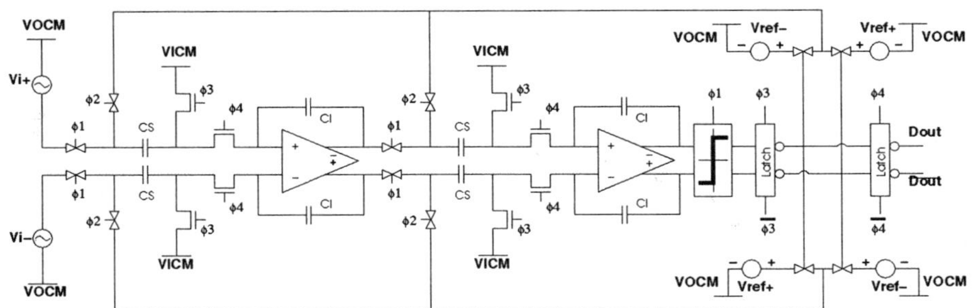

Figure 10.1 Σ-Δ A/D Architecture

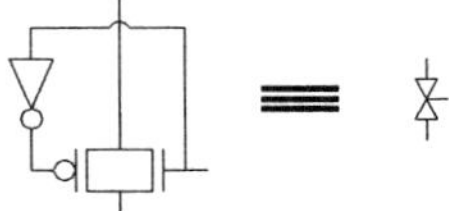

Figure 10.2 Definition of the Symbol for a Transmission Gate

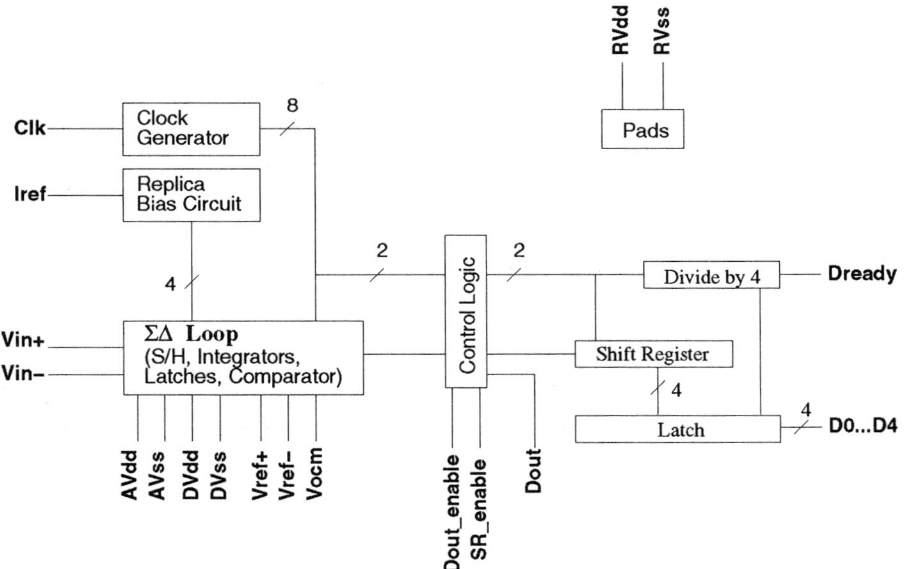

Figure 10.3 Σ-Δ system

Type	Specifications	Value
Performance	Minimum SNR	74 dB
	Nyquist Frequency (f_x)	250 kHz
Operation	Supply voltage	5 V
Technology	Design rules	SCMOS
	SPICE parameters	HP CMOS26B
Constants	Low-level circuit design constants. See Table 10.6	
Layout	Input/output terminal locations, placement, etc.	
Optimization	Non-default flexibility function	default
	Objective for project	minimize area

Table 10.1 Σ-Δ A/D Design Specifications

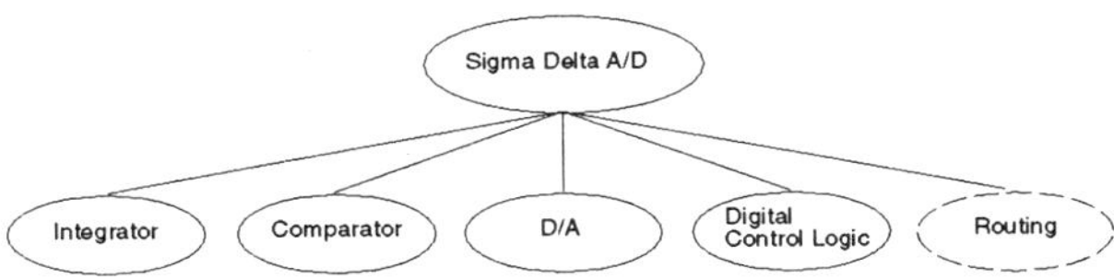

Figure 10.4 Σ-Δ Hierarchy

The specifications along with the specific design values used for the designed and fabricated chip are listed in Table 10.1. The performance specifications for our example design were a minimum SNR of 74 dB and a Nyquist input frequency of 250 kHz. The operating supply voltage was chosen to be 5V. The MOSIS SCMOS design rules were used. The target technology was the HP CMOS26B (0.8 μm minimum gate length) process. The bonding pad locations were specified since the A/D was a stand-alone part. The overall objective for the project was to minimize layout area.

10.3 SYNTHESIS PATH

The two-level hierarchy used for synthesis is shown in Figure 10.4. During high-level synthesis, specifications for the Σ-Δ A/D are mapped onto the integrator, comparator, D/A, digital control logic, and routing. Then in the low-level synthesis phase constraints on these sub-blocks are used to size the schematics.

Parameter	Description
M	Oversampling ratio
V_{FS}	Full scale voltage of Σ-Δ system
O_{off}	Offset due to the OTA
O_{range}	Output range of the OTA relative to V_{FS}
O_{th}	Thermal noise due to the OTA
$O_{1/f}$	$1/f$ noise due to the OTA
O_{gain}	Open loop gain of the OTA
I_{τ}	Time constant for the integrator
I_{SR}	Slew rate of the integrator
$I_{kT/C}$	kT/C noise of the integrator
I_{Amis}	Integrator gain mismatch
I_{gain}	Integrator gain
C_{off}	Offset due to the comparator
$C_{hysteresis}$	Hysteresis due to the comparator
C_{noise}	RMS noise due to the comparator
D_{off}	Offset in the D/A
D_{noise}	Noise contribution for the D/A
ΔSNR	Non-characterized degradations to SNR

Table 10.2   Parameters for the Σ-Δ A/D Block

10.3.1   High-Level Synthesis

The *intermediate level parameters* for the high-level synthesis phase are shown in
Table 10.2. Given values for each of these parameters along with f_x from Table 10.1,
the SNR for the Σ-Δ A/D can be computed using a behavioral model developed for
this particular system and the behavioral simulation, MIDAS [20]. Letting x_i be the ith
parameter ($x_1 \equiv M$, $x_2 \equiv V_{FS}$, $\cdots x_n \equiv \Delta SNR$), n be the number of parameters
(n=18), and c, be a vector containing the additional parameters from Table 10.1, the
behavioral simulator can be represented by the function, g, in Equation 10.1.

$$SNR = g(x_1, x_2, \cdots x_n, c) \tag{10.1}$$

To perform the design decomposition, values for each of these parameters must be
chosen. To simplify the problem, reasonable values for the less critical parameters
were selected by hand, rather than by optimization. This does not invalidate our
approach, since we emphasize designer interaction when required. The values chosen

Parameter	Value
M	100
V_{FS}	2 V
O_{off}	25 mV
I_{Amis}	1%
I_{gain}	0.5
C_{off}	100 mV
D_{noise}	20 μV_{rms}
D_{off}	100 mV
ΔSNR	3.5 dB

Table 10.3   Constant Parameters chosen for the Σ-Δ A/D Block

for these parameters will be constraints just as the parameters chosen by the optimizer will be constraints during the low-level synthesis phase.

Table 10.3 lists the parameters that were chosen by hand. To simplify the problem, M, a fairly critical parameter, was chosen *a priori*, because M usually only takes on particular quantized values. These values are determined by how difficult the decimation filter following the A/D converter is to implement. For a given SNR, a reasonable value for M is easily obtainable using Equation 10.2 and knowing the quantized values. The dynamic range (DR) bounds the SNR. L is the order of the Σ-Δ converter. Had it turned out that M was selected to be too low, M would have been increased during the following design pass.

$$DR = 1.5 \frac{2L+1}{\pi^{2L}} M^{2L+1} \tag{10.2}$$

As before, the objective of high-level synthesis is to *maximize flexibility* of design. Flexibility functions of the form in Equation 4.1 were developed. The coefficients were determined using the method presented in Section 4.1. Table 10.4 lists the moderate and difficult values along with the coefficients for the flexibility functions. The overall flexibility function is shown in Equation 10.3 where m represents the number of variables actually being optimized, and where i represents the ith parameter.

$$flex(x_1, x_2 \cdots x_m) = \sum_{i=1}^{m} flex_i(x_i) \tag{10.3}$$

Parameter	Design Difficulty		Coefficients		
	Moderate	Hard	a	b	c
O_{range}	1.8 V/V	2.0 V/V	13 $(V/V)^{-1}$	0	43 V/V
O_{th}	2.0 μV_{rms}	0.9 μV_{rms}	0	16 μV_{rms}^2	8.2 μV_{rms}
$O_{1/f}$	15 μV_{rms}	6.7 μV_{rms}	0	120 μV_{rms}^2	8.1 μV_{rms}
O_{gain}	250 V/V	900 V/V	1.4×10^{-5} $(V/V)^{-1}$	0	0.84 V/V
I_{τ}	1 ns	0.5 ns	0	10 ns^2	10 ns
I_{SR}	500 V/μs	2500 Vμs	1.7×10^{-6} $(V/\mu s)^{-1}$	0	0.42 V/μs
$I_{kT/C}$	20 μV_{rms}	6.5 μV_{rms}	0	96 μV_{rms}^2	4.8 μV_{rms}
$C_{hysteresis}$	50 mV	20 mV	0	330 mV^2	6.7 mV
C_{noise}	300 μV_{rms}	134 μV_{rms}	0	2400 μV_{rms}^2	8.1 μV_{rms}

Table 10.4   Flexibility Coefficients for the Σ-Δ A/D Parameters

Equation 10.4 shows the high-level optimization problem, where u_i and l_i represent the upper and lower bounds on the parameters, x_i.

$$\max \quad \sum_{i=1}^{m} flex_i(x_i) \qquad (10.4)$$
$$\text{s.t.} \quad \text{SNR}(x_1, x_2, \cdots x_n, c) \geq 74 \text{ dB}$$
$$l_i \leq x_i \leq u_i \qquad \forall i = 1..m$$

As with other optimization problems encountered in these design examples, MINOS [216] could not solve this problem directly, so the problem had to be solved by another approach. The principal difficulties were (1) when computing SNR, MIDAS has a large standard error, ϵ, and (2) SNR is extremely nonlinear with respect to its input variables. The large ϵ is due in part to the fact that SNR is computed using short time domain simulations in which random number generators are used to approximate white noise sources. ϵ is approximately 2 dB, significant compared to the SNR requirement of 74 dB. A typical plot of SNR versus an input variable, in this case O_{range}, is shown in Figure 10.5. Because of the unsmooth response surface, MINOS would either get trapped in local minima or iterate excessively while it reduced the step size. The step size can be adjusted to a fixed value in MINOS, but because of the nonlinear nature, the appropriate step size changed depending on the region. Averaging multiple simulations with a varying sine wave input amplitude between 0.37-0.005 to 0.37+0.005, we reduced ϵ to approximately 0.5 dB.[1]

The following is the optimization algorithm that we used:

1. Start with feasible solution, x^*. Let $x_o = x^*$, $j = 1$.

---

[1] This reduction was not enough to prevent MINOS from getting trapped.

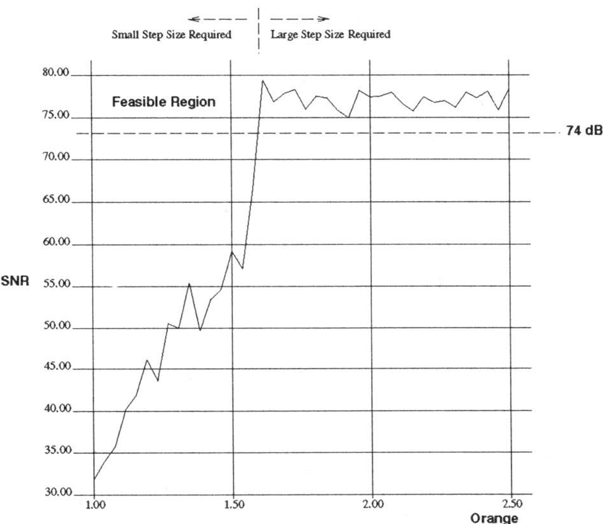

Figure 10.5   SNR versus O_{range}

2. Replace the SNR constraint function with a linearized SNR constraint around x^*.

$$SNR(x) = SNR(x^*) + \left[\frac{\partial SNR(x^*)}{\partial x_1}, \frac{\partial SNR(x^*)}{\partial x_2}, \cdots \frac{\partial SNR(x^*)}{\partial x_m} \right] (x - x^*)$$

3. Impose bounds

$$(1 - d_i)x_i^* \le x_i \le (1 + d_i)x_i^* \quad \forall\, i = 1...m$$

where d_i is a preset value set by the user (typically 0.1) depending on how nonlinear SNR is.

4. Solve this problem using MINOS.[2] Let the solution be x_j'.

5. For each i, if $x_{i,j}'$ is at the upper bound and $x_{i,j-1}'$ was at the lower bound, or if $x_{i,j}'$ is at the lower bound and $x_{i,j-1}'$ was at the upper bound, then reduce d_i by a factor of two.

6. If $\|x_j - x_{j-1}\| \le \epsilon$, exit with x_j' as the solution.

7. Set $j = j + 1$. Goto 2.

---

[2] Since the objective and constraints are now analytic and C^2 smooth within the bounds of interest, MINOS easily solves this problem.

Parameter	Value
O_{range}	1.9 (V/V)
O_{th}	87.4 μV_{rms}
$O_{1/f}$	1.05 mV_{rms}
O_{gain}	250 (V/V)
I_τ	1.5 ns
I_{SR}	1045 V/μs
$I_{kT/C}$	191 μV_{rms}
$C_{hysteresis}$	101 mV
C_{noise}	9.35 mV_{rms}

Table 10.5   Parameter Selection for the Σ-Δ A/D Block

The derivatives in Step 2 are computed using finite differences. However, because of ϵ, a derivative was only considered valid when the finite difference caused a 1 dB change (with a certain tolerance) in the SNR. Binary search was used to find this point. 1 dB was chosen because Δx_i had to be small to be able to evaluate the constraint accurately, but the change in SNR had to be greater than ϵ to obtain the necessary accuracy for the gradient. The idea behind this approach is to dynamically control the step size, using the performance change as the measure of whether or not the step size is valid. This allows the optimizer to see the overall shape of the curve rather than the shape of the noise.

The solution to the high-level decomposition problem is shown in Table 10.5. The value for the flexibility function is 28.5 indicating that the overall difficulty of this design should be easier than what was described to be "moderate" in Table 10.4.

10.3.2   Low-Level Synthesis

With the parameters for the high-level design having been chosen, these now become constraints for the low-level synthesis phase. The following sections describe the design of the basic building blocks.

Integrator/OTA

Figure 10.6 shows the schematic of the switched capacitor integrator. Bottom plate sampling is employed to reduce the charge injection from the switches into the signal. This arrangement also allows for level shifting from the output common mode voltage

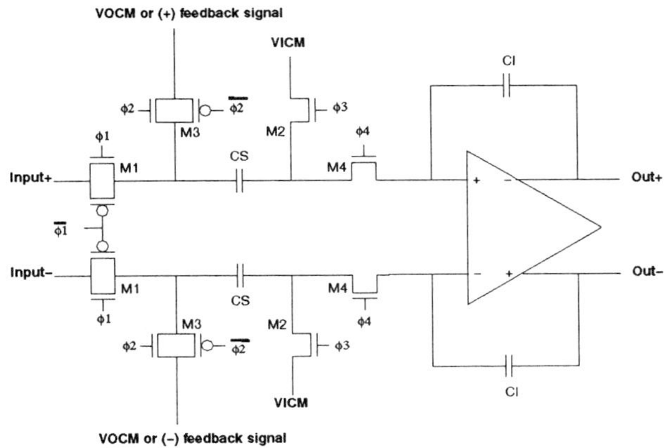

Figure 10.6 Schematic of the Integrator

to the input common mode voltage. C_S is the input sampling capacitor. C_I is the integrating capacitor. This integrator works in two phases. During the first phase, ϕ_1 and ϕ_3 are asserted. The input signal is sampled onto C_S. ϕ_3 is de-asserted two inverter delays before ϕ_1 to allow for the bottom plate sampling. The clock phases are shown in Figure 10.10. Following the de-assertion of ϕ_1, ϕ_2 and ϕ_4 are asserted. The feedback signal is subtracted from C_S and the remaining charge is integrated onto C_I. Again, ϕ_4 is de-asserted two inverter delays before ϕ_2.

A telescopic or unfolded cascode OTA was selected for the integrator. The OTA schematic is shown in Figure 10.7. Its design along with the bias circuitry is similar to the ones in [227]. In this differential OTA, M_1 and M_2 are the input devices. They provide the gain. M_3 through M_6 function as cascode devices to boost the output resistance. M_7 and M_8 in conjunction with M_9 and M_{10} set up the bias current for the OTA based on external bias voltages generated from a bias generator. M_{50} through M_{57} provide a fixed voltage drop to set up the gate voltage for M_3 and M_4 in order to keep M_1 and M_2 saturated. M_{11}, M_{12}, and M_{30}-M_{33} are sized to allow a certain percentage of the tail current to flow down these two bias paths. The transistors driven by the clock lines, ϕ_1, $\overline{\phi_1}$, ϕ_2, $\overline{\phi_2}$, and the capacitors, C_1 and C_2 form the dynamic common mode feedback (CMFB) circuitry [35]. These switches and capacitors set up the output common mode voltage as well as gate voltage for M_9 through M_{12}.

A replica bias circuit, shown in Figure 10.8 was used to provide reference and bias voltages for the OTA and integrator. Given an external reference current, the bias circuit generates the gate voltage (NBIAS) for the NMOS tail current devices, the gate

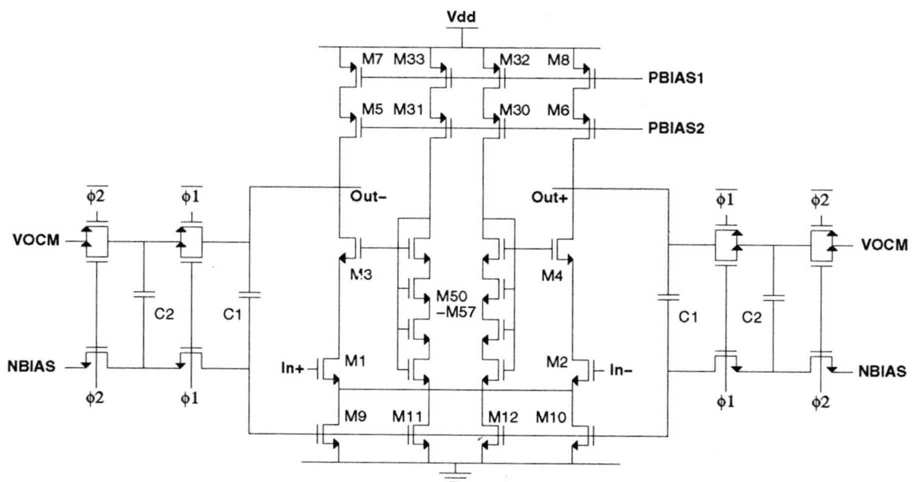

Figure 10.7   Schematic of the OTA

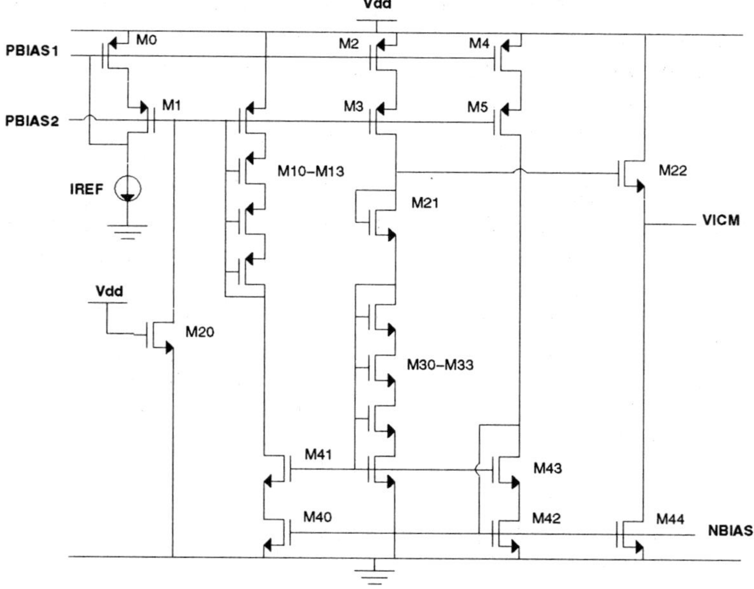

Figure 10.8   Schematic for the bias circuitry

Parameter	Description	Value
V_{safety}	Voltage above V_{dsat} where cascode device will be biased	250 mV
f_{min}	Minimum frequency for flicker noise calculation	1 Hz
I_{ratio}	Current ratio between M_{11} to M_9 of OTA	0.1
$R_{c1,m9gate}$	Ratio between C_1 and M_9 gate capacitance of OTA	0.3
$R_{c1,c2}$	Ratio between C_1 and C_2 of CMFB circuit	5
p_{leak}	Percent leakage current M_{20} of the bias circuit	0.01

Table 10.6 Design constants for the OTA

voltages (PBIAS1, PBIAS2) for the PMOS current sources in the OTA, and the input common mode voltage (VICM) for the OTA. I_{ref} pulls current from M_0 and M_1 and sets up the voltage, PBIAS1. $M_2 - M_5$ mirror the current for the NMOS devices, $M_{40} - M_{43}$, to allow them to set up the voltage, NBIAS. $M_{10} - M_{13}$ set up the proper gate voltage, PBIAS2, in order to keep the top PMOS transistors driven by PBIAS1 in saturation. Likewise, $M_{30} - M_{33}$ keep M_{40} and M_{42} saturated by setting up a proper gate voltage for M_{41} and M_{43}. M_{21} and M_{22} provide a diode drop up followed by a diode drop down to VICM to give it a low impedance output. Finally, since there are two quiescent solutions for this circuit, M_{20} provides a leak current from PBIAS2 to insure that the proper state is selected. For this prototype design, the leakage current device is left on. This alters the reference current by a small percentage. Since the reference current is controlled externally, the offset caused by M_{20} can be compensated externally.

Having chosen the architecture for the integrator and OTA, the task now is to choose component values for the transistors and capacitors in these circuits subject to the *constraints* in Table 10.5 and the design constants shown in Table 10.6. These were derived from the high-level synthesis results (Tables 10.3 and 10.5). Because analytic methods [19] for determining these constraints differed from the MIDAS results, we chose to tighten some of the constraints, taking the more conservative results of the two approaches. This is an example of why we do not propose a fully automatic synthesis process. The designer is required to catch potential design difficulties. Three constraints were changed—the thermal noise bound was lowered, kT/C noise bounds was lowered, and the OTA output range bound was raised. Except for the O_{range}, whose flexibility was lowered from -3.9 to -9.0, the other two constraints' flexibilities remained greater than or equal to 0. Thus, we should not have increased the design difficulty by much, but by taking this more conservative step, we better insured a working chip. Table 10.7 lists the complete set of constraints for the integrator. The only constraint not found in either Table 10.3 or 10.5 is the common-mode feedback time constant constraint, $O_{\tau,CMFB}$. This was arbitrarily set at 4/3 the value of I_{τ}.

Parameter	Value
O_{off}	25 mV
O_{range}	2.0 (V/V)
O_{th}	20.0 μV_{rms}
$O_{1/f}$	1.05 mV$_{rms}$
O_{gain}	250 (V/V)
I_τ	1.5 ns
I_{SR}	1045 V/μs
$I_{kT/C}$	20 μV_{rms}
$O_{\tau,CMFB}$	2.0 ns

Table 10.7   Design Constraints for the Integrator

Parameter	Description
L_{nmos}	Length of transistors $M_1 - M_4$, $M_{50} - M_{57}$ of OTA Length of transistors $M_{21} - M_{22}$, $M_{30} - M_{33}$
L_{pmos}	Length of transistors $M_5 - M_8$, $M_{30} - M_{33}$ of OTA Length of transistors $M_0 - M_5$, $M_{10} - M_{13}$ of the bias circuit
L_{bias}	Length of transistors $M_9 - M_{12}$ of the OTA Length of transistors $M_{40} - M_{43}$ of the bias circuit
L_{M44}	Length of transistor M_{44} of the bias circuit
$V_{gs} - V_T$	$V_{gs} - V_T$ voltage for the transistors in the gain stage
I_{leg}	Current flowing through the transistors in the gain stage
C_S	Size of the sampling capacitor

Table 10.8   Design Parameters for the Integrator

Since it is derived from the high-level results, the constraint-driven nature of this design is maintained.

Seven variables were selected to describe the design of the integrator. They are listed is Table 10.8.[3] This set is sufficient to generate all component transistor and capacitor values, and is sufficient to characterize each of the design constraints listed in Table 10.7. However, there are other possible sets. Section 10.7 shows how these variables translate to devices sizes and performance specifications.

---

[3] Note that in hindsight, $M_{30} - M_{33}$ probably should have been set to scale with L_{bias} rather than L_{nmos} since it is actually setting up the V_{ds} for the bias transistors.

In keeping with the Σ-Δ A/D design specifications, the *objective* during this phase is to minimize the active area of the integrator. Active area is computed using Equation 10.24. Note that for this particular circuit, minimizing area is equivalent to minizing power.

To simplify the integer nature of determining L, the problem was solved in two steps. First, the values of L were determined. L_{nmos}, L_{pmos}, and L_{bias} were minimized subject to the constraint on gain and on common-mode rejection ratio (CMRR) issues. Note that it is critical that L_{nmos} be as small as possible to maximize the speed of the OTA. Enumerating the Ls for L_{nmos} and L_{pmos} for $L = 0.8~\mu m, 1.3~\mu m, 1.8~\mu m, 2.3~\mu m$ it was determined that in order to satisfy the gain requirements with allowance for a safety margin, $L_{nmos} = 1.3~\mu m$ and $L_{pmos} = 1.8~\mu m$. L_{bias} was chosen to have a transistor length of $1.8~\mu m$ based on CMRR issues, and L_{M44} was determined to require an even longer channel length, $2.3~\mu m$, to boost its output resistance because of the larger V_{ds} across M_{44}.

$$\min \quad area(V_{gs} - V_T, I_{leg}, C_S) \tag{10.5}$$

$$\begin{aligned}
\text{s.t.} \quad & O_{off}(V_{gs} - V_T, I_{leg}, C_S) \leq 25~mV \\
& O_{range}(V_{gs} - V_T, I_{leg}, C_S) \geq 2.0~(V/V) \\
& O_{th}(V_{gs} - V_T, I_{leg}, C_S) \leq 20.0~\mu V_{rms} \\
& O_{1/f}(V_{gs} - V_T, I_{leg}, C_S) \leq 1.05~mV_{rms} \\
& O_{gain}(V_{gs} - V_T, I_{leg}, C_S) \geq 250~(V/V) \\
& I_\tau(V_{gs} - V_T, I_{leg}, C_S) \leq 1.5~ns \\
& I_{SR}(V_{gs} - V_T, I_{leg}, C_S) \geq 1045~V/\mu s \\
& I_{kT/C}(V_{gs} - V_T, I_{leg}, C_S) \leq 20~\mu V_{rms} \\
& O_{\tau,CMFB}(V_{gs} - V_T, I_{leg}, C_S) \leq 2.0~ns \\
& V_{gs} - V_T \geq 100~mV \\
& I_{leg} \geq 100~\mu A \\
& C_S \geq 100~fF
\end{aligned}$$

$$\tag{10.6}$$

Equation 10.5 shows the *optimization problem* for the integrator. All performance specifications are evaluated analytically (see Section 10.7), resulting in a problem easily solved by standard optimization techniques. MINOS solved the problem in less than 1 CPU second. The solution is shown in Table 10.9. Table 10.10 shows the performance specifications for the solution indicating which of the constraints were

Parameter	value
$V_{gs} - V_T$	330 mV
I_{leg}	1.22 mA
C_S	207 fF

Table 10.9   Low-level optimization problem solution

Parameter	Value	Active
O_{off}	20 mV	
O_{range}	2.0 (V/V)	√
O_{th}	1.2 μV_{rms}	
$O_{1/f}$	9.9 μV_{rms}	
O_{gain}	1336 (V/V)	
I_τ	0.63 ns	
I_{SR}	1045 V/μs	√
$I_{kT/C}$	20 μV_{rms}	√
$O_{\tau,CMFB}$	0.98 ns	

Table 10.10   Constraint values at low-level solution

active (limiting factors). The active constraints are highly dependent on the high-level specifications and the flexibility function.

Using the values determined for $(V_{gs} - V_T)$, I_{leg}, C_S, gate lengths, and the design constants found in Table 10.6, the component values for the OTA, integrator, and bias circuitry are computed. A program was written to generate a complete SPICE deck for the integrator, OTA, and bias circuit once the component values were calculated. Tables 10.11, 10.12, and 10.13 list the results.[4]

Comparator

Figure 10.9 shows the design of the comparator used in this design [20]. During the non-latching phase, when ϕ_1 is low, transistors M_4 and M_{14} are used to reset the comparator while a voltage difference is set up by M_0 and M_{10}. When ϕ_1 goes high,

---

[4]The size of transistors $M_{40} - M_{43}$ of the bias circuitry should be but are not identical to the size of transistors M_9 and M_{10} of the OTA. They was due to a layout generation error. $M_{40} - M_{43}$ were divided into stacks of six each while M_9 and M_{10} were divided into stacks of ten. This was caught too late in the design cycle to fix, but it should not have a very large effect. The CMFB circuitry compensates for this small difference.

Component	Value
$M_1 - M_4$	W=225 μm, L=1.3 μm
$M_5 - M_8$	W=935 μm, L=1.8 μm
$M_9 - M_{10}$	W=310 μm, L=1.8 μm
$M_{11} - M_{12}$	W=31 μm, L=1.8 μm
$M_{30} - M_{33}$	W=93.5 μm, L=1.8 μm
$M_{50} - M_{57}$	W=12 μm, L=1.3 μm
All Switches	W=3 μm, L=0.8 μm
C_1	213 fF
C_2	43 fF

Table 10.11 Component values for the OTA

Component	Value
NMOS Switches	W=4 μm, L=0.8 μm
PMOS Switches	W=9 μm, L=0.8 μm
C_S	207 fF
C_I	414 fF

Table 10.12 Component values for the Integrator

Component	Value
$M_0 - M_5$	W=935 μm, L=1.8 μm
$M_{10} - M_{13}$	W=459 μm, L=1.8 μm
M_{20}	W=2 μm, L=163 μm
$M_{21} - M_{22}$	W=225 μm, L=1.3 μm
$M_{30} - M_{33}$	W=120 μm, L=1.3 μm
$M_{40} - M_{43}$	W=309 μm, L=1.8 μm
M_{44}	W=396 μm, L=2.3 μm

Table 10.13 Component values for the Bias Circuit

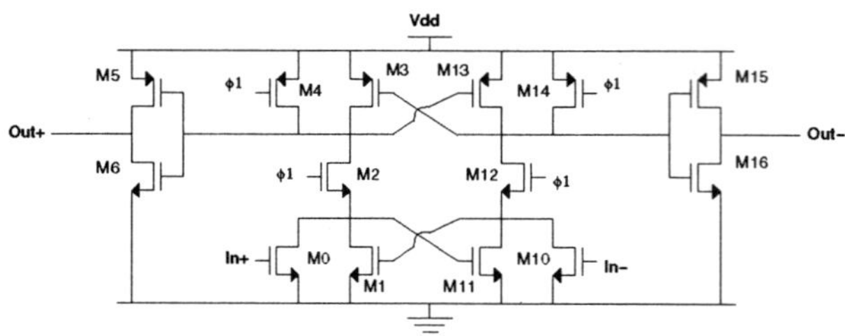

Figure 10.9 Schematic of the comparator used in the Σ-Δ A/D

Parameter	Value
C_{off}	100 mV
$C_{hysteresis}$	100 mV
C_{noise}	9.35 mV$_{rms}$

Table 10.14 Design Constraints for the Comparator

Component	Value
M_0, M_2, M_{10}, M_{12}	W=6 μm, L=0.8 μm
M_1, M_{11}	W=4 μm, L=0.8 μm
M_3, M_4, M_{13}, M_{14}	W=12 μm, L=0.8 μm
M_5, M_{15}	W=9 μm, L=0.8 μm
M_6, M_{16}	W=3 μm, L=0.8 μm

Table 10.15 Component values for the Comparator

M_1, M_{11}, M_3, and M_{13} form a positive feedback gain stage to boost the input to full logic levels. M_5, M_6, M_{15}, and M_{16} form inverting output buffers.

The set of *constraints* or specifications required for the design of the comparator is specified in Table 10.14. As with the constraints generated for the integrator and OTA, these are either from Table 10.3 or from Table 10.5.

Since the design constraints are fairly lax for this comparator, device sizes similar to the switch sizes were chosen for the design. Table 10.15 shows the transistor sizes used in the comparator. These sizes were verified with SPICE and by analytic equations to meet the design constraints.

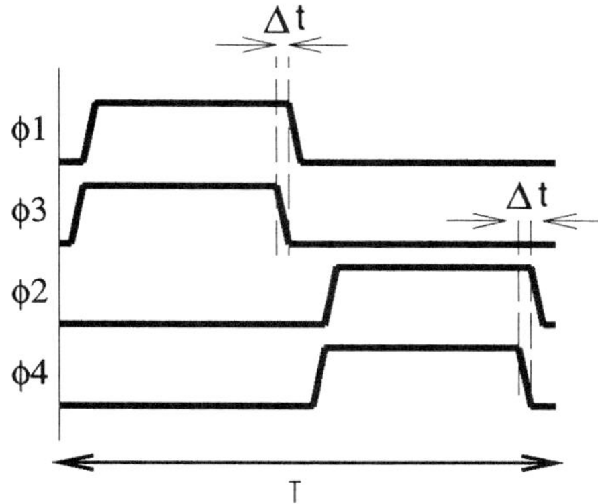

Figure 10.10 Clock Timing Diagram

D/A

The only components that have to be sized for the 1-bit D/A converter are the switches. These were selected to be the same as the integrator.

Digital

There are three digital blocks in this circuit. They are the clock generator, the two latches which follow the comparators, and the shift register at the output for the Σ-Δ A/D.

For the clock generator, four phases and their inverses are required for proper operation of the Σ-Δ system. They are shown in Figure 10.10. ϕ_1 and ϕ_2 are non-overlapping clocks, and ϕ_3 and ϕ_4 are early versions of ϕ_1 and ϕ_2 respectively. ϕ_3 and ϕ_4 are required for the bottom-plate sampling.

The circuit used for the clock generator is shown in Figure 10.11. Given a single phase clock at the desired frequency at node "CLK," the appropriate phases are produced by this circuit. The "1x" sized inverters provide the delays between ϕ_3 and ϕ_1 and between ϕ_4 and ϕ_2. It also provides for the non-overlap time between ϕ_1 and ϕ_2.

The basic elements of the clock circuit are shown in Figure 10.12. The "1x" sized inverter is shown on the left, and the "NOR" gate is shown on the right. The NMOS device size for the inverters were sized in such a way that the "4x" inverter is the same size as the sum of the sizes of all of the switches that the clock phase with the maximum

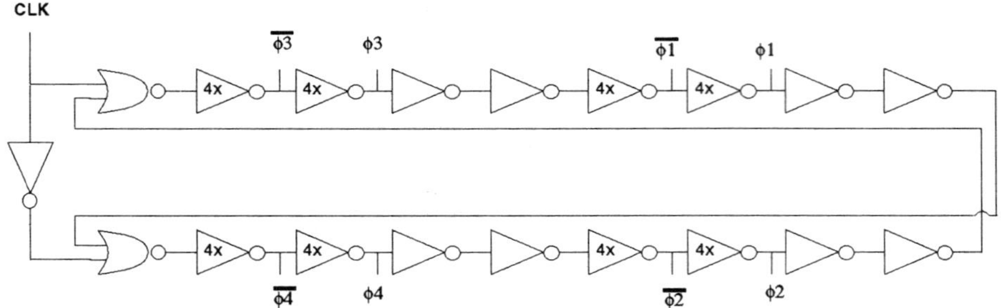

Figure 10.11   Schematic of clock generator

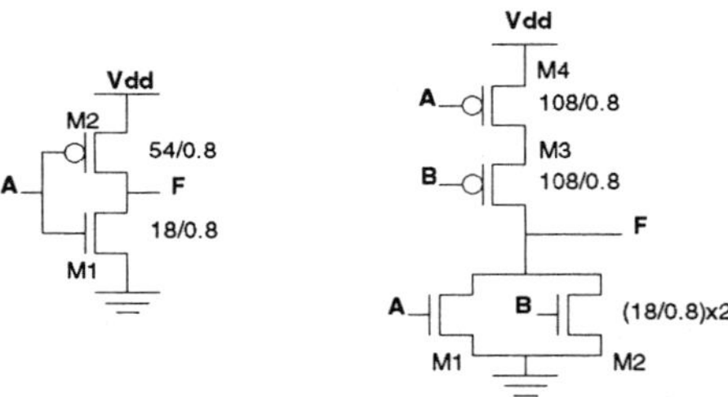

Figure 10.12   Schematic of clock generator elements

load drives. The PMOS device size is then scaled by k'_p/k'_n. The NOR gates were sized to have the same driving capacity as the "1x" sized inverter.

The second of the three digital elements is the latch. It is shown in Figure 10.13. It is a basic eight transistor latch. The transmission gate formed by M_1 and M_2 allows the new input to enter. The other transmission gate formed by M_3 and M_4 closes the two inverter feedback loop which holds the latch's state.

The last of the digital elements is the shift register. This was added so that the output would not have to be read off chip at the full speed of the Σ-Δ system. A four-bit shift register was used. Because the chip was pad limited, adding any more bits would have increased the layout area significantly. This gave a $4x$ lower frequency requirement at the output. The clock divider and shift register system is shown in Figure 10.14.

In the schematic, "Dout" represents the actual data coming out of the Σ-Δ A/D at the full clock rate. It can be shut off by de-asserting "Dout_enable." Because, it was unclear how this digital system would affect substrate noise, a shift-register enable

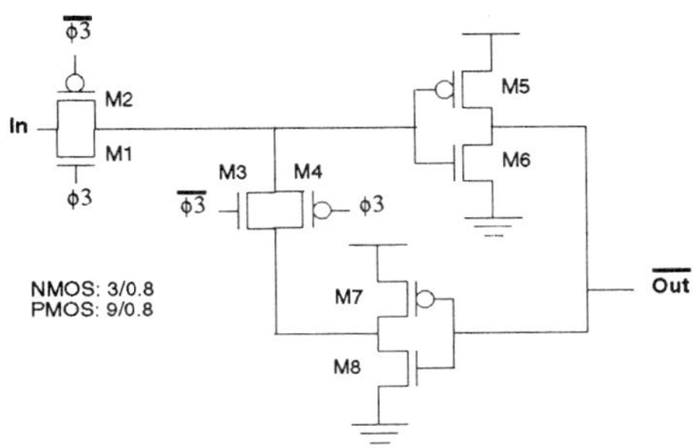

Figure 10.13   Schematic of Latch

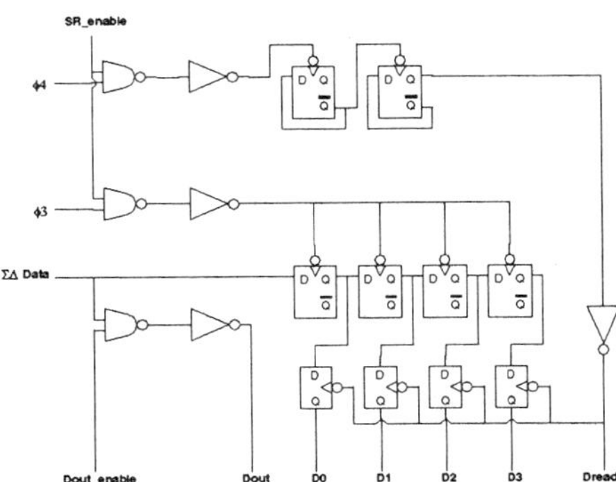

Figure 10.14   Schematic of the Shift Register and Clock Divider

line, "SR_enable," was added to turn off the shift register. The shift register latches its input on the de-asserting edge of ϕ_3. The output buffer, the set of four flip-flops directly underneath the shift-register, latches its output on a clock at $f_{clk}/4$ derived from ϕ_4. The shift register data outputs are D_0 through D_3. The assertion of "D_ready" indicates that the data outputs are valid. This cell was implemented with standard cells and was verified to meet timing requirements.

10.3.3 Physical Design

Physical assembly for the A/D converter will be discussed in this section emphasizing the design flow used for leaf (transistor-level) cell generation. Most of the layout was automatically synthesized. Issues involving why we had to resort to manual layout for some of the circuits will also be discussed.

We describe the steps necessary for leaf cell generation. A consistent example for an OTA synthesis is presented throughout the steps to better illustrate the concepts.

1. Enter the initial schematic in the form of a SPICE deck. A partial SPICE deck with the input transistors for the OTA is shown below (schematic in Figure 10.7; component sizes in Table 10.11):

```
* OTA
...
M1 n304 inp n308 0 CMOSN W=181u L=1.2u
M3 outm n305 n304 0 CMOSN W=181u L=1.2u
...
```

2. Add analog constraints to the SPICE deck either automatically using PARCAR or manually. The extra lines added: (i) indicate transistor matching, (ii) indicate transistor symmetry, and (iii) split transistors into smaller parallel transistors for decreasing drain capacitance and for canceling linear gradient effects. An example for the OTA is below:

```
...
*SYM M1 M2
*SYM M3 M4
...
*SPLIT M1 6
*SPLIT M3 6
...
```

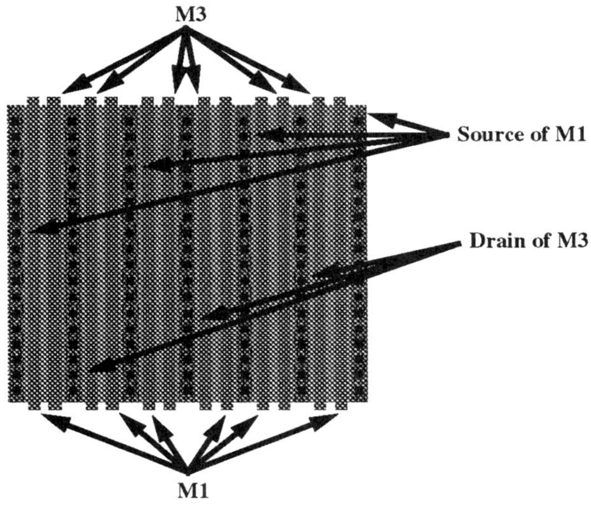

Figure 10.15   Example of a Transistor Stack

```
*MATCH  M1  M2  M3  M4

. . .
```

The "SPLIT" commands ask that transistors M_1 and M_3 be "split" into 6 parallel W=30.2μm transistors. The "SYM" commands ask that the two halves of the differential circuit be made symmetric. M_1 will be placed symmetrically to M_2 and M_3 will be placed symmetrically to M_4. The "MATCH" command asks that the transistors listed be placed close together.

3. Generate the *transistor stacks* using MKSTACK [40]. If transistors have the same gate width and share diffusion contacts, they are combined to form stacks of parallel transistors. The M_1/M_3 stack is shown in Figure 10.15.

4. Generate the capacitors using an available layout tool.

5. Implement the nets of the parallel transistors which are in stacks using another available layout tool. This routing is highly organized. Figure 10.16 shows this for the M_1/M_3 stack. This extra routing step is necessary, because ROAD cannot find this solution.

6. Instantiate objects in the layout and connect the nets symbolically using BD-NET [125][126].

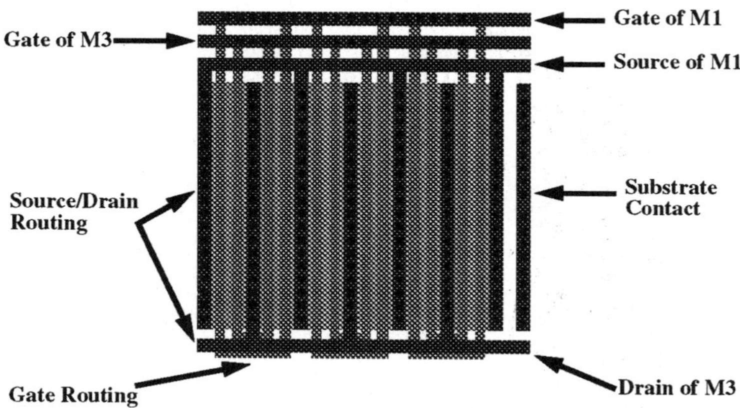

Figure 10.16 Example of Transistor Stack Routing

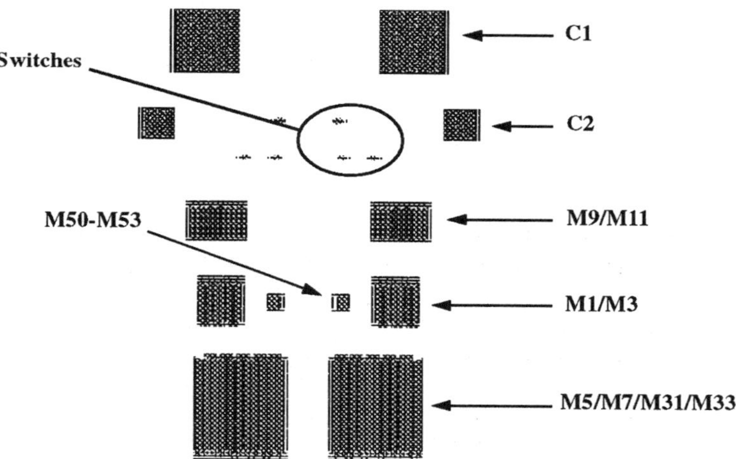

Figure 10.17 Placement by PUPPY-A of the Transistors in the OTA

7. Generate the layout placement using PUPPY-A. Figure 10.17 shows the placement result for the OTA.

8. Put down I/O vias for the leaf cell using PADPLACE.

9. Route the leaf cell using ROAD [191][194].

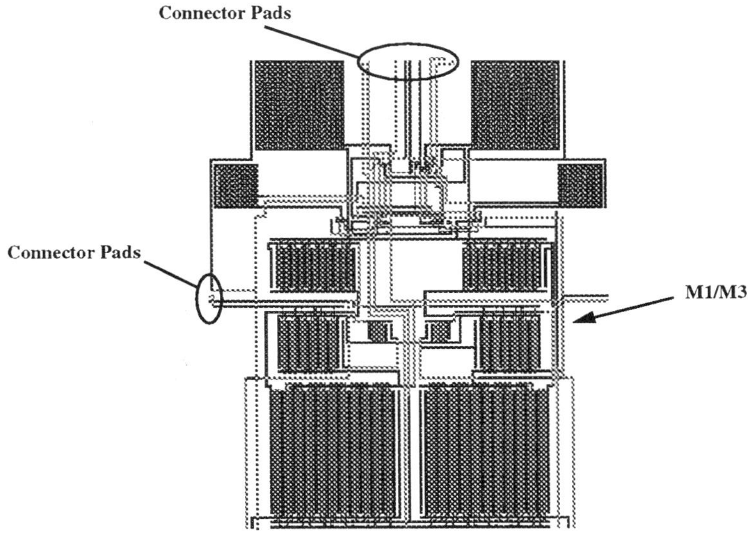

Figure 10.18　Placed, Routed, Compacted OTA

10. Compact the leaf cell using SPARCS-A [79][80] Figure 10.18 shows the placed, routed and compacted result for the OTA.

The level of automation possible depends greatly on the level of organization in the circuits. The automatic tools often fail to find desirable solutions for *highly organized* circuits. One such circuit is the clock generator. Its schematic is shown in Figure 10.11, and the hand layout is shown in Figure 10.21. The layout matches the schematic closely. There is a great deal of organization in the placement. Each inverter is placed one after the other in the chain in almost the exact manner as was drawn in the schematic. The routing is also very organized. The four power lines run horizontally across the cell so that the inverters and NOR gates can be directly tiled. Four power lines were required to separate the substrate and well contacts from the supply and ground lines to reduce noise coupling.

To compare this result with automatic techniques, Figure 10.19 shows the result from automatic placement. Since ROAD failed to route this cell, an exact area comparison could not be made. However, assuming that it could have been routed with *no* expansion (the transistor stacks seem too close together, so this is a conservative estimate), this cell is already 42% larger than the hand layout. A second attempt was made in which hand placement was followed by ROAD. Figure 10.20 shows the result. SPARCS-A

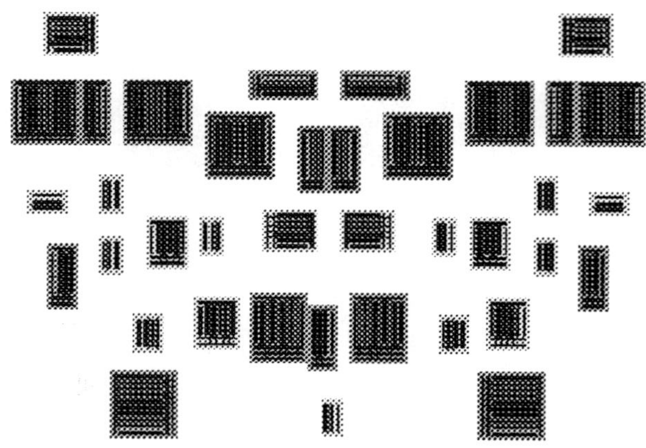

Figure 10.19   Clock Layout: Automatic Placement

was unable to compact this due to detected overconstraints. However, even if SPARCS-A were to succeed, there are portions which are not compressible. These are areas where ROAD created a non-optimal topology where an obvious optimal existed. For example, rather than routing two parallel wires which turn a corner in parallel, vias and extra jogs have been inserted, so that one wire overlaps the other. SPARCS-A can only push the vias closer to the other wire. It cannot remove them. These sections will cause the layout to be larger and more irregular than necessary. Without compression, the layout is 250% larger than the hand layout.

A layout for a circuit which requires *little organization* is the latch layout shown in Figure 10.22. The schematic is shown in Figure 10.13.

A circuit with a *medium level of organization* is the OTA shown in Figure 10.18. The transistors have certain matching and symmetry requirements, however, these can be satisfied using a variety of placement topologies. In this layout, we did intervene by hand during the placement phase. For aesthetic reasons, a couple of the transistor stacks were swapped.

Table 10.16 lists the various circuits in the Σ-Δ converter with the various layout steps enumerated. The tool required for each layout step for each layout block is indicated. An asterisk (*) indicates that slight hand modification were required after the tool was invoked. Custom tools were developed for this project. This is indicated by a $\sqrt{}$. "Step" refers the step in the layout flow described previously. "n/a" indicates that for the circuit in question, that step was not applicable. The last column gives

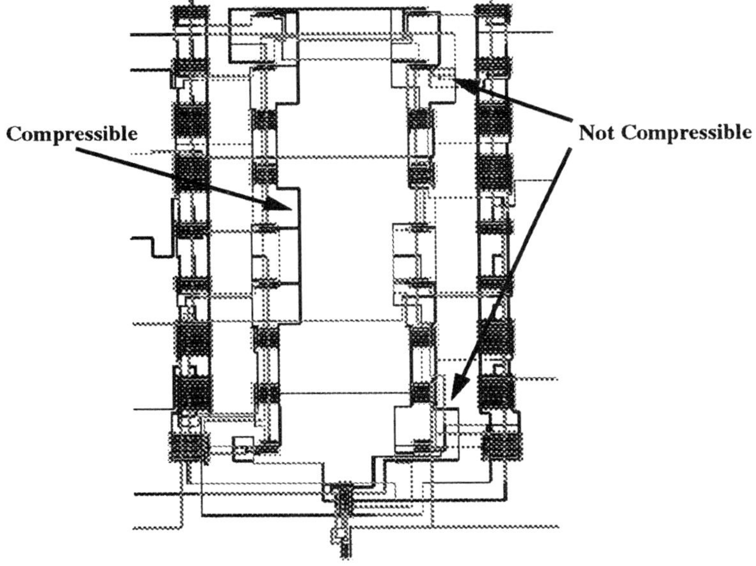

Figure 10.20 Clock Layout: Manual Placement/Automatic Routing

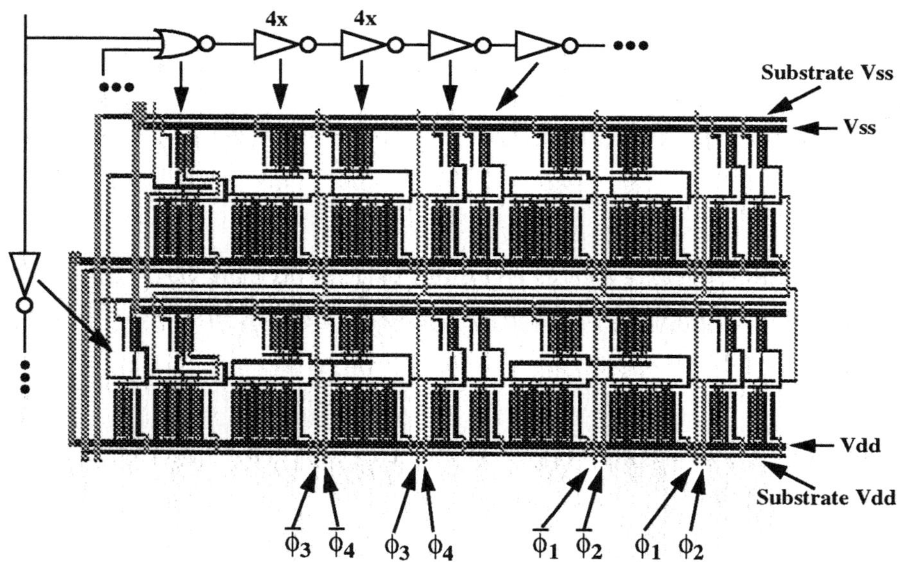

Figure 10.21 Layout of the Clock Generator

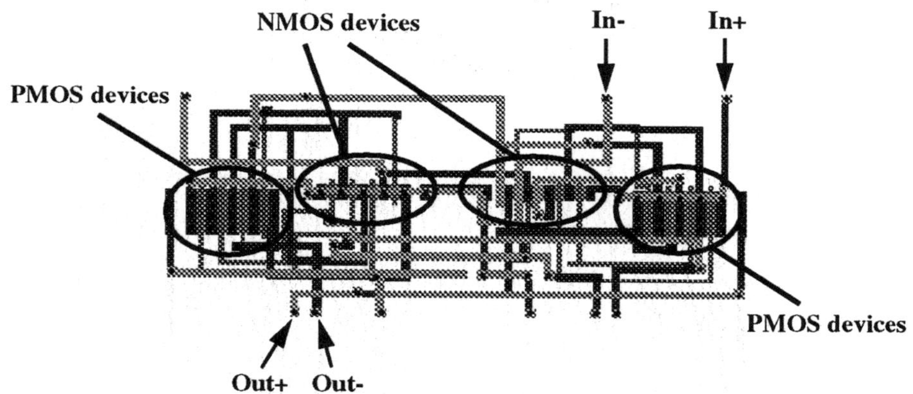

Figure 10.22 Layout of the Latch

Circuit	Element Generation Step 3,4	Stack Routing Step 5	Placement Step 7	Routing Step 9	Compaction Step 10	Time (days)
Bias	MKSTACK	√	PUPPY-A	ROAD	SPARCS-A	0.5
OTA	MKSTACK	√	PUPPY-A*	ROAD	SPARCS-A	1.0
Integrator	MKSTACK	√	PUPPY-A	ROAD	SPARCS-A	0.5
Comparator	MKSTACK	√	PUPPY-A	hand done		0.5
Latch	MKSTACK	√	PUPPY-A	ROAD	SPARCS-A	0.5
D/A	MKSTACK	√	PUPPY-A	ROAD	SPARCS-A	0.5
Clock	hand done					0.5
Shift Reg.	digital standard cells		WOLFE [269]			0.5
Pads	<Module Generator from D/A project>					0.1
Chip	n/a	n/a	hand done	MOSAICO [30]	SPARCS	1.5

Table 10.16 Level of Automation in Σ-Δ Converter Layout

the approximate *total* design time required for that layout. The D/A is a combination of the comparator and the latch plus the switches necessary for selecting the proper reference voltages.

The layout for the final chip is shown in Figure 10.23. The major components of the integrated circuit are labeled. At the top is the clock generator and the shift register. On the right is the bias circuitry along with the latches and the D/A. The two integrators are in the lower left hand corner of the chip. The area not including the bonding pads (active area), is fairly small, 1.1 mm^2. The total area is approximately 3.1 mm^2.

10.4 EXTRACTION AND VERIFICATION

We used hierarchical simulation to verify circuit performance, and we used non-hierarchical simulations to verify circuit functionality. In the hierarchical method, each sub-block was extracted. Using SPICE simulations, each sub-block's performance specifications were determined. These results were fed into MIDAS which returned the SNR for the Σ-Δ A/D. This verified the performance.

In the non-hierarchical method, we extracted the entire circuit, and performed pad-to-pad SPICE transient simulations for a number of clock periods, given a fixed amplitude sine wave at the input. The bit stream result was fed into MIDAS which decimated the output and returned an SNR value which was greater than the one returned from the hierarchical simulations. This was expected, since there were no noise sources in the flat SPICE simulation. The higher SNR value verified the circuit's functionality.

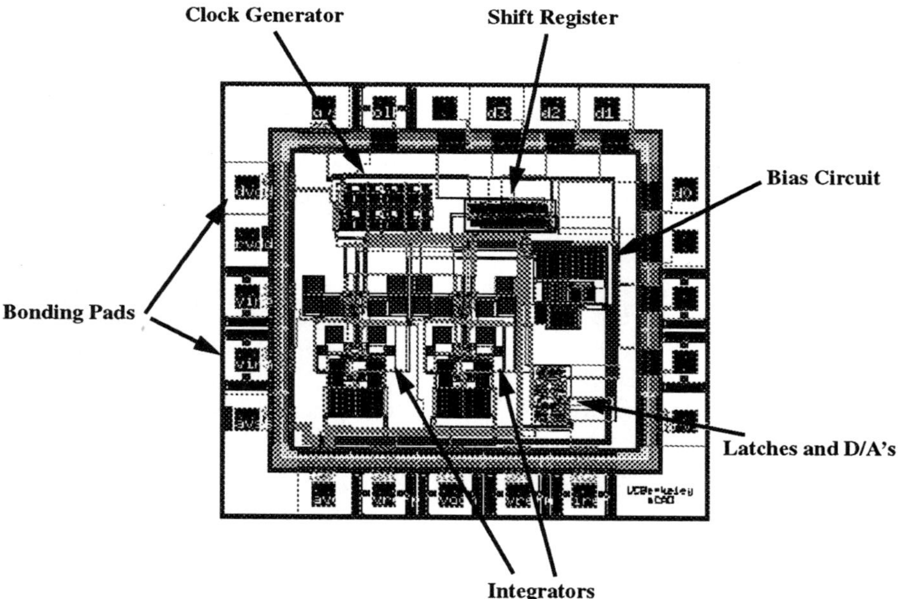

Figure 10.23 Layout of Σ-Δ A/D Converter

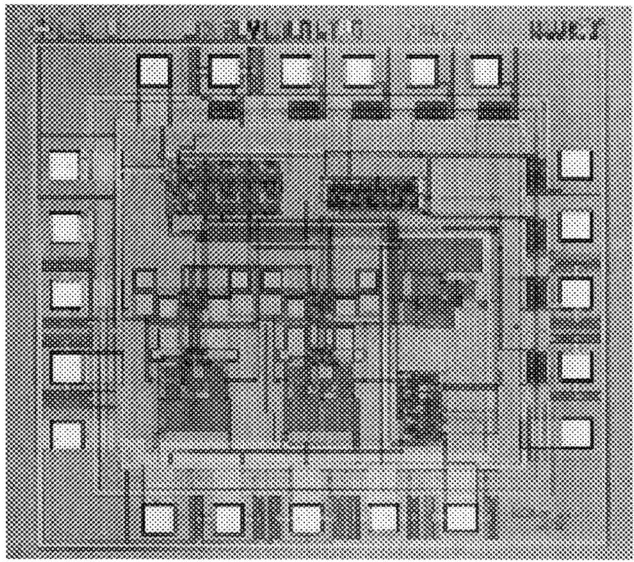

Figure 10.24 Σ-Δ A/D Converter Die Photo

10.5   EXPERIMENTAL RESULTS

This chip was fabricated on an HP 0.8 μm process by MOSIS. A die photo is shown in Figure 10.24. A test circuit board was built consisting of coaxial cable inputs, voltage regulators, by-pass capacitors, etc. Using a sine wave generator fed to a transformer to create a differential signal, a logic analyzer is used to measure the bit stream output from the Σ-Δ A/D converter. This data is then sent to MIDAS which performs the decimation and returns the SNR. Preliminary results are shown in Figure 10.17. It is believed that these results are pessimistic, because (1) we have had trouble increasing the analog input signal's SNR above 73 dB, and (2) we cannot acquire enough sample points using the logic analyzer to get a precise FFT plot. We believe that not meeting the Nyquist frequency constraint was due to the digital logic, which was automatically synthesized, being too slow.

Specifications	Value
SNR	69 dB
f_x	165 kHz

Table 10.17 Σ-Δ A/D Experiment Results

Design Phase	Time Required
Behavioral model development	4 days
High-level optimization	15 days
Low-level synthesis	5 days
Layout generation	10 days
Extraction/Verification	5 days

Table 10.18 Design Times

10.6 DESIGN TIMES

The total elapsed time for this project, design synthesis and physical synthesis phase, was approximately two months. Table 10.18 shows how this time is broken down. These times are intended to give an accurate account of the length of this project from its inception to "taping-out" to illustrate the *real time required* for design when using this methodology.

10.7 ANALYTIC EQUATIONS FOR Σ-Δ A/D

Given the values for the parameters found in Tables 10.19, 10.20, 10.22, and 10.21, a variety of performance approximations and design characteristics can be determined for the integrator described in Section 10.3.2 using analytic equations. These need to be solved during the design process of the integrator/OTA. The flicker noise parameters, K_{fn}, K_{fp}, A_{fn}, and A_{fp}, were determined by fitting curves to device measurements of flicker noise obtained directly from HP. $\Delta k'_n$, $\Delta k'_p$, and a_{VFB} are estimated values.

This section will show how these characteristics can be calculated. Most of the equations are taken directly from or derived using [105]. These characteristics will be summarized at the end of this section with their corresponding equation numbers for reference.

Parameter	Description
V_{dd}	Supply Voltage
f_x	Nyquist Frequency
M	Oversampling ratio
V_{FS}	Full scale voltage of Σ-Δ system
I_{gain}	Integrator gain
L_{nmos}	Length of transistors $M_1 - M_4$, $M_{50} - M_{57}$ of OTA Length of transistors $M_{21} - M_{22}$, $M_{30} - M_{33}$
L_{pmos}	Length of transistors $M_5 - M_8$, $M_{30} - M_{33}$ of OTA Length of transistors $M_0 - M_5$, $M_{10} - M_{13}$ of the bias circuit
L_{bias}	Length of transistors $M_9 - M_{12}$ of the OTA Length of transistors $M_{40} - M_{43}$ of the bias circuit
L_{M44}	Length of transistor M_{44} of the bias circuit
W_{M20}	Width of transistor M_{20} of the bias circuit
$V_{gs} - V_T$	$V_{gs} - V_T$ voltage for the transistors in the gain stage
I_{leg}	Current flowing through the transistors in the gain stage
C_S	Size of the sampling capacitor
V_{safety}	Voltage above V_{ds} where cascode device will be biased
f_{min}	Minimum frequency for flicker noise calculation
I_{ratio}	Current ratio between M_{11} to M_9 of OTA
$R_{c1,m9gate}$	Ratio between C_1 and M_9 gate capacitance of OTA
$R_{c1,c2}$	Ratio between C_1 and C_2 of CMFB circuit
p_{leak}	Percent leakage current M_{20} of the bias circuit

Table 10.19 Integrator Design Parameters

Parameter	Description	NMOS device	PMOS device
$\Delta W_n, \Delta W_p$	Lithography width variation	0.25 μm	0.39 μm
L_{Dn}, L_{Dp}	Gate Lateral Diffusion	0.10 μm	0.02 μm
V_{Tn}, V_{Tp}	Zero-bias Threshold Voltage	0.608V	0.838V
k'_n, k'_p	Transconductance Parameters	130 $\mu A/V^2$	43 $\mu A/V^2$
γ_n, γ_p	Bulk Threshold Parameter	0.448 $V^{\frac{1}{2}}$	0.497 $V^{\frac{1}{2}}$
t_{ox}	Oxide Thickness	16 nm	16 nm
$C_{jo,n}, C_{jo,p}$	Zero Bias Bulk Junction Cap.	$0.11962 \cdot 10^{-3}$ F/m^2	$0.53093 \cdot 10^{-3}$ F/m^2
$m_{j,n}, m_{j,p}$	Bulk Junction Exponential	0.4398	0.5074
phi	Surface Potential	0.6 V	0.6 V
K_{fn}, K_{fp}	Flicker Noise Coefficient	$6.5 \cdot 10^{-24} V^2$F	$2.1 \cdot 10^{-24} V^2$F
A_{fn}, A_{fp}	Flicker Noise Exponential	1.2	1.2
$\Delta k'_n, \Delta k'_p$	Transconductance Matching	4%	4%
a_{VFB}	Threshold Voltage Matching	0.0623 $\mu V m$	0.0623 $\mu V m$

Table 10.20   Relevant HP CMOS26B Process Parameters

Parameter	Description	value
L_{drain}	Length of Drain	3.0 μm
L_{source}	Length of Source	2.5 μm

Table 10.21   Relevant MOSIS SCMOS Design Rules

Parameter	Description	Value
k	Boltzmann's constant	$1.38 \times 10^{-23} J/K$
T	Temperature	300 K
ϵ_o	Dielectric constant of vacuum	$8.85 \times 10^{-12} F/m$
κ_{ox}	Silicon Oxide dielectric ratio	3.97

Table 10.22   Physical Constants

There are nine basic performance parameters that need to be calculated. These are listed in Table 10.7.

The *open-loop gain*, O_{gain}, of the OTA is

$$O_{gain} = g_m \left[\frac{1}{r_{on}(1 + g_m r_{on})} + \frac{1}{r_{op}(1 + g_m r_{op})} \right]^{-1} \tag{10.7}$$

where g_m, the transconductance, is $\frac{2I_{leg}}{V_{gs} - V_T}$, $r_{on} = \frac{1}{I_{leg}\lambda_n}$, and $r_{op} = \frac{1}{I_{leg}\lambda_p}$. The channel length modulation parameters, λ_n and λ_p, were determined by SPICE simulation using transistors with L_{nmos} and L_{pmos} for transistor lengths respectively. λ_n and λ_p are extremely strong functions of L_{nmos} and L_{pmos} for short channels.

The *kT/C noise* of the integrator, $I_{kT/C}$, is

$$I_{kT/C} = \left(\frac{2kT}{MC_S} \right)^{\frac{1}{2}} \tag{10.8}$$

The factor of 2 is present because the circuit is differential with two sampling capacitors.

The equation for *thermal noise* for this OTA (shown in Figure 10.7), O_{th}, is

$$O_{th} = \left[2 \cdot 4kT \frac{2}{3g_{m1}} f_x + 2 \cdot \left(\frac{g_{m7}}{g_{m1}} \right)^2 \cdot 4kT \frac{2}{3g_{m7}} f_x \right]^{\frac{1}{2}} \tag{10.9}$$

In this design since $V_{gs} - V_T$ is the same for all of transistors in the main current legs, $g_{m1} = g_{m7} = g_m$, thus Equation 10.9 reduces to:

$$O_{th} = \left(4 \cdot 4kT \frac{2}{3g_m} f_x \right)^{\frac{1}{2}} \tag{10.10}$$

The *output voltage range* of the OTA is

$$V_{range} = 2(V_{o,max} - V_{o,min})$$

where $V_{o,max} = V_{dd} - 2(V_{gs} - V_T) - 2V_{safety}$ and $V_{o,min} = 3(V_{gs} - V_T) + 3V_{safety}$. Again, the factor of 2 is because the system is differential. The output voltage range relative to V_{FS}, then, is

$$O_{range} = \frac{V_{range}}{V_{FS}} \tag{10.11}$$

Also, the output common voltage is

$$V_{ocm} = \frac{V_{o,max} + V_{o,min}}{2} \qquad (10.12)$$

In order to calculate the remaining performance parameters, some device sizes must be calculated using

$$W_{M1-M4} = \frac{2I_{leg}(L_{nmos} - L_{Dn})}{k'_n(V_{gs} - V_T)^2} + \Delta W_n \qquad (10.13)$$

$$W_{M5-M8} = \frac{2I_{leg}(L_{pmos} - L_{Dp})}{k'_p(V_{gs} - V_T)^2} + \Delta W_p \qquad (10.14)$$

$$W_{M9-M10} = \frac{2I_{leg}(L_{bias} - L_{Dn})}{k'_n(V_{gs} - V_T)^2} + \Delta W_n \qquad (10.15)$$

The *flicker noise* for this type of OTA is

$$v_{1/f,n}^2 = \begin{cases} 2 \cdot \dfrac{K_{fn}}{W_{M1}(L_{nmos}-L_{Dn})C_{ox}} \ln \dfrac{f_x}{f_{min}} & \text{if } A_{fn} = 1 \\[2ex] 2 \cdot \dfrac{K_{fn}}{W_{M1}(L_{nmos}-L_{Dn})C_{ox}} \cdot \dfrac{f_x^{1-A_{fn}} - f_{min}^{1-A_{fn}}}{1-A_{fn}} & \text{otherwise} \end{cases}$$

$$v_{1/f,p}^2 = \begin{cases} 2 \cdot \left(\dfrac{g_{m7}}{g_{m1}}\right)^2 \cdot \dfrac{K_{fp}}{W_{M7}(L_{pmos}-L_{Dp})C_{ox}} \ln \dfrac{f_x}{f_{min}} & \text{if } A_{fp} = 1 \\[2ex] 2 \cdot \left(\dfrac{g_{m7}}{g_{m1}}\right)^2 \cdot \dfrac{K_{fp}}{W_{M7}(L_{pmos}-L_{Dp})C_{ox}} \cdot \dfrac{f_x^{1-A_{fp}} - f_{min}^{1-A_{fp}}}{1-A_{fp}} & \text{otherwise} \end{cases}$$

$$O_{1/f} = \left(v_{1/f,n}^2 + v_{1/f,p}^2\right)^{\frac{1}{2}} \qquad (10.16)$$

where $C_{ox} = \frac{\kappa_{ox}\epsilon_o}{t_{ox}}$. Again, since the g_m for all of the transistors are the same, $\frac{g_{m7}}{g_{m1}} = 1$ and the equations are slightly simplified.

To calculate the *slew rate*, I_{SR}, the capacitance at the output must be computed. One component is the drain capacitances of M_3-M_6. These computations assume that the transistors are interleaved in the layout, reducing the junction capacitance by approximately a factor of two and virtually eliminating the sidewall capacitance. The design rule, L_{drain} is the gate-to-gate distance of interleaved transistors. The drain capacitances are

$$C_{d,M3} = \frac{W_{M3}}{2} L_{drain} \cdot \frac{C_{jo,n}}{(1 + \frac{V_{ocm}}{\phi})^{m_{j,n}}}$$

$$C_{d,M5} = \frac{W_{M5}}{2} L_{drain} \cdot \frac{C_{jo,p}}{(1 + \frac{V_{ocm}}{\phi})^{m_{j,p}}}$$

Another component is the series capacitance of C_1 and the capacitance at the gate of M_9. The capacitance at the gate of M_9 is $C_{g,M9} = \frac{2}{3}C_{ox}W_{M9}(L_{bias} - L_{Dn})$. C_1 is taken as

$$C_1 = R_{c1,m9gate}C_{g,M9} \tag{10.17}$$

There is also a fairly large parasitic capacitance associated with C_1. Since this is only a single poly process, the parasitic capacitance between the bottom plate of the capacitor to the substrate is approximately the same as the capacitor itself. To reduce the drain capacitance, the bottom plate for C_1 was selected to be the node at the gate of M_9. Thus, $C_{g,M9}$ has in parallel with it approximately C_1. The total series capacitance then is

$$C_{series} = \left[\frac{1}{C_1} + \frac{1}{C_{g,M9}(1 + I_{ratio}) + C_1} \right]^{\frac{1}{2}}$$

The $(1 + I_{ratio})$ scaling term is necessary to include the gate capacitance of M_{11}. The total output capacitance is

$$C_{out} = C_{d,M3} + C_{d,M5} + C_{series} + C_I + C_S$$

The slew rate, then, is finally

$$I_{SR} = \frac{I_{leg}}{C_{out}} \tag{10.18}$$

where

$$C_I = \frac{C_S}{I_{gain}} \tag{10.19}$$

Because each capacitor carries with it a parasitic capacitance of almost the same value as the capacitor itself, this must be considered when choosing which side (top or bottom) to connect to the OTA. Traditionally, top plates are connected to the inputs of the OTA because the inputs are very sensitive. This minimizes the capacitance at the gate and improves the feedback factor. The bottom plate also acts as a shield from substrate noise for the sensitive node. However, in this design, as an experiment, since the optimization at the low-level showed that slew rate was a limiting factor (see Table 10.10) and not I_τ, the bottom plates were connected to the input of the OTA, thus minimizing the drain capacitance, and as such, improving the slew rate. Also, consider that the gain of this integrator is only $1/2$. For this low amount of closed loop gain, an increase in the parasitics at the input of the OTA, does not affect the system too greatly (see Equation 10.20).

Figure 10.25 shows the position of the parasitics. The side in bold on the capacitors indicates the bottom plate. The extra capacitor adjacent to C_S and C_I depicts where the parasitic capacitance is.

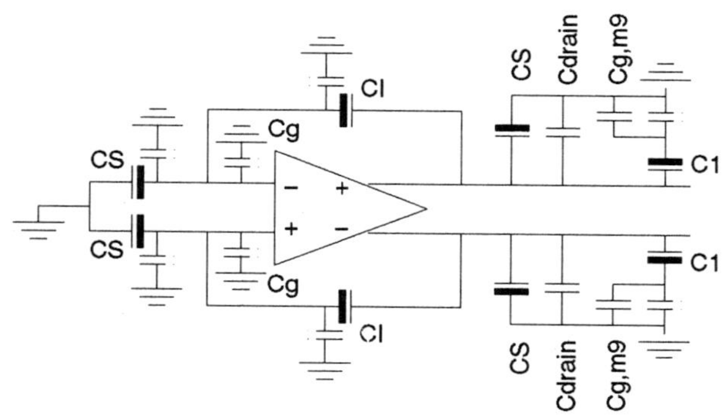

Figure 10.25 Location of Parasitics introduced by capacitors, C_S and C_I

To calculate the *time constant for the integrator*, I_τ, first, the feedback factor is computed. It is

$$f = \frac{C_I}{C_I + C_{I,par} + C_S + C_{S,par} + C_{g,M3}} \tag{10.20}$$

where $C_{g,M3} = 2/3 \cdot C_{ox} W_{M3}(L_{nmos} - L_{Dn})$ and $C_{I,par}$ and $C_{S,par}$ are the parasitic capacitances from C_I and C_S respectively. Assuming that $C_{I,par} \approx C_I$ and $C_{S,par} \approx C_S$, then

$$f = \frac{C_I}{2C_I + 2C_S + C_{g,M3}}$$

From this,

$$I_\tau = \frac{C_{out}}{g_m} \cdot \frac{1}{f} \tag{10.21}$$

The feedback factor for the common mode feedback circuit can be calculated in a similar fashion. Again assuming the parasitic capacitor associated with C_1 is about the same as C_1, the feedback factor is

$$f_{cmfb} = \frac{C_1}{2C_1 + C_{g,M9}(1 + I_{ratio})}$$

The *time constant for the common mode feedback circuit* is

$$O_{\tau,CMFB} = \frac{C_{out}}{g_m} \cdot \frac{1}{f_{cmfb}} \tag{10.22}$$

The final of the nine performance parameters is the OTA *input offset voltage, O_{off}*. From Equation 9.1,

$$\sigma_{vt,n} = \frac{\sqrt{a_{VFB}}}{W_{M1}(L_{nmos} - L_{Dn})}$$

$$\sigma_{vt,p} = \frac{\sqrt{a_{VFB}}}{W_{M7}(L_{pmos} - L_{Dp})}$$

From this

$$O_{off} = \sigma_{vt,n} + \frac{V_{gs} - V_T}{2\Delta k'_n} + \frac{g_{m7}}{g_{m1}}\left(\sigma_{vt,p} + \frac{V_{gs} - V_T}{2\Delta k'_p}\right) \tag{10.23}$$

To obtain the rough approximation of the *active area* necessary for use as the objective function in the low-level optimization step in Section 10.3.2, the area is taken as

$$area = W_{M1}L_{nmos} + W_{M7}L_{pmos} \tag{10.24}$$

Note that all of the capacitors scale with this area, so this is a valid approximation, even when the capacitors are large.

The above is enough to set up and solve the low-level optimization problem. However, in order to generate the full schematic/spice deck for the OTA, integrator, and bias circuit, there are a few more expressions that need to be evaluated. Most of these device sizes are direct consequences of the formulas previously developed. For example,

$$W_{M11-M12} = I_{ratio}W_{M9} \tag{10.25}$$

$$W_{M30-M33} = I_{ratio}W_{M5} \tag{10.26}$$

$$W_{M0-M5,bias} = W_{M5} \tag{10.27}$$

$$W_{M21-M22,bias} = W_{M1} \tag{10.28}$$

$$W_{M40-M43,bias} = W_{M9} \tag{10.29}$$

$$W_{M44} = \frac{2I_{leg}L_{M44}}{k'_n(V_{gs} - V_T)^2} + \Delta W_n \tag{10.30}$$

The other capacitor in the CMFB circuit, C_2 was chosen to be

$$C_2 = \frac{C_1}{R_{c1,c2}} \tag{10.31}$$

The value for $R_{c1,c2}$ of was simply chosen as a rule of thumb.

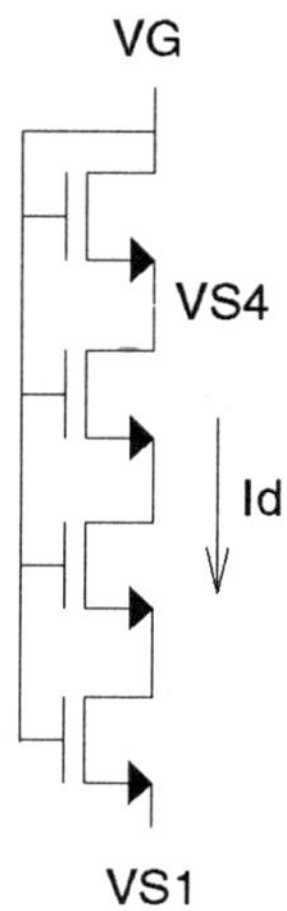

Figure 10.26   Diode bias chain

Finally, the widths of the diode connected transistors need to be determined. Figure 10.26 shows the schematic with node names and currents for the diode chain. The transistor widths can be approximated by solving simultaneously, Equations 10.32 through 10.35 for W and V_{S4}.

$$V_{t1} = V_{T0} + \gamma(\sqrt{\phi + V_{S1}} - \sqrt{\phi}) \tag{10.32}$$

$$V_{t4} = V_{T0} + \gamma(\sqrt{\phi + V_{S4}} - \sqrt{\phi}) \tag{10.33}$$

$$I_D = \frac{k'(W - \Delta W)}{2(n-1)(L - L_D)}[2(V_G - V_{S1} - V_{t1})(V_{S4} - V_{S1}) - (V_{S4} - V_{S1})^2] \tag{10.34}$$

$$I_D = \frac{k'(W - \Delta W)}{2(L - L_D)}(V_G - V_{S4} - V_{t4})^2 \tag{10.35}$$

n is the number of diodes in the chain. $n = 4$ in this case. For the NMOS chain, $W_{M50-M57}$, let

$$I_D = I_{leg}I_{ratio}$$
$$\gamma = \gamma_n$$
$$k' = k'_n$$
$$L = L_{nmos}$$
$$V_{T0} = V_{Tn}$$

$$V_{S1} = V_{gs} - V_T + V_{safety}$$
$$V_G = 3(V_{gs} - V_T) + 2V_{safety} + V_{tx}$$
$$\text{where} \quad V_{tx} = V_{Tn} + \gamma_n(\sqrt{\phi + 2(V_{gs} - V_T) + 2V_{safety}} - \sqrt{\phi})$$

For the PMOS chain, $W_{M10-M13}$, let

$$I_D = I_{leg}$$
$$\gamma = \gamma_p$$
$$k' = k'_p$$
$$L = L_{pmos}$$
$$V_{T0} = V_{Tp}$$
$$V_{S1} = 0$$
$$V_G = 2(V_{gs} - V_T) + V_{safety} + V_{Tp}$$

Solving twice, then, Equations 10.32 to 10.35, one can obtain $W_{M50-M57}$ and $W_{M10-M13}$. Finally,

$$W_{M30-M33} = \frac{W_{M50-M57}}{I_{ratio}} \tag{10.36}$$

Note that this only yields an approximate solution. These values had to be increased 10% to 20% to achieve the exact headroom voltage. This was verified through SPICE simulation.

The last transistor length, L_{M20} to be determined is the one for M_{20}, the leak transistor in the bias circuit. For this,

$$L_{M20} = \frac{(W_{M20} - \Delta W)k'_n}{2p_{leak}I_{leg}}(V_{dd} - V_{Tn})^2 + L_{Dn} \tag{10.37}$$

These equations characterize the critical performance parameters for the integrator and OTA in this Σ-Δ A/D design example. They also define all of the device sizes. Tables 10.23 and 10.24 summarize the two sets of results.

10.8 CONCLUSION

During the span of two months a Σ-Δ A/D was synthesized following the methodology and sent for fabrication. The total design time was reasonable, and tests indicate that

Performance	Description	Equation
O_{off}	Offset voltage due to the OTA	10.23
O_{range}	Output range of the OTA relative to V_{FS}	10.11
O_{th}	Thermal noise due to the OTA	10.10
$O_{1/f}$	Flicker noise for the OTA	10.16
O_{gain}	Open loop gain of the OTA	10.7
I_τ	Time constant for the integrator	10.21
I_{SR}	Slew Rate of the integrator	10.18
$I_{kT/C}$	kT/C noise of the integrator	10.8
$O_{\tau,CMFB}$	Time constant for the CMFB circuit of the OTA	10.22
area	Active Area approximation	10.24

Table 10.23 Calculated Integrator/OTA Performance Parameters

Parameter	Description	Equation
V_{ocm}	Output common mode voltage	10.12
W_{M1-M4}	Width of transistors $M_1 - M_4$	10.13
W_{M5-M8}	Width of transistors $M_5 - M_8$	10.14
W_{M9-M10}	Width of transistors $M_9 - M_{10}$	10.15
$W_{M11-M12}$	Width of transistors $M_{11} - M_{12}$	10.25
$W_{M30-M33}$	Width of transistors $M_{30} - M_{33}$	10.26
W_{M44}	Width of transistor M_{44}	10.30
$W_{M50-M57}$	Width of transistors $M_{30} - M_{33}$	10.32-10.35
$W_{M0-M5,bias}$	Width of transistors $M_0 - M_5$ in the bias circuit	10.27
$W_{M10-M13,bias}$	Width of transistors $M_{10} - M_{13}$ in the bias circuit	10.32-10.35
L_{M20}	Length of transistor M_{20} in the bias circuit	10.37
$W_{M21-M22,bias}$	Width of transistors $M_{21} - M_{22}$ in the bias circuit	10.28
$W_{M30-M33,bias}$	Width of transistors $M_{30} - M_{33}$ in the bias circuit	10.36
$W_{M40-M43,bias}$	Width of transistors $M_{40} - M_{43}$ in the bias circuit	10.29
C_1	First of two CMFB capacitors	10.17
C_2	Second of two CMFB capacitors	10.31
C_I	Integrating capacitor	10.19

Table 10.24 Determined Integrator/OTA Voltages and Device Sizing

the A/D converter is close to meeting performance specifications. Future work for the Σ-Δ A/D consists of further tests and the development of a testing methodology which will attempt to accelerate the testing process. One possible approach is to find an optimal set of sine wave input amplitudes and frequencies in order to quickly characterize SNR.

11

VIDEO DRIVER DESIGN EXAMPLE

11.1 INTRODUCTION

The design of a video driver system is the last example in our current set. New behavioral modeling, optimization, and layout techniques have been developed or extended from existing ones, in order to provide a full set of tools supporting the design of a class of similar mixed-signal systems. This description focuses on the "critical path" of the design. At the high-level synthesis phase, emphasis is given at the PLL behavioral models and simulation techniques. The setup of the PLL optimization problem that performs the constraint mapping, together with the appropriate optimization algorithm are also described. The novelty of the design approach includes a jitter constraint which is set at the system level and mapped onto circuit level constraints. Typically, simulation (if any, and usually only at the circuit level) is used to verify the performance at the bottom of the hierarchy, thus causing expensive design iterations. Following the "critical path" of the design, the VCO synthesis phase is depicted, with focus on the optimization approach that takes into account layout parasitics. The layout constraints generated at the circuit level are enforced during the VCO layout synthesis phase. Finally, detailed extraction of the sub-blocks and behavioral system-level simulation is used for the verification of the system performance for the PLL.

11.2 SYSTEM DESCRIPTION

Video driver systems such as RAMDAC [29] or similar [319][46][173][251] have become indispensable modules in the PC and workstations graphics. The function of a RAMDAC is to convert a set of colors, 2^n, into another set of colors, 2^m, using a RAM look-up table to convert the output word into an analog signal for the display.

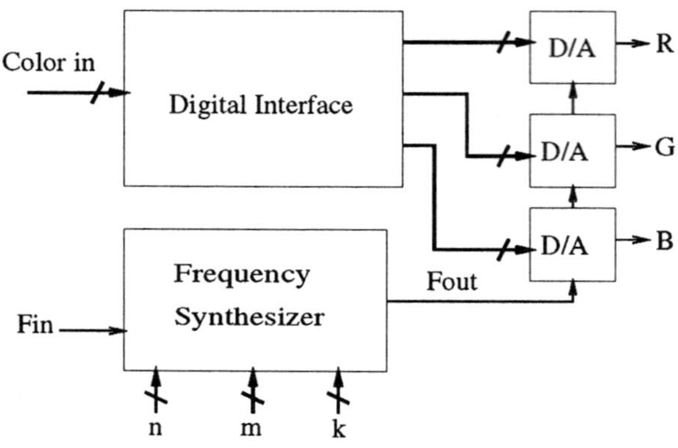

Figure 11.1   Display Driver System Diagram

Typically $n = 8$, $m = 24$, so 256 out of 16 million possible colors can be viewed at the same time. An external oscillator can be used to provide the clocks required. However, integration of a programmable frequency synthesizer on the same die with the RAMDAC [54] reduces the cost and gives the flexibility of using the same chip in different systems.

The video driver system we chose to implement includes three basic sub-systems: first is the frequency synthesizer PLL, second are the three D/A converters, and third is a digital interface file register for loading the D/A converters and programming the frequency synthesizer. This system is similar to a RAMDAC except that the SRAM lookup table is not implemented. A general block diagram of the system is shown in Figure 11.1. In this system, the PLL functions as a programmable clock generator, synthesizing the various dot clocks required for different screen resolutions. The D/A converters generate the output currents required for the red, green, and blue signals to the monitor.

This system is more complex than the previous design examples. It contains two large analog components and requires an extra layer of hierarchy. It also more closely resembles large commercial systems. Its level of complexity is shown in Figure 9.1.

11.3   DESIGN SPECIFICATIONS

Crucial for the performance of such a system is the speed and resolution of the D/A converters, the frequency range and the timing jitter of the frequency synthesizer. The

Type	Specification	Value
Performance	Output Frequencies	25 to 135 MHz
	Timing jitter	$\leq$ 1%
	Video signal INL	$\leq$ 1 LSB
	Video signal DNL	$\leq$ 0.5 LSB
	D/A resolution	8 bits
Operation	Supply voltage	5 V
Technology	Spice models	HP CMOS34
	Design rules	SCMOS

Table 11.1 Video Driver System Specifications

specifications for the system are given in Table 11.1. A number of discrete frequencies need to be generated that correspond to different display standards. A 1 μm MOSIS HP process and a 5 volt supply will be used.

11.4 SYNTHESIS PATH

The design hierarchy for the system is shown in Figure 11.2. The first level of hierarchy divides the video driver into its basic system components—three D/A converters, a PLL, and a data and control path to handle the incoming pixel data as well as registers to control the frequency of the PLL output. The system-level constraints are mapped onto constraints for the PLL and the D/A converters. The description of the methodology will focus on the path highlighted in Figure 11.2. Behavioral modeling and optimization is used to map PLL performance constraints onto constraints for its building blocks. The most critical building block is the VCO. Its low-level design and physical synthesis are described in detail. Other, less critical paths in the design hierarchy, will be described briefly. At every level of the hierarchy, constraints are also imposed onto routing parasitics, in order to account for the performance degradation after the final layout is done.

11.4.1 High-Level Synthesis

Unlike the previous design examples, three high-level decompositions are required in this problem. (1) The video driver system is broken down its subcomponents. (2) The

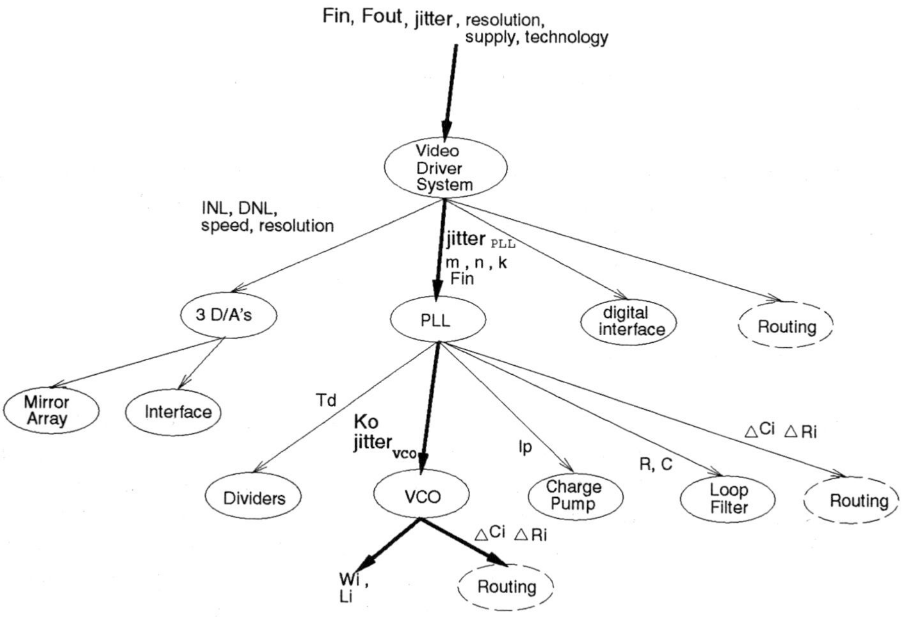

Figure 11.2   Video Driver System Hierarchy

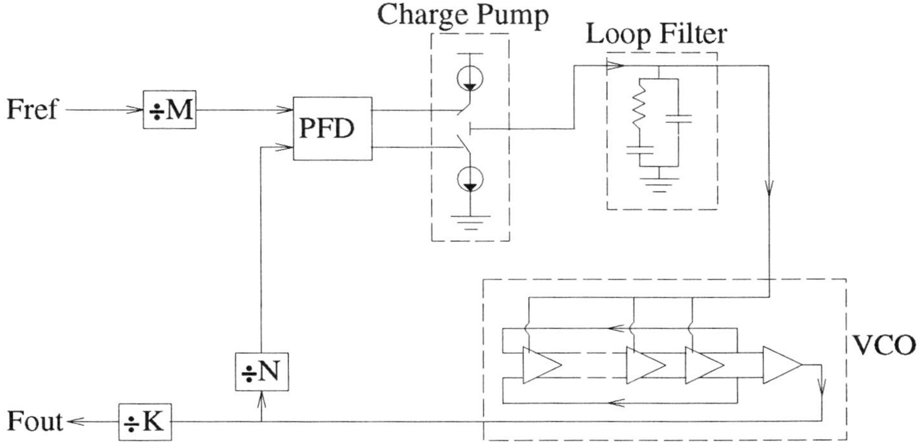

Figure 11.3 PLL Programmable Frequency Synthesizer

PLL system is broken down into its subcomponents. And (3) the D/A converters are mapped into their subcomponents.

Video Driver System

The constraints of Table 11.1 can be immediately decomposed into D/A and frequency synthesizer constraints. The D/A synthesis hierarchy stops after the D/A constraints are given, since a module generator [220] is used for their design and layout. [220] provides more details about the constraints used in the layout synthesis tool. Custom cells were used for the design of the file register.

Phase-Locked Loop

The architecture selected for the PLL is a charge-pump PLL using a ring oscillator VCO (Figure 11.3). This is one of the most popular architectures for CMOS frequency synthesizers and is also used in [251][322][54][140]. The main advantage of this architecture is that does not require any external components and can be easily integrated. It consists of a PFD which compares the input reference frequency (F_{ref}) that has been divided by m and the output from the VCO which has been divided by n. The PFD controls the charge pump. The current from the charge pump goes through a low pass filter ("loop filter"). A lead-lag RC filter is used most commonly in order to have a wide range of operating frequencies. The loop forces the two compared signals in

$F_{in}(MHz)$	F_{out}	N	M	K
	25.175	176	25	4
	30.24	169	20	4
	31.5	88	10	4
	36	101	20	2
	40	95	17	2
	44.9	69	11	2
14.318	50	70	10	2
	64	152	17	2
	65	118	13	2
	75	89	17	1
	80	95	17	1
	86	60	10	1
	127	133	15	1
	135	132	14	1

Table 11.2   Frequency Synthesizer Divider Values

the PFD to have equal phase. By changing the divider values, various integer fractions
of the input clock can be realized:

$$F_{out} = \frac{N}{M \cdot K} \cdot F_{ref} \qquad (11.1)$$

The last divide by K is used so that the PLL operating frequency range required is
reduced and the design is made easier. As a reference, an external crystal oscillator is
used and the divider values needed for the various output frequencies are selected so
that the desired operating frequencies are generated precisely.

Table 11.2 summarizes the divider values that correspond to the desired output frequencies.

PLL Behavioral Models

For the high-level mapping, a behavioral description for the PLL is used. It is important that the behavioral models are *implementation independent* and capture all the
important second-order effects determining the performance of analog circuits. This is
different than macromodels and first-order functional behavioral models that are often

used. A behavioral simulator for mixed-signal systems was developed and is described in Section 3.6. The PLL is described by a set of differential equations:

$$\frac{\partial \theta_i}{\partial t} = 2\pi f_i(t) \tag{11.2}$$

$$\frac{\partial V_c}{\partial t} = \frac{I_{p_{eff}}}{C_2} - \frac{1}{RC_2}V_c + \frac{1}{RC_2}V_x \tag{11.3}$$

$$\frac{\partial V_x}{\partial t} = \frac{1}{RC}V_c - \frac{1}{RC}V_x \tag{11.4}$$

$$\frac{\partial \theta_j}{\partial t} = 2\pi F(V_c(t))n_d \qquad \forall j, j = 1 \ldots n_d \tag{11.5}$$

where:

$$I_{p_{eff}} = ST \cdot I_p(1 + \frac{\Delta I_p}{I_p} \cdot ST) - ST\frac{\Delta V}{R_{out}} \tag{11.6}$$

$$F(V_c(t)) = F_0 + K_0 V_c \tag{11.7}$$

The state variables θ_i, V_c, V_x, θ_j represent the phase of the input clock, the VCO control voltage, the voltage on capacitor C, and the phases of the n_d stages of the VCO delay stages respectively. $ST = 0, -1, 1$, depending on the state of the PFD. The Jacobian of the above equations with respect to the state variables is also given analytically to the simulator. $F(V_c(t))$ is the instantaneous VCO frequency. A linear model is used for the VCO voltage-to-frequency characteristic.

The *behavioral representation of timing jitter* is crucial for our design since it was used as an initial system constraint. Timing jitter is defined as the standard deviation of the period of the output of the PLL:

$$\Delta \tau_{rms} = \sqrt{E\left[(T - E\left[T\right])^2\right]} \tag{11.8}$$

where T is the output period. The most fundamental timing jitter noise source is the thermal noise of the electronic circuits, which causes uncertainty in the digital waveform transitions. Assuming that the other timing jitter sources have been dealt with, timing jitter of circuit components can be attributed exclusively to thermal noise and be modeled as a white Gaussian random process. In addition to that, we can assume that when the system is in lock, the noise is small enough so that it does not drive the system out of lock. The overall PLL jitter is then predicted by adding random noise at the time of each VCO transition and subsequently processing the resulting waveform [66]. A complete listing of the behavioral parameters used is given in Table 11.3.

Module	Parameter name	Description
VCO	K_o	Voltage to frequency gain
	F_o	Center frequency
	$\frac{\Delta T}{T}$	Timing jitter
	V_{min}	Lower saturation voltage
	V_{max}	Upper saturation voltage
Charge-pump	I_p	Nominal charge-pump current
	$\frac{\Delta I_p}{I_p}$	Up and down current mismatch
	R_{lin}	Linear region resistance
	R_{sat}	Saturation region resistance
	V_{satup}	Upper source saturation voltage
	V_{satlo}	Lower source saturation voltage
Loop Filter	R	Resistor value
	C_1, C_2	Capacitor values
PFD		State transition table
	t_{dead_zone}	Dead zone
	t_d	Delay
Divider	t_d	Delay

Table 11.3   PLL High-Level Parameters

Module	Parameter name	Value
VCO	F_o	100 MHz
	V_{min}	1.8 V
	V_{max}	5 V
Charge-pump	$\frac{\Delta I_p}{I_p}$	5%
	R_{sat}	2 MΩ
	V_{satup}	4.8 V
	V_{satlo}	0.2 V
PFD	t_{dead_zone}	0
	t_d	2 ns
Divider	t_d	2 ns

Table 11.4   Constant High-Level PLL Parameters

Each one of the performance constraints can be interpreted as a function of the above variables. However, an optimization problem including all the above parameters would be unnecessarily complicated since many of them have little effect on the system performance and given a particular technology, values close to optimal could be chosen manually.

In the case of the VCO, given the topology described in Section 11.4.2, typical values can be easily chosen for V_{min}, V_{max}. The VCO center frequency F_o, can be set in the middle of the desired frequency range. We can also manually select the parameters $\frac{\Delta I_p}{I_p}$, R_{lin}, R_{sat}, V_{satup}, V_{satlo} of the charge pump, since small variations of these parameters have little effect on the PLL performance, as shown by behavioral simulation. Digital delays do not affect the system performance and the PFD dead zone problem can be eliminated, as shown subsequently.

Selecting manually the parameters which do not affect optimality does not invalidate the top-down approach. Those values will be constraints for the next levels of hierarchy. Table 11.4 summarizes the parameters that were manually chosen.

High-Level Optimization Objective

In order to determine the remaining parameters, crucial to the PLL performance, optimization needs to be used. The behavioral models used are, again, high-level implementation independent. The area or power consumption of the low-level blocks cannot be determined yet. Therefore, to perform the high-level optimization an alter-

flexibility type	Parameter name	"Easy"	"Hard"
parabolic	$\sqrt{\Delta\tau_{VCO}^2}$ @140 MHz	7 ps	1.4 ps
hyperbolic	K_0	50 MHz/V	80 MHz/V
	I_p	20 μA	200 μA
	R, C_1, C_2	0.05 mm^2	0.1 mm^2

Table 11.5 Flexibility Coefficients

nate measure of optimality has to be selected. As in the previous design examples, we chose to maximize the *flexibility* of the design. The objective function for the high-level optimization is the sum of the flexibility functions for each of the remaining design parameters. Two types of functions were used, one hyperbolic and one parabolic:

$$flex_1(x) \quad = \quad -\alpha x^2 + c \qquad (11.9)$$

$$flex_2(x) \quad = \quad -\frac{\beta}{x} + c \qquad (11.10)$$

The hyperbolic type is used when a higher value of the parameter is harder to meet while the parabolic is used when the opposite is true. In the case of the parabolic function, the intuitive criterion is that a zero parameter value is almost impossible to achieve, while for higher parameters it becomes easier and for high enough values the flexibility "increase" is insignificant.

The criterion used to build the flexibility function was attributing a value of $flex(x) = -10$ for a parameter value that is "hard" to obtain and $flex(x) = 0$ for a parameter value that is "easy" to obtain. The coefficients are subsequently calculated by solving Equations 11.9, 11.10. Table 11.5 summarizes the values used for the coefficient selection for the high-level parameters of the PLL. The flexibilities for the R, C elements of the loop filter are expressed in terms of the area they occupy in the layout.

The flexibility functions for parameters $\Delta\tau_{VCO_{rms}}$ and K_0 are shown in Figure 11.4 and 11.5, respectively. Due to its heuristic nature, such an approach can cause loss of the overall optimality. However, if a solution is found, its feasibility is guaranteed. The use of flexibility functions accelerates significantly the design process at the high level, allowing the fast selection of a feasible design solution.

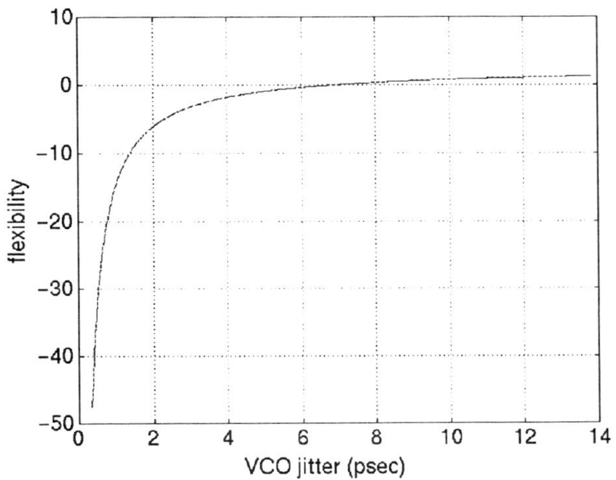

Figure 11.4 Jitter flexibility function

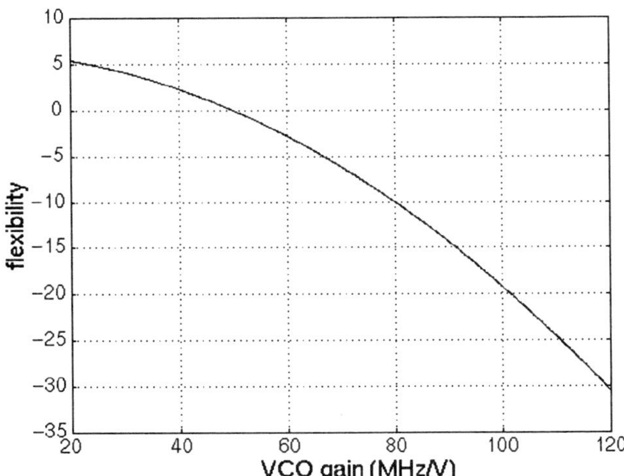

Figure 11.5 VCO gain flexibility function

Type	PLL input frequency	N-divider	VCO frequency	Value
Stability	0.56 MHz	100 250	56 MHz 140 MHz	
$\frac{\Delta T}{T}$	0.56 MHz	250	140 MHz	≤ 0.007
Phase Margin				$\geq 45^\circ$

Table 11.6   PLL Specifications

High-Level Optimization Problem Formulation

The constraints for the PLL performance can be derived directly from Table 11.1 combined with Table 11.2. However, if the performance had to be checked for each one of the operating frequencies the optimization problem would be unnecessarily complex. From a linear analysis, we can guarantee stability at all intermediate operating frequencies if we guarantee stability at the lowest frequency. Also at high frequencies, VCO overload can affect the stability of the system. Consequently, after verifying those assumptions using behavioral simulation, only two extreme frequencies for the VCO were used in the optimization that were determined by the divider values. A frequency slightly lower than the lowest PLL operating frequency was used, which is 0.572 MHz from Table 11.2.

Following the analysis presented in Section 11.7, only the worst case for jitter accumulation was considered, which is at the lowest value of the VCO gain (highest divider n value).

It is very critical for any circuit design optimization problem to yield a result that is *tolerant to parameter variations*. For example, typical R, C variations due to the fabrication process can be as high as 30% of the nominal value. In the high-level optimization problem this was taken into account indirectly by adding a *phase margin constraint*. This can be calculated by considering the linearized open-loop equation for the PLL (Equation 11.29). A phase margin of 45° was found to be adequate to verify stability under typical parameter variations.

All PLL constraints are summarized in Table 11.6. Constraints have been tightened with respect to the ones of Table 11.1 in order to provide for an extra safety margin.

The high-level optimization problem can be expressed as:

$$max \; \sum_{i=1}^{n} flex_i(x_i) \tag{11.11}$$

$$s.t. \; PM(x_1, x_2, \ldots x_n, c) \; \geq \; 45^o \tag{11.12}$$

$$\Delta \tau_{PLL\,140\,MHz}(x_1, x_2, \ldots x_n, c) \; \leq \; 50 \, \text{ps} \tag{11.13}$$

where n is the number of parameters used in the optimization: K_o, $\Delta\tau_{VCO}$, I_p, R, C_1, C_2. A number of other parameters, such as charge-pump output resistances, digital delays, and current mismatches, which have a minimal effect to the PLL performance and can be manually selected, are represented by vector c.

High-Level Optimization Algorithm

To solve the above optimization problem, a nonlinear optimization method should be used. There are several methods in the literature dealing with nonlinear optimization. Typical are gradient methods that use first or second derivative information of the constraint and objective functions. Such methods are used in nonlinear optimization software like MINOS [216].

Disadvantages of these nonlinear optimization algorithms are:

- the values of the first and second-order derivatives must be very accurate for the method to converge. When simulators are used to calculate the constraint functions, finite accuracy can cause large errors in the first and second-order derivatives.

- sometimes, they can only find local extremes of the objective function

- often the gradients of the constraint function are not defined outside the feasible region, as in the case of the PLL where we cannot define timing jitter when the system is unstable.

One solution is to use simulated annealing as in [123]. The drawback is that such a solution can be computationally too expensive and does not use the inherent gradient behavior that circuit performance parameters have, thus discarding useful information about the circuit contained in the gradients.

A quite efficient method for such problems has been proved to be the supporting hyperplane algorithm, used in [220][36].

The supporting hyperplane method can solve any optimization problem of the form:

$$minimize \ \ \mathbf{c}^T \mathbf{x}$$
$$subject \ to \ \ \mathbf{g}(\mathbf{x}) \leq \mathbf{0} \tag{11.14}$$

where:

- $\mathbf{x} \in E^n$,

- $\mathbf{g}(\mathbf{x}) \in E^p$,

- $g_i(\mathbf{x})$ are continuously differentiable functions, and

- the space defined by the constraints $\mathbf{g}(\mathbf{x})$ is convex.

The algorithm works as follows: after an initial feasible point is given, an unconstrained optimization is performed. Then, the nonlinear constraints are checked and if the solution point $\mathbf{P_0}$ is feasible then the algorithm stops and the solution is a global minimum. If a constraint g_i is violated, then the point $\mathbf{u_{k+1}}$ is found on the line joining the initial feasible point $\mathbf{P_0}$ and the last solution $\mathbf{P_{k+1}}$, that lies on the boundary of the feasible region S. Then a linear constraint is added such as:

$$\nabla g_j(\mathbf{u_{k+1}})(\mathbf{x} - \mathbf{u_{k+1}}) \leq 0 \tag{11.15}$$

Following that, the linearized constrained optimization problem is solved again. This process is repeated until a global minimum is found that meets the nonlinear constraints. The algorithm is depicted graphically in Figure 11.6. The key idea is to approximate the feasible space with a polytope, using the tangent hyperplanes at the border points of the feasible space. Then, the linear problem is solved using linear optimization.

Principal advantages of this method are:

- only first derivative information is needed to compute the equations for the tangent plane. Thus the problem created by simulator inaccuracies is smaller than methods using second derivative information.

- derivatives are only needed in the feasible space, where the constraint functions are well defined. In the case of the PLL a great problem is eliminated, since the jitter constraint is not defined when the system is unstable.

- the linear subproblem can be solved easily and efficiently by employing some linear optimization tool, in this case MINOS [216].

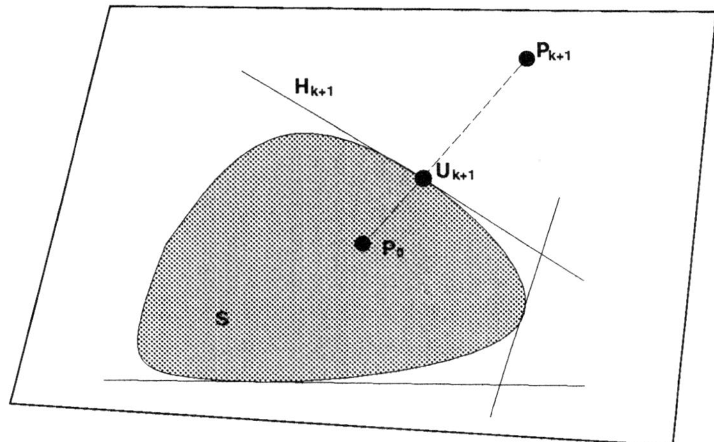

Figure 11.6   Supporting Hyperplane Method

■   provided that the feasible space is convex, a global optimum is computed.

Those advantages are mitigated by the assumption about the convexity of the problem. In most circuit design problems, convexity cannot be guaranteed. Sometimes this can cause the algorithm to fail, but in most cases, even in non-convex problems, a local minimum can be found.

Another source of failure of the algorithm can be the inexact computation of the gradients due to simulators' finite accuracy. This can cause the new added constraint not to exclude the last non-feasible optimal, putting the algorithm in an infinite loop. It can also cause the complete elimination of the feasible region as in the case of non-convex problems.

Great care has been taken to avoid as much as possible the above problems. Since the derivatives are calculated using finite differences, the step for the finite difference is large enough to eliminate the effect of simulator inaccuracy. However, the steps must be taken always towards the feasible direction otherwise the algorithm may wander in the infeasible region where the constraint functions are undefined. The step of the algorithm starts with a high default value and is reduced if the problem is infeasible. If the step is reduced beyond the limit where the difference is higher than the simulator inaccuracy, a different point is used for drawing the tangent plane which is more "inside" the feasible region. This process may cause loss of optimality but makes the algorithm more robust.

```
set_initial_feasible_point;
repeat

    do_linear_optimization;
    foreach non_linear_constraint

        check_constraint;
    if not_feasible

        find_border_point_binary_search;
        compute_gradients;
        add_new_linear_constraint;
until optimal_is_feasible
```

Figure 11.7 Optimization Algorithm

The pseudocode for the optimization algorithm is depicted in Figure 11.7. This algorithm was used in both optimizations done for the PLL design, the high-level optimization, and the circuit optimization.

Equation 11.12 expresses the phase margin constraint, necessary to guarantee a result tolerant to parameter variations. Phase margin is computed using Equation 11.29. The timing jitter constraint is calculated using the behavioral simulation tool described in Section 3.6. The stability constraints of Table 11.6 are not explicitly set as constraints in the optimization problem described in Equations 11.11- 11.13, since stability and jitter are detected in only one simulation for a given set of parameters. Stability detected by the behavioral simulator is not a continuous constraint so it would be hard to compute a measure for it. If instability is detected during the timing jitter simulation, then the simulated point is in the infeasible region so no gradients need to be computed. Higher PLL bandwidth results in better jitter performance according to Equations 11.32, 11.33. This would result in the optimization algorithm pushing the solution close to the instability region. This is partially avoided again by adding a tight phase margin constraint. In order to speed up the optimization, the phase margin constraint is always computed first and if it is satisfied the timing jitter constraint is calculated using behavioral simulation. To find the border point in the supporting hyperplane optimization algorithm, phase margin constraint has priority over the jitter constraint, since it is computationally much less expensive. The parameter values for the initial feasible point needed by the supporting hyperplane algorithm were selected with the help of behavioral simulation.

		Optimization results			
Parameters	K_0 (MHz/V)	35	40	50	60
	$\Delta\tau_{VCO_{rms}}$ (ps)	3.1	3.33	3.46	3.33
	I_p (μA)	18.7	15.8	14.58	11.44
	R (KΩ)	258.2	200.5	198.7	179.6
	C_1 (pF)	40	57.8	51.95	66.9
	C_2 (pF)	5	5	5	5
Constraints	$\Delta\tau_{PLL_{rms}} \leq 50$ (ps)	48.08	50.42	51.05	50.32
	Phase margin ≤ 45 ($^\circ$)	43.5	43.6	43.17	44.55
Objective	Flexibility	4.34	2.79	1.14	-3.11
CPU Time (sec)			7606.1	9198	11529
Iterations		11	8	8	9

Table 11.7 Optimization Results

The supporting hyperplane algorithm needs a convex feasible space in order to guarantee convergence. In the case of the PLL, using the parameters of Table 11.5 and the constraints of Equation 11.13 results in a non-convex feasible space causing the algorithm to fail. To overcome this problem, the value of parameter K_0 was held constant and optimizations were done with different values of K_0. Since the VCO parameters suffer from large temperature and process variations and a heuristic measure of optimality is used, we can claim that the optimality of the solution is not affected.

High-Level Optimization Results

Table 11.7 presents the optimization results for different values of K_0. The optimizations were done in a DEC Alpha-Server 2100 5/250 with 256Mb of memory and 4 CPUs. In every case the optimization converges in not more than 11 iterations and within reasonable computation time. Table 11.8 shows the initial points used in each optimization. In every case, the results of Table 11.7 show a great improvement over the initial solution.

The optimal results appear to sometimes violate slightly the constraints. This is due to the numerical calculation of the border point used in the supporting hyperplane algorithm: small violations are allowed in order to detect if a point belongs to the "border" region. The violations are minimal and have insignificant effect in the performance.

		Initial solutions			
Parameters	K_0 (MHz/V)	35	40	50	60
	$\Delta\tau_{VCO_{rms}}$ (ps)	1.03	1.03	1.03	1.03
	I_p (μA)	5	5	5	5
	R (KΩ)	220	220	220	220
	C_1 (pF)	200	200	220	200
	C_2 (pF)	5	5	5	5
Objective	Flexibility	-38.6	-39.5	-48.9	-44.67

Table 11.8   Initial Solutions for Optimization Algorithm

From the results of Table 11.7 the one that corresponds to $K_0 = 35$ (MHz/V) yields the highest flexibility. However the result for $K_0 = 40$ (MHz/V) was selected since behavioral simulation showed that it is less affected by parameter variations. This by no means reduces the value of the methodology since it also involves designer's judgment in the constraint propagation process. Optimization is used as a tool to propose various solutions while the designer makes the final decision.

Figure 11.8 shows the output of the behavioral simulator for the selected parameters. The graph presents the transient simulation of the VCO control voltage for a step in the n divider value.

In Figure 11.9 a graph of the input and output timing jitter is shown from the Monte Carlo simulation for the jitter computation. The graph shows the input and output jitter events in every transition. The open loop characteristic for gain and phase is depicted in Figure 11.10 as calculated from Equation 11.29.

Since parameters variations were only taken into account indirectly, through the phase margin constraint, it was necessary to verify the performance of the PLL at the worst case variations of the top-level parameters. For the R, C parameters those worst-case variations were given by the process and for I_p, K_0 typical parameters were selected. The PLL was found to meet the constraints for the variations of Table 11.9.

11.4.2   Low-Level Synthesis

After the selection of the high-level parameters, the next step of the methodology is building the low-level blocks. Following the methodology, the high-level parameters

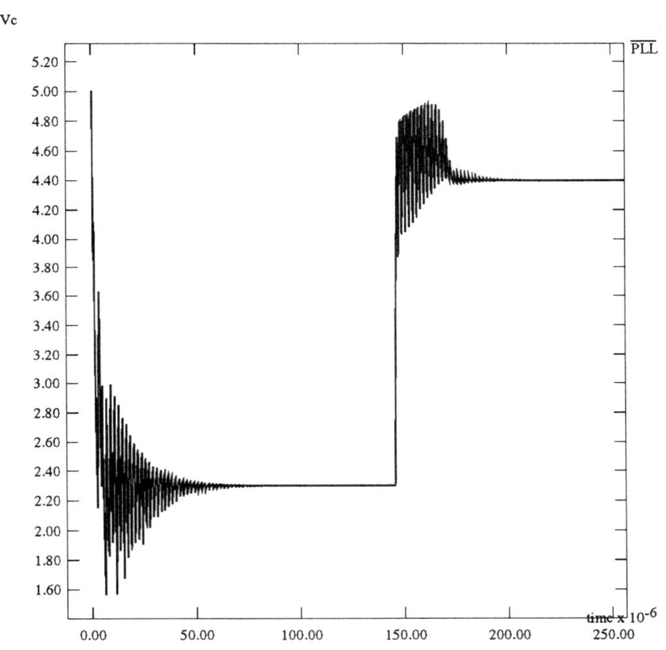

Figure 11.8   PLL Behavioral Simulation

Parameter	Value	Tolerance
K_0	40 MHz/V	$\pm$ 10 MHz/V
F_0	100 MHz	$\pm$ 30 MHz
I_p	16 μA	$\pm 1\ \mu$A
R	200 KΩ	± 60 KΩ
C_1	57.8 pF	± 6 pF
C_2	5 pF	± 1 pF

Table 11.9   PLL Parameter Tolerances

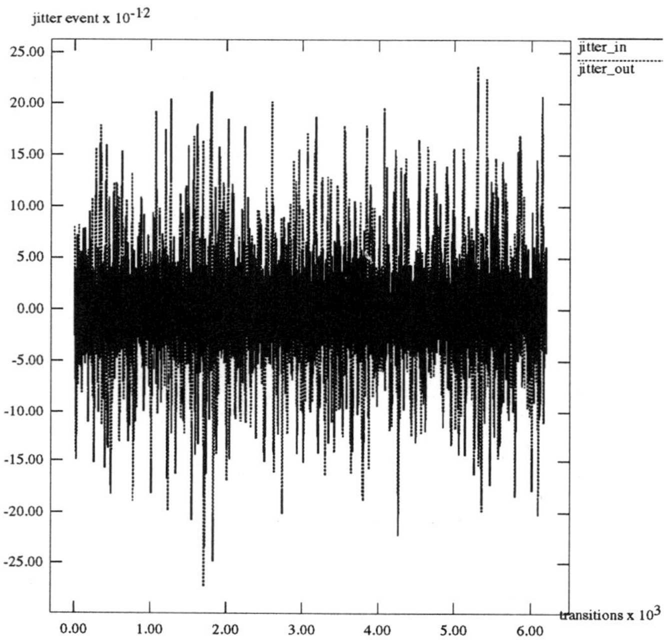

Figure 11.9   PLL Jitter Simulation

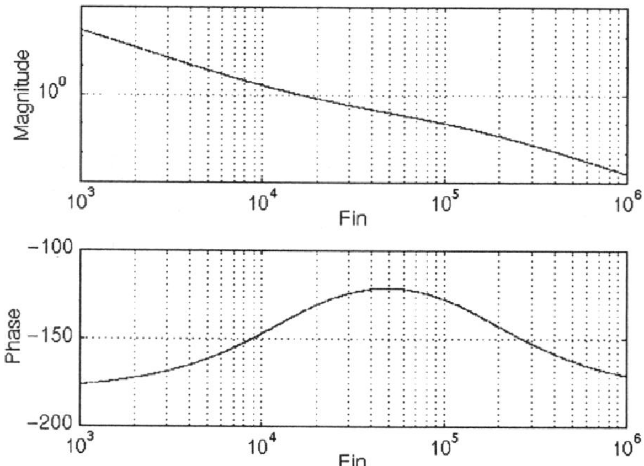

Figure 11.10   PLL Open Loop Characteristic

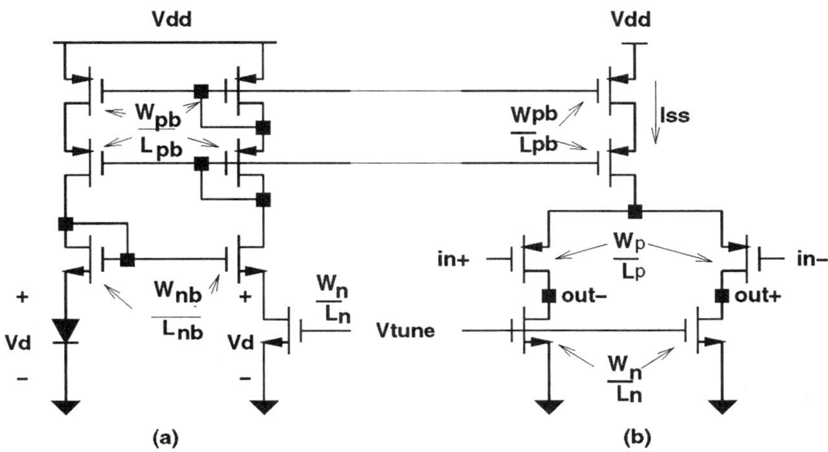

Figure 11.11   VCO Delay Cell and Bias Circuit

will become the *performance constraints* for the low-level building blocks and will be mapped onto an architecture of transistors, device sizes and layout parasitics. The most important block that affects the PLL performance is the VCO and will be discussed extensively.

Voltage Controlled Oscillator

As mentioned before, a ring oscillator VCO topology was selected. The best topology to decrease the effect of power supply and substrate coupling is the differential cell with CMOS loads in triode region, as used in [251][322][154].

A modified version of the cell topology described in [251] was used, since it was the simplest way our requirements could be met. The topology of the cell with the bias circuit is shown in Figure 11.11. In the bias circuit, the voltage across the NMOS transistor in triode is forced by the feedback loop to be equal to the voltage across the diode. Assuming a simple triode model and ignoring short channel effects, the bias current can be given by:

$$I_{ss} = \mu_n C_{ox} \frac{W_n}{L_n}(V_{tune} - V_{THN} - \frac{V_d}{2})V_d \qquad (11.16)$$

I_{ss} is mirrored by the top mirror to the delay cell current. The NMOS loads are made equal to the NMOS bias transistor, so when the current fully switches to one branch,

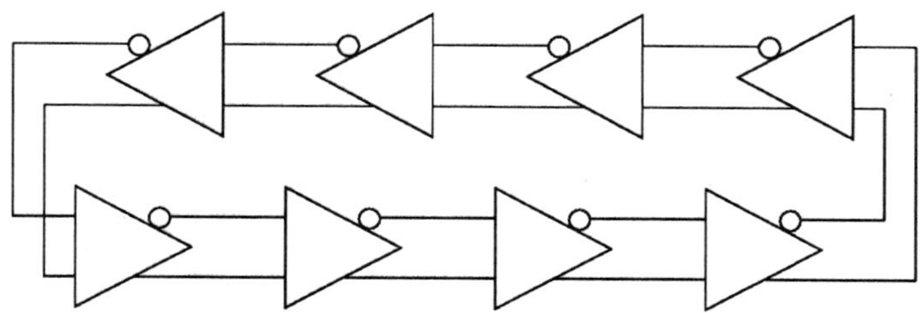

Figure 11.12　Ring Oscillator VCO

the maximum swing across the output of the delay cell is limited to the voltage across the diode V_d, which is approximately a constant.

Using a first order approximation, the delay of one cell can be given by:

$$t_d = \frac{V_d C_L}{I_{ss}} \tag{11.17}$$

where C_L is the capacitive load of the output node. Thus the frequency of oscillation for a N-stage ring oscillator would be:

$$F_o = \frac{I_{ss}}{2N V_d C_L} \tag{11.18}$$

which shows at first order a linear dependency with V_{tune} as desired. For the given HP 1μ technology, it was found by simulation that 8 cells were needed to give the desired VCO gain. The VCO frequency-to-voltage characteristic is linear at first order, provided that the NMOS bias transistor is in the triode region of operation. For low values of V_{tune} the transistor is in saturation and Equation 11.16 is not valid.

For the ring oscillator to oscillate, it is necessary that the gain a_v of each cell is high enough. Typically $a_v \simeq 1.5$ is enough to guarantee oscillation.

The bias circuit parameters $W_{nb}, L_{nb}, W_{pb}, L_{pb}$ do not affect the VCO performance as long as the current mirrors work properly. The parameters that affect the VCO performance are W_n, L_n, W_p, L_p. Since the objective of the optimization is to minimize power dissipation, parameter L_p can also be set to the minimum possible value so that the load capacitance is minimized and the I_{ss} needed to achieve a certain delay is smaller according to Equation 11.18. The transistor sizes selected for the bias circuit and L_p are given in Table 11.10.

The remaining parameters were obtained by optimization. The objective was to minimize power. The VCO constraints of Tables 11.9, 11.7 can be converted to

W_{nb}	L_{nb}	W_{pb}	L_{pb}	L_p
10 μm	1 μm	20 μm	1 μm	1 μm

Table 11.10 VCO Pre-selected Parameters

constraints for a minimum and a maximum VCO frequency. The corresponding minimum and maximum applied voltages are almost constants for the specific topology, given our technology.

Optimization Taking into Account Parasitics

To ensure that the performance constraints are met after the layout is done, it is critical that layout parasitics are taken into account during the optimization phase. In our specific optimization problem, the optimization algorithm will try to adjust I/C_{out} in order to achieve the desirable frequency performance, while minimizing power dissipation. This will result in the use of minimum size transistors, so the final circuit will be extremely sensitive to layout parasitic capacitances, which was observed in preliminary optimizations.

Let $\mathbf{P}$ denote a performance vector, $\mathbf{C}$ the parasitics vector, and $\mathbf{\Delta P_{max}}$ the corresponding maximum allowed performance degradation due to those parasitics. Assuming a linear model around the nominal performance and small parasitics, the performance degradation ΔP_i can be given by:

$$\Delta P_i = \left[\mathbf{S_C^{P_i}}\right]^T \cdot \mathbf{\Delta C} \tag{11.19}$$

where

$$\left[\mathbf{S_C^{P_i}}\right]^T = \left[\left.\frac{\partial P_i}{\partial C_1}\right|_{C_1=C_{1NOM}} , \left.\frac{\partial P_i}{\partial C_2}\right|_{C_2=C_{2NOM}} , \cdots , \left.\frac{\partial P_i}{\partial C_{Nc}}\right|_{C_{Nc}=C_{Nc\,NOM}} \right] \tag{11.20}$$

and $\mathbf{\Delta C}$ is the deviation from the nominal value of the parasitics. Given an estimate $\mathbf{\Delta C_{max}}$ of the maximum deviations from the nominal parasitics, we can include the following constraint to our optimization problem:

$$\left[\mathbf{S_C^{P}}\right]^T \cdot \mathbf{\Delta C_{max}} \leq \mathbf{\Delta P_{max}} \tag{11.21}$$

This will force the optimization result to have a reduced sensitivity to parasitics.

Parameter name	Parameter description	Constraint Value
F_{max}	output at $V_{in} = 4.8$ V	$\leq$ 173 MHz $\geq$ 163 MHz
F_{min}	output at $V_{in} = 1.8$ V	$\leq$ 58 MHz $\geq$ 38 MHz
$\Delta\tau_{VCOrms}$	rms VCO jitter at 140 MHz	$\leq$ 3 ps
V_{max}	maximum output voltage	$\geq$ 0.5 V
V_{min}	minimum output voltage	$\leq$ 0.1 V
$\frac{\Delta F_{max}}{F_{max}}$	maximum variation of output due to layout parasitics	$\leq$ 10%

Table 11.11 VCO Optimization Constraints

VCO Low-Level Optimization Problem Formulation and Results

For the case of the VCO optimization, the parasitic capacitances at the outputs of the differential gates were used, since they have a significant effect to the output frequency. The layout effect was based on an *a priori* estimation of the maximum allowed deviation from the nominal value of layout parasitics, which is subsequently used as a constraint for the VCO layout generator. The constraints were evaluated at a maximum deviation of 50% from a nominal estimate of 15 pF for the parasitics, which was found to be a reasonable value for the layout synthesis phase.

Another two circuit specific constraints were added regarding the minimum and maximum output voltages. Those constraints are necessary for the level restoring circuit to work. Table 11.11 summarizes the constraints for the VCO optimization problem.

The overall optimization problem for the VCO can be expressed as:

$$
\begin{aligned}
minimize \ \ &\text{Power}_{VCO}(W_n, L_n, W_p) &&&& (11.22)\\
subject \ to \ \ &F_{maxVCO}(W_n, L_n, W_p) &\leq& \ \ F_{maxmax}\\
&F_{maxVCO}(W_n, L_n, W_p) &\geq& \ \ F_{maxmin}\\
&F_{minVCO}(W_n, L_n, W_p) &\leq& \ \ F_{minmax}\\
&F_{minVCO}(W_n, L_n, W_p) &\geq& \ \ F_{minmin}\\
&\Delta\tau_{VCOrms}(W_n, L_n, W_p) &\leq& \ \ \Delta\tau_{max}
\end{aligned}
$$

		Initial	Final
Parameter name	W_n	3.8 μm	2.6 μm
	L_n	4 μm	4 μm
	W_p	55 μm	36 μm
Performance variable	F_{max}	165 MHz	163 MHz
	F_{min}	50 MHz	48 MHz
	$\Delta T_{VCO_{rms}}$	1.25 ps	1.5 ps
	V_{max}	0.67 V	0.66 V
	V_{min}	0 V	0 V
	$\frac{\Delta F_{max}}{F_{max}}$	6.8%	10%
Objective	Power	8.6 mW	6.2 mW
Iterations		4	
CPU time		2069 CPU sec	

Table 11.12 VCO Optimization Results

$$V_{max}(W_n, L_n, W_p) \geq V_{max}$$
$$V_{min}(W_n, L_n, W_p) \leq V_{min}$$
$$\frac{\Delta F_{max}}{F_{max}} \leq P_{tol}$$

The optimization problem was again solved using the supporting hyperplane algorithm. All constraints were evaluated using SPICE simulations except for the timing jitter constraint that was evaluated using Equation 11.34. Since transistor sizes can only take discrete values, the results were truncated to the nearest feasible value, taking into account the technology specific parameters ΔL, ΔW, $\lambda = 0.6$ μm. Table 11.12 summarizes the optimization results. The initial point satisfying the constraints is provided for the supporting hyperplane algorithm to work.

Level Restoring Circuit

The specific VCO topology used produces a signal of V_d maximum amplitude differential signal. This signal has to be converted to full CMOS logic levels for the digital parts of the chip. This task is performed by the level restoring circuit shown in Figure 11.13.

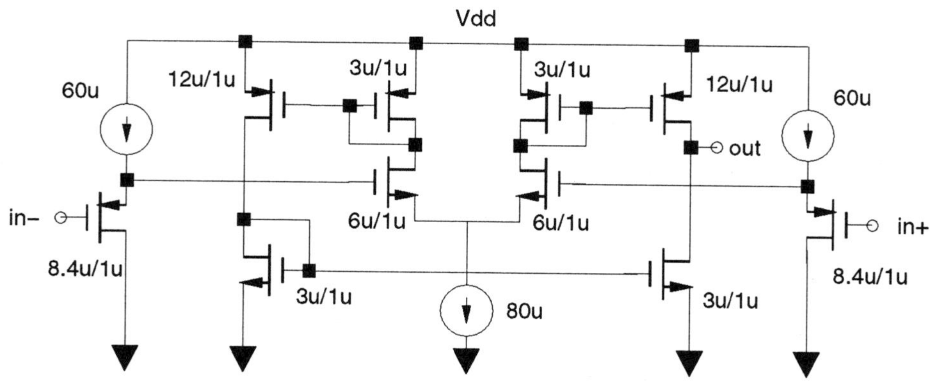

Figure 11.13   Level Restoring Circuit

The differential signal is level shifted by two level shifters and then applied to a unity gain differential pair. The output of the differential pair is amplified and restored to singled ended CMOS levels. The signal is buffered by a buffer chain not shown in Figure 11.13. The total static power dissipation of the level restorer is 1 mW.

Phase-Frequency Detector

The PFD is a sequential circuit that generates a digital pulse of duration equal to the phase difference between the two compared pulses. If the upper pulse comes earlier an UP pulse is generated; if the lower pulse comes earlier a DOWN pulse is generated. When the edges of the two pulses are perfectly synchronized no pulse is generated. The function of the PFD can be shown in Figure 11.15.

The PFD used is the one of Figure 11.14. The dead zone problem is avoided by adding delays in the loop using buffers. The PFD generates two minimum length UP and DOWN pulses even when the compared waveforms are perfectly synchronized. Those pulses are identical in length, so the up and down currents cancel each other assuming good matching. When the pulses are not perfectly synchronized, the difference between the pulses equals the phase difference between the compared clocks. The first order function of the PFD is identical to the one described in Figure 11.15.

The PFD layout was automatically synthesized from the input schematic using OCT-TOOLS [125] and is shown in Figure 11.16.

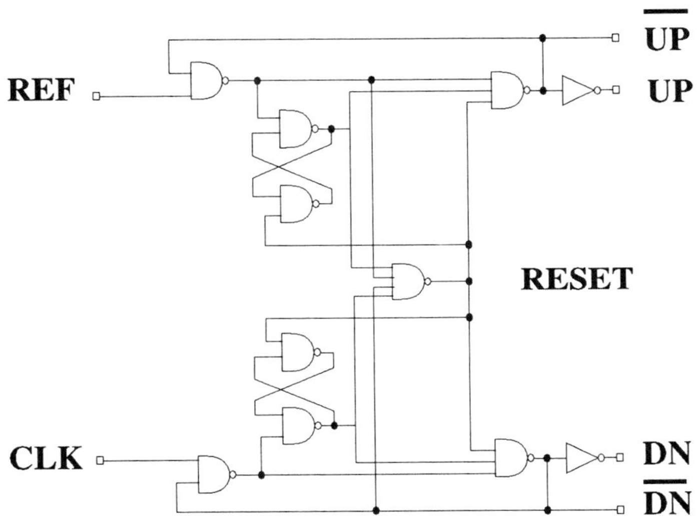

Figure 11.14   Phase-Frequency Detector

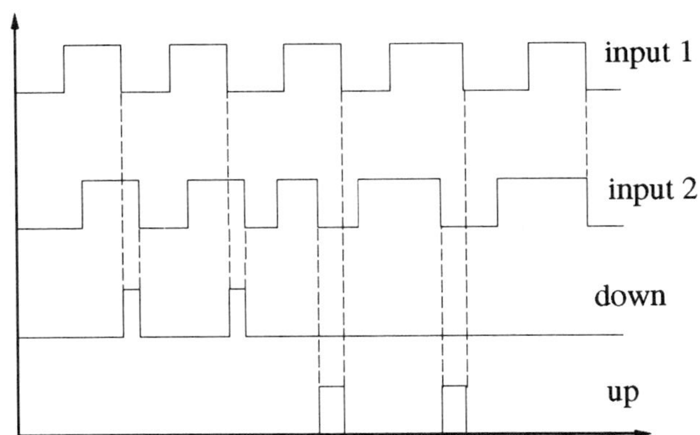

Figure 11.15   PFD I/O Relationship

Figure 11.16   PFD Layout

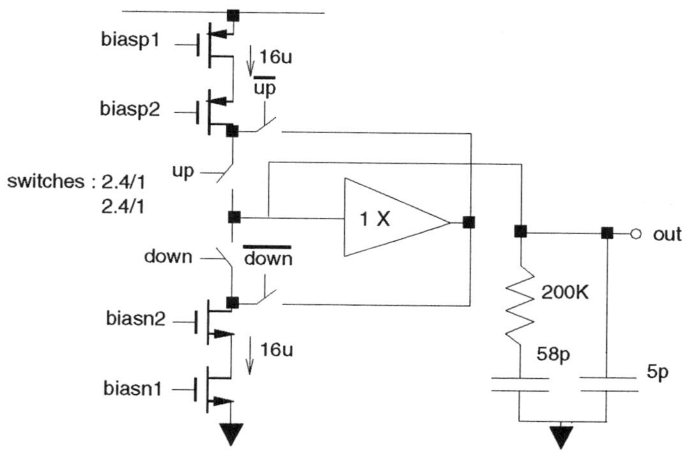

Figure 11.17   Charge-Pump Circuit

Charge Pump

The charge-pump used is the one shown in Figure 11.17 and is based on the one used in [322]. It consists of two cascode current sources with equal currents. The sources were cascoded in order to make the two currents match as much as possible. Since the PFD eliminates the dead zone by turning both sources on at the same time and for the same period, a mismatch could inject extra noise to the output node. The charge pump is biased from a high-swing cascode bias circuit shown in Figure 11.18 and is based on the one in [167].

In order to avoid the problem of turning on and off the sources every time they switch, which could cause charge injection and slow response of the charge pump when any of the UP or DOWN pulses is low, the corresponding current source is redirected to the

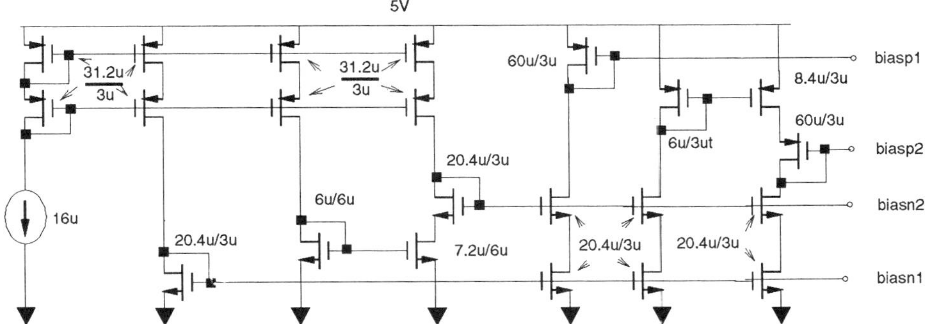

Figure 11.18   Charge-Pump Bias Circuit

output of a unity gain buffer kept at the same voltage as the output of the loop filter. This prevents the current fluctuations which are due to finite output resistance of the current sources.

Dividers

The divider used was the synchronous programmable divider shown in Figure 11.19. It is basically a down counter which, after reaching zero, loads the programming word. The logic family used was true single phase clock family (TSPC), since fast dividers operating at around 150 MHz were needed. The basic D flip-flop element used is shown in Figure 11.20.

11.4.3   Physical Synthesis

Voltage Controlled Oscillator

The hierarchical, constraint-driven approach was also used in the layout phase for the VCO. The constraints set in the optimization problem of Equation 11.22 were used in the layout generation. In addition to that, constraints for all parasitics were generated using the constraint generation techniques described in [50]. The sensitivities of every performance parameter with respect to every parasitic resistance and capacitance are calculated automatically. Then, given a maximum allowable performance deviation, bounds are imposed on every parasitic using quadratic optimization.

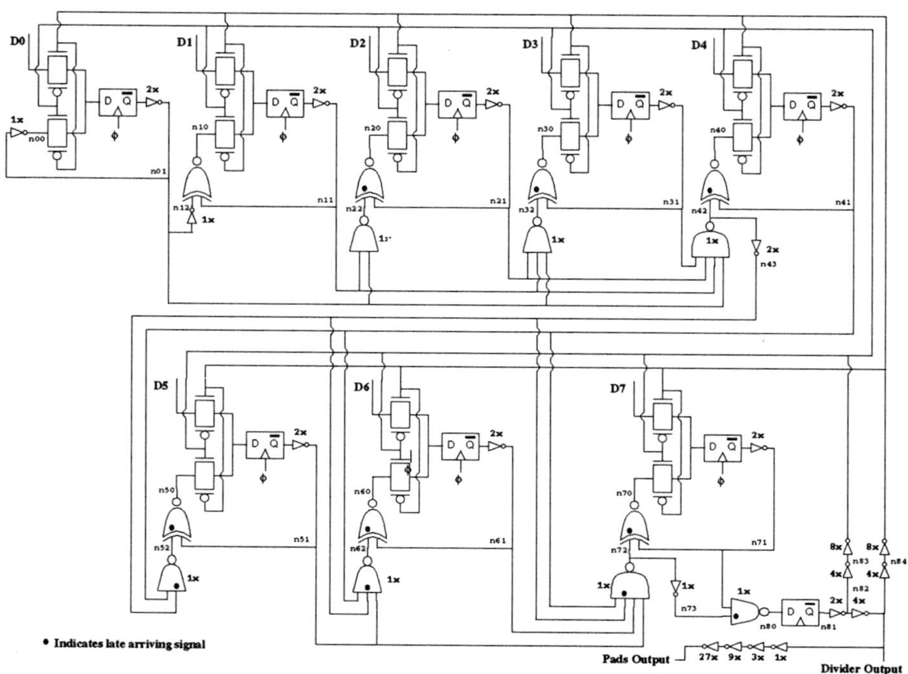

Figure 11.19   Programmable Divider

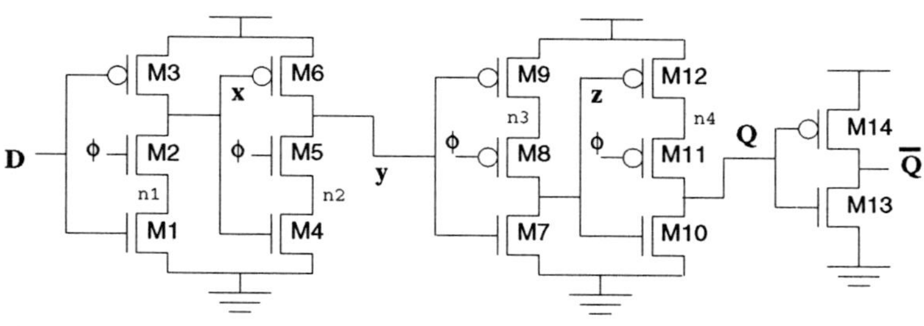

Figure 11.20   TSPC Flip-Flop

```
begin{
     foreach(P_j)
          foreach(R_i, C_i) calculate( ∂P_j/∂C_i, ∂P_j/∂R_i );
100: calculate(R_imax, C_imax); /* quadratic optimization*/
     foreach(i) {
     set W_i = W_imin and L_i = L_imin  ⟹  C_i = C_0 W_min L_min ;
       do {
          evaluate R_i = ρ W_i/L_i;
          if (R_i < R_imax) then exit;
          else W_i = W_i + ΔW ;
       } while (C_i < C_imax);
       if ((C_i > C_imax) or (R_i > R_imax)) then "infeasible" ; ⟹ goto 100;
       else "constraint enforced" ;
     }
}end
```

Figure 11.21   Layout Generation Algorithm

Figure 11.22   Ring Oscillator VCO Layout

A parametric layout generator was written for the specific VCO topology. It uses a fixed floorplan and takes as parameters the number of delay cells, the device sizes, and the parasitic constraints. Additional parasitic constraints are generated for the parasitics that were not accounted for in the circuit optimization.

The algorithm for the layout generation is shown in Figure 11.21. ΔW is the minimum increment (λ) allowed by the process design rules, P_j is the performance j, and i is the number of the parametric wires.

The final VCO layout is shown in Figure 11.22.

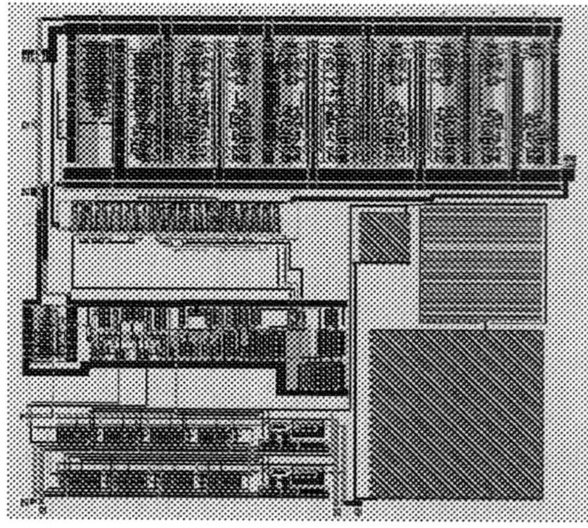

Figure 11.23   PLL Layout

PLL

Since parasitic constraints were not as critical for the other PLL sub-blocks, the charge-pump, dividers, bias and level restoring circuits described in the previous sections were done in a custom mode. On-chip decoupling capacitors were used to minimize the effect of power supply coupling to the overall jitter performance. The final PLL system was routed using MOSAICO [30]. The result is shown in Figure 11.23. The PLL area is approximately 0.45 mm^2.

11.4.4   D/A Converters

A module generator [220] was used to directly synthesize the D/A layout from specifications. The result is shown in Figure 11.24.

11.4.5   Video Driver System

The final layout for the Video Driver System was also put together using MOSAICO. It is shown in Figure 11.25. The die is approximately 3.4 mm × 3.9 mm = 13.26 mm^2. Different analog and digital supplies were used and additional supplies were provided for the VCO in order to avoid as much as possible supply coupled noise which can significantly contribute to timing jitter.

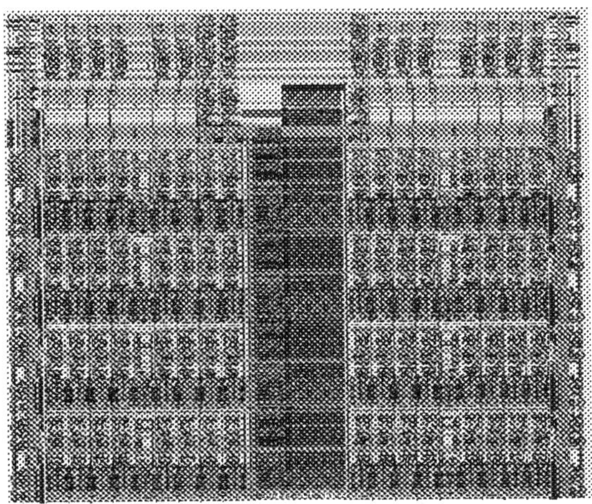

Figure 11.24   D/A Layout

Figure 11.25   Video Driver System Layout

Module	Parameter name	Value
VCO	K_o	39 MHz/V
	F_o	110 MHz/V
	$\Delta \tau_{VCOrms}$	4 ps
	V_{min}	1.4 V
Charge-pump	I_p	16 μA
	$\frac{\Delta I_p}{I_p}$	10% (est)
	R_{linup}	20 KΩ
	R_{satup}	500 MΩ
	V_{satup}	4.7 V
	R_{linlo}	13.3 KΩ
	R_{satlo}	2000 MΩ
	V_{satlo}	0.33 V
Loop Filter	R	200 KΩ
	C_1	58 pF
	C_2	5 pF
PFD	t_{dead_zone}	0
	t_d	2 ns
Divider	t_d	2 ns

Table 11.13   PLL Extracted Parameters

11.5   BOTTOM-UP VERIFICATION

The overall design was verified using hierarchical simulation. First the functionality of the lowest level sub-blocks, such as VCO, dividers, charge-pump, was verified using SPICE and the parameters needed for the higher level blocks were extracted. The timing jitter performance of the VCO was extracted using the non-Monte Carlo, nonlinear noise simulator described in Chapter 6.3. Then behavioral simulation was used to verify the performance of the whole system.

The extracted parameters of the blocks used in the PLL behavioral simulation are given in Table 11.13. To illustrate the value of bottom-up verification, the result of a flat circuit simulation is compared to the result of the behavioral simulation shown in Figure 11.26. The waveform simulated is the control voltage of the VCO when the PLL is in acquisition mode and is done to detect stability in the worst case divide ratio. Waveforms from both simulations are almost identical.

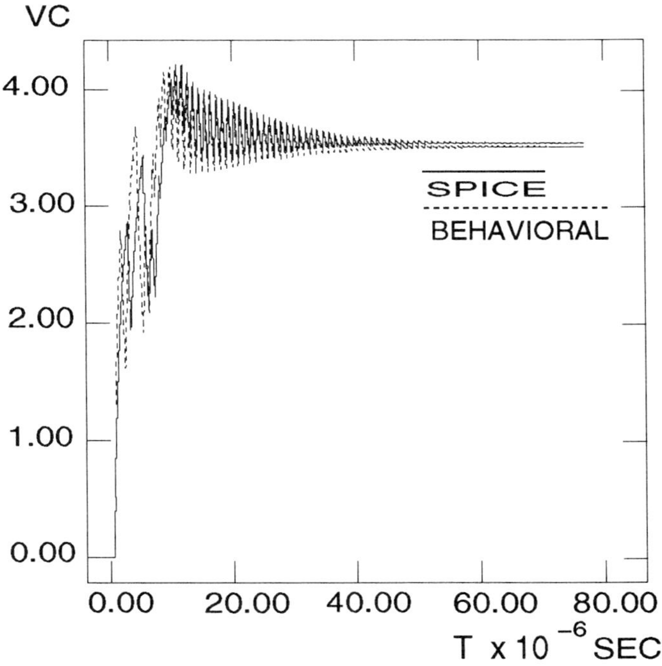

Figure 11.26   PLL Verification

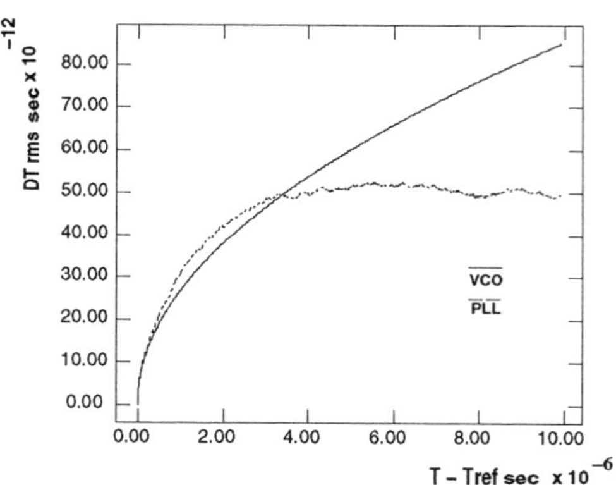

Figure 11.27 ΔT_{rms}

Type	PLL input Frequency	N-divider	VCO frequency	Constraint	Result
Stability	0.56 MHz	100	56 MHz	ok	ok
		250	140 MHz	ok	ok
$\frac{\Delta T}{T}$	0.56 MHz	250	140 MHz	≤ 50 ps	47 ps
Phase Margin				$\geq 45°$	$45°$

Table 11.14   PLL Verification Results

For the behavioral simulation, 560 CPU seconds were needed, while the full circuit simulation took 20 CPU hours (using macromodels for the dividers). Both simulations were done in a DEC Alpha Server 2100 5/250 with 256 MB of memory and 4 CPUs.

Figures 11.27 and 11.28 show the results of a behavioral simulation for the timing jitter using the extracted parameters for $F_{out} = 100$ MHz. Figure 11.27 shows the PLL and VCO rms timing jitter as a function of the distance from the reference transition. The PLL timing jitter increases and finally converges to its final value, as expected. Figure 11.28 shows the square of the PLL and VCO rms timing jitter. As expected, ΔT_{VCO}^2 accumulates linearly to infinity, since there is no correction from the PLL loop.

The results of the bottom-up verification are summarized in Table 11.14. The projected performance is based only in the calculation of the thermal jitter of the VCO, which sets the fundamental performance bound. Still, the performance of the actual system is expected to be close to the one predicted since care has been taken to reduce as much as possible all other jitter sources.

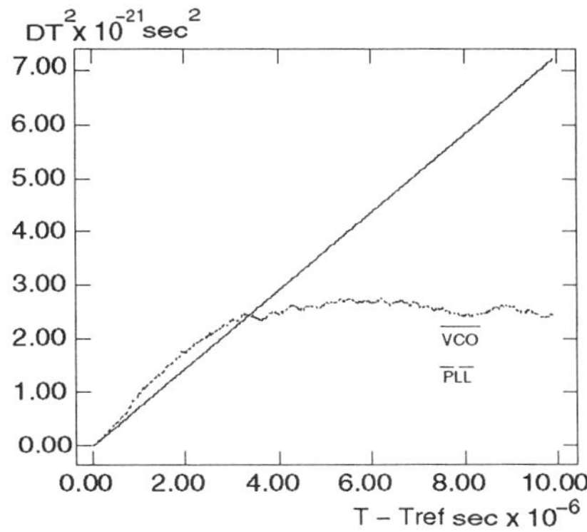

Figure 11.28 ΔT^2

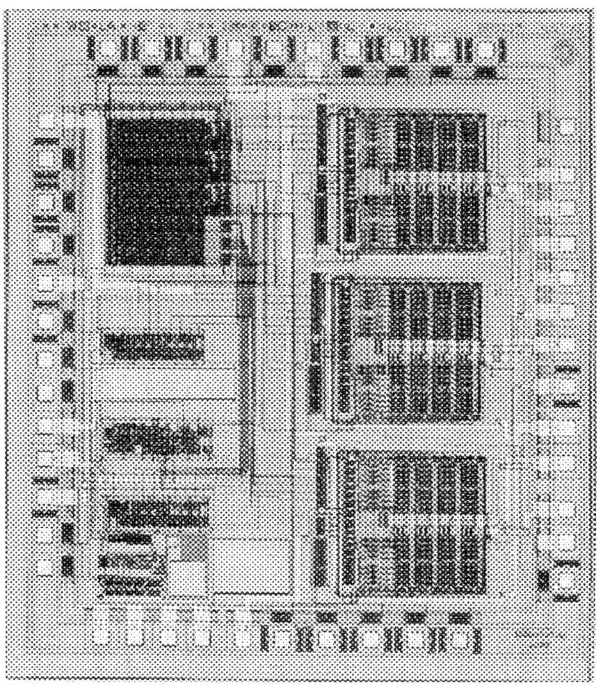

Figure 11.29 Video Driver System Die Photo

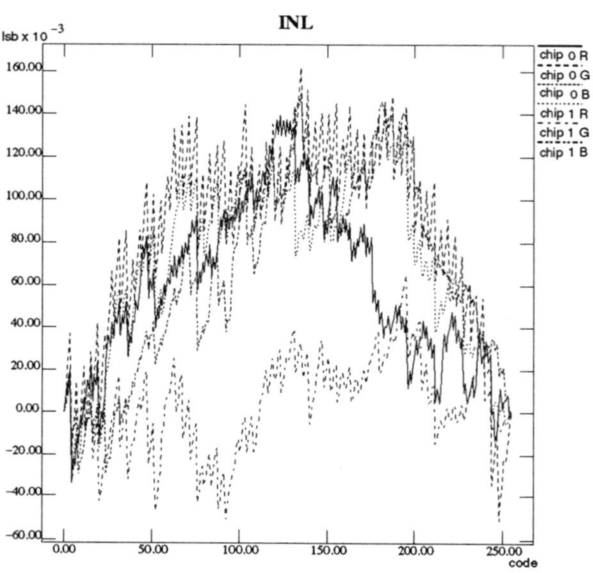

Figure 11.30   INL measurement results

11.6   EXPERIMENTAL RESULTS

The chip was fabricated on a MOSIS HP 1.0 μm technology. A die photo is shown in
Figure 11.29. The 17,000 transistor system occupies an area of 3.4 mm $\times$ 3.9 mm =
13.26 mm^2. A printed circuit board was designed and manufactured in order to
measure the performance of the chip. Experimental results show that the D/A INL and
DNL performance is 0.16 LSB and 0.05 LSB, respectively, and that the settling speed
requirements are satisfied (T_{set} = 6 ns). Figures 11.30 and 11.31 show experimental
INL and DNL measurement results as functions of the input code for various chips.

The PLL frequency generator meets the specifications for generating frequencies from
25 MHz to 130 MHz. Figure 11.33 shows the output waveform at 130 MHz. A
small deviation from the expected speed in the upper edge of the specifications is due
to an error in the parameters file used in the synthesis phase. Detailed timing jitter
measurements were done using a Tektronix 11801B high-bandwidth digitizing oscil-
loscope. Figure 11.32 shows an output waveform at 100 MHz and the corresponding
jitter histogram at a transition edge 7 μs from the reference, so that the accumulated
jitter converges to its final value. The rms jitter at 100 MHz is 65 ps (0.65%), which
is close to the specifications. The results match the predictions within 30% for the
worst case chip. Component process variations affecting the PLL bandwidth, simpli-

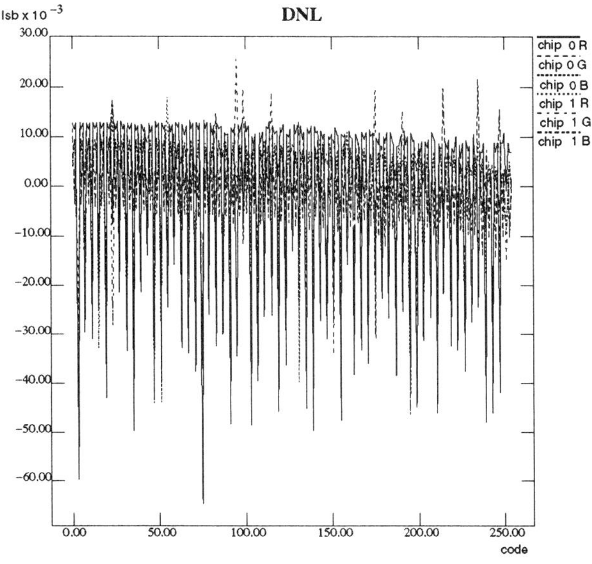

Figure 11.31 DNL measurement results

fied noise models for the devices, and power supply and substrate coupling can cause the measured value to deviate from the predicted value. Reflections and coupling from the testing board could have also affected the measurements. Considering these factors, the agreement between the measured results and the simulation results is quite satisfactory.

11.7 PLL EQUATIONS

The PLL is a feedback system that exhibits a strong nonlinear behavior, especially when the phase difference of the two compared signals is large and the system is trying to null it. Analyses of the nonlinear problem have been done [90], but even a linearized solution can give a lot of information. A classical linearized analysis can be found also in [90]. Assuming that:

- the PD output for small phase differences is given by

$$v_d = K_d \cdot (\theta_i - \theta_o) \tag{11.23}$$

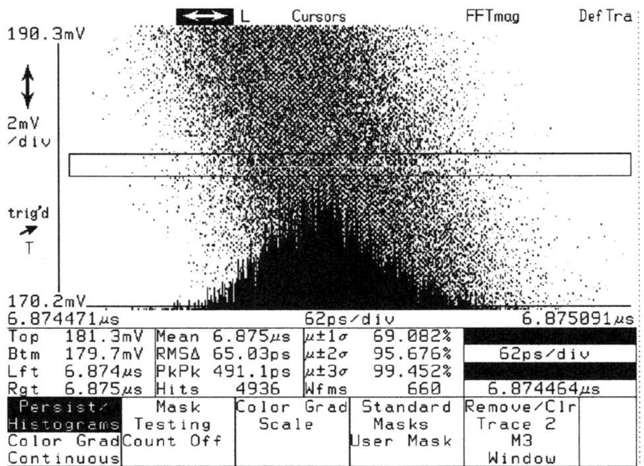

Figure 11.32   Jitter Histogram

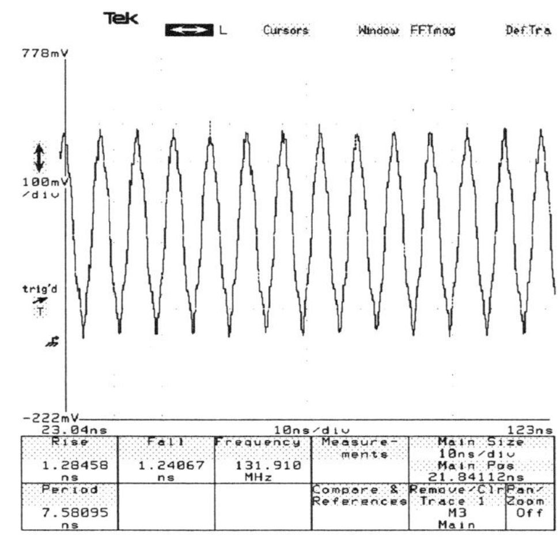

Figure 11.33   Frequency Synthesizer Output

where θ_i is the input phase, θ_o is the VCO phase and K_d is a constant

- the transfer function of the loop filter is $F(s)$

- The output frequency of the VCO is given by:

$$\omega_{vco}(s) = \omega_o + K_o \cdot V_c(s) \tag{11.24}$$

it can easily be proved that the PLL transfer function is given by:

$$\frac{\theta_o}{\theta_i} = H(s) = \frac{K_o \cdot K_d \cdot F(s)}{s + K_o \cdot K_d \cdot F(s)} \tag{11.25}$$

The quantity

$$BW = K_o \cdot K_d \tag{11.26}$$

is also referred to as *PLL bandwidth*. The precise form of the transfer function depends on the type of the loop filter used. In this analysis it was also assumed that the type of phase detector used was an ideal multiplier.

11.7.1 Charge Pump PLLs

Based on the assumption that the phase differences between the pulses are small and the bandwidth of the PLL is much smaller than the input frequency, a continuous time approximation can be done to quantify the behavior of a charge pump PLL [91]. Assuming that:

- the charge pump current is $\pm I_p$

- the filter impedance is $Z_F(s)$

- the VCO gain is K_o

- θ_i, θ_o are the input and output phases respectively

- R, C_1, C_2 are the loop filter components

the loop transfer function can be expressed as:

$$\frac{\theta_o}{\theta_i} = H(s) = \frac{K_o \cdot I_p \cdot Z_F(s)}{2 \cdot \pi \cdot s + K_o \cdot I_p \cdot Z_F(s)} \tag{11.27}$$

Assuming a second-order loop filter as in Figure 11.3 the transfer function becomes:

$$\frac{\theta_o}{\theta_i} = H(s) = \frac{G(s)}{1 + G(s)} \tag{11.28}$$

$$G(s) = \frac{K_o \cdot I_p}{2 \cdot \pi \cdot (C_1 + C_2) \cdot s^2} \cdot \frac{1 + s \cdot R \cdot C_1}{1 + s \cdot R \cdot \frac{C_1 \cdot C_2}{C_1 + C_2}} \tag{11.29}$$

where G(s) is the open loop transfer function. From Equation 11.29 the *phase margin* of the PLL can be calculated, which can give a measure of the stability of the system. A large phase margin is desirable to ensure stability under variations of the parameters of the building blocks.

11.7.2 Timing Jitter

Very critical to the performance of a PLL used as a clock generator is the *timing jitter*. In a display driver system this can cause patterns on the screen. This is defined as the standard deviation of the period of the output of the PLL:

$$\overline{\Delta\tau^2} = E\left[(T - E[T])^2\right] \tag{11.30}$$

where T is the output period. The *phase jitter* can be accordingly defined as:

$$E\left[\Theta_{TOT}^2\right] = \left(\frac{2\pi}{T}\right)^2 \cdot \overline{\Delta\tau^2} \tag{11.31}$$

In the case where the reference clock is a crystal oscillator, which is typical in clock synthesizer PLLs, the incoming clock is assumed to be "clean" so, the main source of timing jitter is the variation $\overline{\Delta\tau_{VCO}^2}$ due to the VCO. The fundamental sources of VCO jitter can be:

- thermal noise

- coupling of digital noise from the power supply

- coupling of digital noise from the substrate

The PLL feedback loop acts to correct the variations of the VCO output period that could accumulate to infinity. Depending on how fast the feedback loop acts, the accumulation factor for the jitter varies. Thus, a wide bandwidth PLL would have a smaller accumulation factor than a narrow bandwidth PLL. It is very common that ring oscillators are used for VCOs. Under the assumption that $T \ll RC_1$ and $C_1 \gg C_2$ an analysis of a charge-pump PLL with a second-order passive RC loop filter is done in [155]. Given that each of the N stages of the VCO contributes $\overline{\Delta \tau_N{}^2}$, an expression is calculated for the PLL phase noise:

$$\sqrt{E\left[\Theta_{TOT}{}^2\right]} = \alpha \frac{2\pi \sqrt{\overline{\Delta \tau_N{}^2}}}{T} \tag{11.32}$$

where α is the PLL accumulation factor:

$$\alpha = \sqrt{\frac{1}{2 I_p K_o R T}} \tag{11.33}$$

The jitter accumulation factor is proportional to the PLL bandwidth $\omega_L \simeq K_o I_p R$, for $T \ll RC_1$ and $C_1 \gg C_2$.

Differential delay cells do not completely eliminate the problem of power supply rejection, since the transistors of the differential pair are in different operating regions during the transition, but they greatly improve the noise performance. The fundamental limit in the noise performance of the differential cells is set by the device generated noise. For a VCO used in a PLL we are mainly interested in the thermal noise limit, since the 1/f noise is mainly low frequency and is rejected by the PLL loop acting as a high pass filter for the VCO generated jitter [155].

In [313] the average jitter in a ring oscillator VCO using differential stages is calculated to be:

$$\frac{\Delta \tau_{rms}}{t_d} = \sqrt{\frac{kT}{C_L} \cdot \frac{1}{V_{GS} - V_T}} \cdot \xi \tag{11.34}$$

where C_L is the capacitive load of each stage of the ring oscillator, t_d is the delay of one stage, and ξ is a design parameter of the cell given by:

$$\xi = \sqrt{1 + \frac{2}{3}\alpha_v \left(1 - e^{-t/\tau}\right) + \frac{2\sqrt{2}}{3}\alpha_v e^{-t/\tau}} \tag{11.35}$$

where α_v is the small-signal gain of the cell, $\tau \simeq t_d$. From the above equations it can be inferred that there is a trade-off between the power consumption of the VCO and the timing jitter.

11.8 CONCLUSION

We have presented the complete design flow for a video driver system, illustrating the value of the top-down, constraint-driven paradigm for more complex systems. The experimental results verify the validity of our design methodology. Fundamental to our approach was the use of behavioral simulation and optimization for hierarchical constraint propagation. Combined with the tools used, this methodology significantly impacts the design of complex systems such as this RAMDAC by reducing design time, over-design, and costly fabrication iterations.

12

CONCLUSIONS

A new design methodology for analog integrated circuit design has been presented. This methodology is aimed at reducing substantially the design cycle of complex analog and mixed analog-digital systems while maintaining or even improving the performance of the final design. A hierarchical approach with early verification and constraint propagation is favored as a way of achieving this goal. The tools, developed within the top-down, constraint-driven paradigm, are the product of the continuing long term effort to meet the design needs of analog integrated circuit design. To show the feasibility of such an approach in doing design, large design examples have been presented. Unlike many other methodologies that have been proposed in the past, we have taken our design examples all the way from the design synthesis path to testing to show that our methodology and tools can actually be used in practice. Most other methodologies have stopped short of fabrication, leaving it up to the end user who wants to use their tools to prove that it can actually be used in practice. One of our goals has been to eliminate this uncertainty.

This work has shown the methodology's potential in its ability to systematize design and speed up the design process. It is hoped from this work with its two completed design examples and the one in progress, from the discussion of the methodology, and from the discussion on its tool set, that this methodology will gain acceptance and be adopted by the design community.

Future work in the the development of this methodology will undoubtly include refinements to this approach. One way to make these refinements is through the pursuit of even larger design examples. Not only should they be larger, but a wide variety of circuits should be examined. If all goes well, it should be easier to demonstrate the effectiveness of this methodology in larger systems. These are systems in which the complexity level can overwhelm traditional hand design, and, therefore, will absolutely require computer-aided designs tools for design management.

Immediate work is the completion of the video driver system. One possible area for the future is in RF circuits. The design of RF systems is being done at high level, where decisions have to be made about the partitioning of the constraints among the various high-level blocks, mostly heuristically, using designer's expertise. We are proposing to fit the design methodology into the requirements of such RF systems, and as an example we can start with a direct conversion CMOS transceiver. The methodology should accelerate and make design more systematic to handle the complexities involved. The tool set will also grow, because in order to work on this RF system, behavioral models will have to be built, and chances are new optimization algorithms will have to be developed as was necessary for the A/D and video driver design.

This methodology will only evolve from doing new designs to constantly challenge it. The field of analog integrated circuit design is constantly changing. In order for the tool set and the methodology to not become obsolete, the designs chosen must follow the state-of-the-art in analog circuit design.

ACKNOWLEDGEMENTS

The authors would like to acknowledge the financial support of the Semiconductor Research Corporation (SRC), the Advanced Research Projects Agency (ARPA), the MICRO Program of the State of California (corporate sponsors: Bell Northern Research, Fujitsu, Harris, Hewlett Packard, Mentor Graphics, Motorola, Philips, Rockwell), Cadence, SGS-Thomson, the Italian National Council of Research, Asea Brown Boveri (Baden, Switzerland), the Swiss Science Foundation (Bern, Switzerland), the National Science Foundation (NSF), the Intel Foundation, and the Armed Forces Communications and Electronics Association. They would also like to thank Professor Paul Gray and his students, especially Robert Neff, for their excellent input and collaboration.

REFERENCES

[1] E.H.L. Aarts and P.J.M. van Laarhoven, *Simulated Annealing: Theory and Applications*, D. Reidel Publishing, 1987.

[2] A. Abidi and R.G. Meyer, "Noise in Relaxation Oscillators," *IEEE Journal of Solid-State Circuits*, vol. SC-18, n. 6, December 1983.

[3] P.E. Allen, "A Tutorial—Computer Aided Design of Analog Integrated Circuits," in *Proc. IEEE Custom Integrated Circuits Conference*, pp. 608–616, 1986.

[4] P.E. Allen and E.R. Macaluso, "AIDE2: An Automated Analog IC Design System," in *Proc. IEEE Custom Integrated Circuits Conference*, pp. 498–501 May 1985.

[5] P.E. Allen and H. Nevarez-Lozano, "Automated Design of MOS Op Amps," in *Proc. IEEE International Symposium on Circuits and Systems*, pp. 1286–1289, 1983.

[6] D.J. Allstot and W.C. Black, Jr., "Technological Considerations for Monolithic MOS Switched-Capacitor Filtering Systems," in *Proc. of the IEEE*, pp. 967–986, 1983.

[7] L. Arnold, *Stochastic Differential Equations: Theory and Applications*, John Wiley & Sons, 1974.

[8] J. Assael, P. Senn and M.S. Tawfik, "A Switched-Capacitor Filter Silicon Compiler," *IEEE Journal of Solid-State Circuits*, vol. SC-23, n. 1, pp. 166–174, February 1988.

[9] *ASTAP—Advanced Statistical Analysis Program*, IBM Program Product Document SH20-1118-0, IBM Data Processing Division, White Plains, NY, 1973.

[10] E. Barke, "Line-to-Ground Capacitance Calculation for VLSI: A Comparison," *IEEE Trans. on Computer-Aided Design of Integrated Circuits and Systems*, vol. CAD-7, n. 2, pp. 295–298, February 1988.

[11] A. Barlow, K. Takasuka, Y. Nambu, T. Adachi and J.I. Konno, "An Integrated Switched-Capacitor Filter Design System," in *Proc. IEEE Custom Integrated Circuits Conference*, pp. 451–455, May 1989.

[12] B. Basaran, R.A. Rutenbar and L.R. Carley, "Latchup-Aware Placement and Parasitic-Bounded Routing of Custom Analog Cells," in *Proc. IEEE International Conference on Computer-Aided Design*, pp. 415–421, November 1993.

[13] E. Berkcan, "MxSICO: a Silicon Compiler for Mixed Analog Digital Circuits," in *Proc. IEEE International Conference on Computer Design*, pp. 33–36, September 1990.

[14] E. Berkcan, M. d'Abreu and W. Laughton, "Analog Compilation Based on Successive Decompositions," in *Proc. IEEE/ACM Design Automation Conference*, pp. 369–375, June 1988.

[15] E. Berkcan and F. Yassa, "Towards Mixed Analog/Digital Design Automation: A Review," in *Proc. IEEE International Symposium on Circuits and Systems*, pp. 809–815, May 1990.

[16] V.M. zu Bexten, C. Moraga, R. Klinke, W. Brockherde and K.-G. Hess, "ALSYN: Flexible Rule-Based Layout Synthesis for Analog ICs," *IEEE Journal of Solid-State Circuits*, vol. SC-28, n. 3, pp. 261–268, March 1993.

[17] P. Bolcato and R. Poujois, "A New Approach for Noise Simulation in Transient Analysis," in *Proc. IEEE International Symposium on Circuits and Systems*, May 1992.

[18] L. Bonet, J. Ganger, J. Girardeu, C. Greaves, M. Pendelton and D. Yatim, "Test Features of the MC145472 ISDN U-Transceiver," *Proc. International Test Conference*, pp. 68–79, 1990.

[19] B.E. Boser, EECS 290Y class notes, Electrical Engineering and Computer Sciences, University of California, Berkeley, Fall 1993.

[20] B.E. Boser, *Design and Implementation of Oversampled Analog-to-Digital Converters*, Ph.D. thesis, Integrated Circuits Laboratory, Stanford University, Stanford, October 1988.

[21] B.E. Boser, et al., "Simulating and Testing Oversampled Analog-to-Digital Converters," *IEEE Trans. on Computer-Aided Design of Integrated Circuits and Systems*, vol. 7, pp. 668–674, June 1988.

[22] R.J. Bowman and D.J. Lane, "A Knowledge-Based System for Analog Integrated Circuit Design," in *Proc. IEEE International Conference on Computer-Aided Design*, pp. 210–212, 1985.

[23] G.E.P. Box and N.R. Draper, *Empirical Model-Building and Response Surfaces*, Wiley, New York, 1987.

[24] R. Brayton, G. Hachtel and A.L. Sangiovanni-Vincentelli, "A Survey of Optimization Techniques for Integrated-Circuit Design," *Proc. of the IEEE*, vol. 69, n. 10, pp. 1334–1364, 1981.

[25] P.A. Brennan, N. Raver and A.E. Ruehli, "Three Dimensional Inductance Computations with Partial Element Equivalent Circuits," *IBM Journal of Research and Development*, vol. 23, pp. 661–668, November 1979.

[26] D.R. Brillinger, "The Identification of Polynomial Systems by Means of Higher Order Spectra," *Journal Sound Vibration*, vol. 12, pp. 301–31, 1970.

[27] D.R. Brillinger, *Time Series Data Analysis and Theory, Expanded Edition*, McGraw-Hill, New York, 1981.

[28] R.W. Brodersen, et al., *LAGER User's Manual*, Department of Electrical Engineering and Computer Sciences, University of California, Berkeley, 1990.

[29] Brooktree Corporation, *Brooktree Graphics and Imaging Product Databook*, 1993.

[30] J. Burns, A. Casotto, M. Igusa, F. Marron, F. Romeo, A.L. Sangiovanni-Vincentelli, C. Sechen, H. Shin, G. Srinath and H. Yaghutiel, "MOSAICO: An Integrated Macro-Cell Layout System," in *VLSI '87*, August 1987.

[31] M. Burstein, "Channel Routing," in *Layout Design and Verification*, ch. 4, pp. 133–167. T. Ohtsuki Ed., North Holland, 1986.

[32] O. Buset, M. Declercq, F. Rahali and P. Vaucher, "Fast Prototyping of Semi-Custom Bipolar Analog ASICs," in *Proc. IEEE Custom Integrated Circuits Conference*, pp. 561–564, May 1991.

[33] G. Casinovi, *Macromodeling for the Simulation of Large Scale Analog Integrated Circuits*, Ph.D. thesis, University of California, Berkeley, 1988.

[34] G. Casinovi and J-M. Yang, "Multi-Level Simulation of Large Analog Systems Containing Behavioral Models," *IEEE Trans. on Computer-Aided Design of Integrated Circuits and Systems*, vol. 13, n. 11, pp. 1391, November 1994.

[35] R. Castello and P. Gray, "A High-Performance Micropower Switched-Capacitor Filter," *IEEE Journal of Solid-State Circuits*, vol. SC-20, pp. 1122–1132, December 1985.

[36] H. Chang, E. Liu, R. Neff, E. Felt, E. Malavasi, E. Charbon, A.L. Sangiovanni-Vincentelli and P.R. Gray, "Top-Down, Constraint-Driven Methodology Based Generation of n-bit Interpolative Current Source D/A Converters," in *Proc. IEEE Custom Integrated Circuits Conference*, pp. 369–372, May 1994.

[37] H. Chang, A.L. Sangiovanni-Vincentelli, F. Balarin, E. Charbon, U. Choudhury, G. Jusuf, E. Liu, E. Malavasi, R. Neff and P. Gray, "A Top-Down, Constraint-Driven Design Methodology for Analog Integrated Circuits," in *Proc. IEEE Custom Integrated Circuits Conference*, pp. 841–846, May 1992.

[38] E. Charbon, *Constraint-Driven Analysis and Synthesis of High-Performance Analog IC Layout*, Ph.D. thesis, University of California, Berkeley, December 1995.

[39] E. Charbon, E. Malavasi, U. Choudhury, A. Casotto and A.L. Sangiovanni-Vincentelli, "A Constraint-Driven Placement Methodology for Analog Integrated Circuits," in *Proc. IEEE Custom Integrated Circuits Conference*, pp. 2821–2824, May 1992.

[40] E. Charbon, E. Malavasi, D. Pandini and A.L. Sangiovanni-Vincentelli, "Imposing Tight Specifications on Analog ICs through Simultaneous Placement and Module Optimization," in *Proc. IEEE Custom Integrated Circuits Conference*, pp. 525–528, May 1994.

[41] E. Charbon, E. Malavasi, D. Pandini and A.L. Sangiovanni-Vincentelli, "Simultaneous Placement and Module Optimization of Analog ICs," in *Proc. IEEE/ACM Design Automation Conference*, pp. 31–35, June 1994.

[42] E. Charbon, E. Malavasi and A.L. Sangiovanni-Vincentelli, "Generalized Constraint Generation for Analog Circuit Design," in *Proc. IEEE International Conference on Computer-Aided Design*, pp. 408–414, November 1993.

[43] B.R. Chawla, H.K. Gummel and P. Kozak, "MOTIS—An MOS Timing Simulator," *IEEE Trans. on Circuits and Systems*, vol. CAS-22, pp. 901–909, December 1975.

[44] D.J. Chen, J.C. Lee and B.J. Sheu, "SLAM: A Smart Analog Module Generator for Mixed Analog-Digital VLSI Design," in *Proc. IEEE International Conference on Computer Design*, pp. 24–27, 1989.

[45] D.J. Chen and B.J. Sheu, "Automatic Custom Layout of Analog ICs using Constraint-Based Module Generation," in *Proc. IEEE Custom Integrated Circuits Conference*, pp. 551–554, May 1991.

[46] K. Chi, et al., "A CMOS Triple 100-Mbit/s Video DA Converter with Shift Register and Color Map," *IEEE Journal of Solid-State Circuits*, vol. SC-21, n. 6, pp. 989–996, December 1986.

[47] M. Chou, M. Kamon, K. Nabors, J. Phillips and J. White, "3-D Extraction Techniques for Signal Integrity Analysis," in *Proc. IEEE Custom Integrated Circuits Conference*, pp. 379–382, May 1995.

[48] U. Choudhury, "Sensitivity Computation in SPICE3," M.S. thesis, University of California, Berkeley, 1988.

[49] U. Choudhury and A.L. Sangiovanni-Vincentelli, "Constraint-Based Channel Routing for Analog and Mixed-Analog Digital Circuits," in *Proc. IEEE International Conference on Computer-Aided Design*, pp. 198–201, November 1990.

[50] U. Choudhury and A.L. Sangiovanni-Vincentelli, "Constraint Generation for Routing Analog Circuits," in *Proc. IEEE/ACM Design Automation Conference*, pp. 561–566, June 1990.

[51] U. Choudhury and A.L. Sangiovanni-Vincentelli, "Use of Performance Sensitivities in Routing of Analog Circuits," in *Proc. IEEE International Symposium on Circuits and Systems*, pp. 348–351, May 1990.

[52] U. Choudhury and A.L. Sangiovanni-Vincentelli, "An Analytical-Model Generator for Interconnect Capacitances," in *Proc. IEEE Custom Integrated Circuits Conference*, pp. 861–864, May 1991.

[53] U. Choudhury and A.L. Sangiovanni-Vincentelli, "Automatic Generation of Analytical Models for Interconnect Capacitances," *IEEE Trans. on Computer-Aided Design of Integrated Circuits and Systems*, vol. CAD-14, n. 4, pp. 470–480, April 1995.

[54] Chrontel, Inc., *CH8398 16-bit Interface ChronDAC^{TM} True-Color RAMDAC + Dual Clocks*, 1994.

[55] G.W. Clow, "A Global Routing Algorithm for General Cells," in *Proc. IEEE/ACM Design Automation Conference*, pp. 45–51, June 1984.

[56] J.M. Cohn, D.J. Garrod, R.A. Rutenbar and L.R. Carley, "New Algorithms for Placement and Routing of Custom Analog Cells in ACACIA," in *Proc. IEEE Custom Integrated Circuits Conference*, pp. 2761–2765, May 1990.

[57] J.M. Cohn, D.J. Garrod, R.A. Rutenbar and L.R. Carley, "KOAN/ANAGRAM II: New Tools for Device-Level Analog Placement and Routing," *IEEE Journal of Solid-State Circuits*, vol. SC-26, n. 3, pp. 330–342, March 1991.

[58] J.D. Conway and G.G. Schrooten, "An Automatic Layout Generator for Analog Circuits," in *Proc. European Design Automation Conference,* pp. 513–519, March 1992.

[59] S.B. Crary, "Optimal Design of Experiments for Sensor Calibration," *Proc. IEEE International Conference on Solid-State Sensors and Actuators,* pp. 404–407, 1991.

[60] J. Cullum, "An Algorithm for Minimizing a Differentiable Function that Uses only Function Values," in *Techniques of Optimization,* pp. 117–127. A.V. Balakrishnan Ed., Academic Press, New York, NY, 1972.

[61] M. Degrauwe, et al., "IDAC: An Interactive Design Tool for Analog CMOS Circuits," *IEEE Journal of Solid-State Circuits,* vol. SC-22, n. 6, pp. 1106–1116, December 1987.

[62] M.G. DeGrauwe, et al., "An Analog Expert Design of Analog Integrated Circuits," in *Proc. IEEE International Solid-State Circuits Conference,* pp. 212–213, 1987.

[63] H. DeMan, G. Arnout and P. Reyneart, "Mixed-Mode Circuit Simulation Techniques and Their Implementation in DIANA," in *Computer Design Aids for VLSI Circuits,* pp. 113–174. P. Antognetti and D.O. Pederson and H. DeMan Eds., Sijthoff & Noordhoff (The Netherlands), 1980.

[64] A. Demir, E. Liu and A. Sangiovanni-Vincentelli, "Time-Domain non-Monte Carlo Noise Simulation for Nonlinear Dynamic Circuits with Arbitrary Excitations," in *Proc. IEEE International Conference on Computer-Aided Design,* November 1994.

[65] A. Demir, E. Liu and A. Sangiovanni-Vincentelli, "Time-Domain non-Monte Carlo Noise Simulation for Nonlinear Dynamic Circuits with Arbitrary Excitations," *IEEE Trans. on Computer-Aided Design of Integrated Circuits and Systems,* vol. 15, n. 5, May 1996.

[66] A. Demir, E. Liu, A.L. Sangiovanni-Vincentelli and I. Vassiliou, "Behavioral Simulation Techniques for Phase/Delay-Locked Systems," in *Proc. IEEE Custom Integrated Circuits Conference,* pp. 453–456, May 1994.

[67] A. Demir and A. Sangiovanni-Vincentelli, "Simulation and Modeling of Phase Noise in Open-Loop Oscillators," in *Proc. IEEE Custom Integrated Circuits Conference,* May 1996.

[68] Analog Devices, *Data Converter Reference Manual, Vol. II,* Norwood, MA, 1992.

[69] S.W. Director and R.A. Rohrer, "The Generalized Adjoint Network and Network Sensitivities," *IEEE Trans. on Circuit Theory*, vol. CT-16, pp. 318–323, August 1969.

[70] S. Donnay, K. Swings, G. Gielen and W. Sansen, "A Methodology for Analog High-Level Synthesis," in *Proc. IEEE Custom Integrated Circuits Conference*, pp. 373–376, May 1994.

[71] R.I. Dowell, *Automated Biasing of Integrated Circuits*, Ph.D. thesis, University of California, Berkeley, April 1972.

[72] A.E. Dunlop, V.D. Agrawal, D.N. Deutsch, M.F. Jukl, P. Kozak and M. Wiesel, "Chip Layout Optimization Using Critical Path Weighting," in *Proc. IEEE/ACM Design Automation Conference*, pp. 133–136, June 1984.

[73] A.E. Dunlop, G.F. Gross, C.D. Kimble, M.Y. Luong, K.J. Stern and E.J. Swanson, "Features in LTX2 for Analog Layout," in *Proc. IEEE International Symposium on Circuits and Systems*, pp. 21–24, 1985.

[74] A.E. Dunlop and B.W. Kernighan, "A Procedure for Placement of Standard-Cell VLSI Circuits," *IEEE Trans. on Computer-Aided Design of Integrated Circuits and Systems*, vol. CAD-4, n. 1, pp. 92–98, January 1985.

[75] G.V. Eaton, D.G. Nairn, W.M. Snelgrove and A.S. Sedra, "SICOMP: A Silicon Compiler for SC Filters," in *Proc. IEEE International Symposium on Circuits and Systems*, pp. 321–324, 1987.

[76] F. El-Turky and E.E. Perry, "BLADES: An A.I. Approach to Analog Circuit Design," *IEEE Trans. on Computer-Aided Design of Integrated Circuits and Systems*, vol. CAD-8, n. 6, pp. 680–692, June 1989.

[77] S.C. Fang, Y.P. Tsividis and O. Wing, "SWITCAP: A Switched Capacitor Network Analysis Program," *IEEE Circuits and Systems*, vol. 5, n. 3, September 1983.

[78] P. Feldmann and R.W. Freund, "Reduced-Order Modeling of Large Linear Subcircuits via a Block Lanczos Algorithm," in *Proc. IEEE/ACM Design Automation Conference*, June 1995.

[79] E. Felt, E. Charbon, E. Malavasi and A.L. Sangiovanni-Vincentelli, "An Efficient Methodology for Symbolic Compaction of Analog ICs with Multiple Symmetry Constraints," in *Proc. European Design Automation Conference*, pp. 148–153, September 1992.

[80] E. Felt, E. Malavasi, E. Charbon, R. Totaro and A.L. Sangiovanni-Vincentelli, "Performance-Driven Compaction for Analog Integrated Circuits," in *Proc. IEEE Custom Integrated Circuits Conference*, pp. 1731–1735, May 1993.

[81] E. Felt, A. Narayan and A. Sangiovanni-Vincentelli, "Measurement and Modeling of MOS Transistor Current Mismatch in Analog ICs," in *Proc. IEEE International Conference on Computer-Aided Design*, pp. 272–277, November 1994.

[82] E. Felt and A. Sangiovanni-Vincentelli, "Testing of Analog Systems Using Behavioral Models and Optimal Experimental Design Techniques," in *Proc. IEEE International Conference on Computer-Aided Design*, pp. 672–678, November 1994.

[83] R.W. Freund and P. Feldmann, "Efficient Small-Signal Circuit Analysis and Sensitivity Computations with the PVL Algorithm," in *Proc. IEEE International Conference on Computer-Aided Design*, November 1994.

[84] A.H. Fung, D.J. Chen, Y.N. Lai and B.J. Sheu, "Knowledge-Based Analog Circuit Synthesis with Flexible Architecture," in *Proc. IEEE International Conference on Computer Design*, pp. 48–51, October 1988.

[85] "The Future Role of the Analog Designer," discussion session at *Solid-State Circuits Conference*, Februrary 24, 1994.

[86] G. Gad-El-Karim, *Sensitivity-Driven Placement of Analog Modules*, Ph.D. thesis, Carnegie-Mellon University, 1992.

[87] G. Gad-El-Karim and R.S. Gyurcsik, "Generation of Performance Sensitivities for Analog Cell Layout," in *Proc. IEEE/ACM Design Automation Conference*, pp. 500–505, June 1991.

[88] G. Gad-El-Karim and R.S. Gyurcsik, "Use of Performance Sensitivities in Analog Cell Layout," in *Proc. IEEE International Symposium on Circuits and Systems*, vol. 4, pp. 2008–2011, June 1991.

[89] C.W. Gardiner, *Handbook of Stochastic Methods for Physics, Chemistry and the Natural Sciences*, Springer-Verlag, 1983.

[90] F.M. Gardner, *Phaselock Techniques*, J. Wiley & Sons, New York, 1979.

[91] F.M. Gardner, "Charge-Pump Phase-Lock Loops," *IEEE Trans. on Communications*, vol. COM-28, n. 11, pp. 1849–1858, November 1980.

[92] W.A. Gardner, *Introduction to Random Processes with Applications to Signals & Systems*, McGraw-Hill, second edition, 1990.

[93] U. Gatti, V. Liberali, F. Maloberti and A. Scianna, "An Expert System for the Design of Analog Building Blocks," in *Proc. 10th International Workshop on Expert Systems and their Applications, Avignon*, pp. 75–86, May 1990.

[94] C.W. Gear, *Numerical Integration of Stiff Ordinary Differential Equations*, Technical Report 221, University of Illinois, Urbana, 1967.

[95] R. Gharpurey, *Modeling and Analysis of Substrate Coupling in ICs*, Ph.D. thesis, University of California, Berkeley, May 1995.

[96] R. Gharpurey and R.G. Meyer, "Modeling and Analysis of Substrate Coupling in Integrated Circuits," in *Proc. IEEE Custom Integrated Circuits Conference*, May 1995.

[97] A. Ghosh, S. Devadas and A. Richard Newton, *Sequential Logic Testing and Verification*, Kluwer, Boston, 1992.

[98] G. Gielen, E. Liu, A. Sangiovanni-Vincentelli and P. Gray, "Analog Behavioral Models for Simulation and Synthesis of Mixed-Signal Systems," in *Proc. European Design Automation Conference*, March 1992.

[99] G. Gielen and W. Sansen, *Symbolic Analysis for Automated Design of Analog Integrated Circuits*, Kluwer Academic Publishers, Boston, MA, 1991.

[100] G. Gielen, H. Walscharts and W. Sansen, "ISAAC: A Symbolic Simulator for Snalog Integrated Circuits," *IEEE Journal of Solid-State Circuits*, vol. SC-24, pp. 1587–1597, 1989.

[101] G. Gielen, H. Walscharts and W. Sansen, "Analog Circuit Design Optimization Based on Symbolic Simulation and Simulated Annealing," *IEEE Journal of Solid-State Circuits*, vol. SC-25, pp. 707–713, 1990.

[102] G. Gielen, Z. Wang and W. Sansen, "Fault Detection and Input Stimulus Determination for the Testing of Analog Integrated Circuits Based on Power-Supply Current Monitoring," in *Proc. IEEE International Conference on Computer-Aided Design*, pp. 495–498, November 1994.

[103] N. Gohar, P. Cheung and C. Pun, "RACHANA: An Integrated Placement and Routing Approach to CMOS Analog Cells," in *Proc. IEEE International Symposium on Circuits and Systems*, vol. 6, pp. 2981–2984, May 1992.

[104] R.S. Graham, "Relay Computer for Network Analysis," *Bell Labs Rec.*, vol. 31, pp. 152–157, April 1953.

[105] P.R. Gray and R.G. Meyer, *Analysis and Design of Analog Integrated Circuits*, J. Wiley & Sons, New York, 1977.

[106] P.R. Gray and R.G. Meyer, *Analysis and Design of Analog Integrated Circuits*, John Wiley & Sons, 2nd edition, 1984.

[107] P.R. Gray and R.G. Meyer, *Analysis and Design of Analog Integrated Circuits*, J. Wiley & Sons, 3rd Edition, 1993.

[108] P.R. Gray and R.G. Meyer, "Future Directions in Silicon ICs for RF Personal Communications," in *Proc. IEEE Custom Integrated Circuits Conference*, May 1995.

[109] R.W. Gregor, "On the Relationship between Topography and Transistor Matching in an Analog CMOS Technology," *IEEE Trans. on Electron Devices*, vol. 39, n. 2, pp. 275–282, February 1992.

[110] G.R. Grimmet and D.R. Stirzaker, *Probability and Random Processes*, Oxford Science Publications, second edition, 1992.

[111] C. Guardiani, A. Tomasini, J. Benkoski, M. Quarantelli and P. Gubian, "Applying a Submicron Mismatch Model to Practical IC Design," in *Proc. IEEE Custom Integrated Circuits Conference*, pp. 297–300, May 1994.

[112] K.C. Gupta, R. Garg and R. Chardha, *Computer Aided Design for Microwave Circuits*, Artech House, Norwood, MA, 1981.

[113] R.S. Gyurcsik and J.-C. Jeen, "A Generalized Approach to Routing Mixed Analog and Digital Signal Nets in a Channel," *IEEE Journal of Solid-State Circuits*, vol. SC-24, n. 2, pp. 436–442, April 1989.

[114] G. Hachtel and A.L. Sangiovanni-Vincentelli, "A Survey of Third-Generation Simulation Techniques," *Proc. of the IEEE*, vol. 69, n. 10, pp. 1264–1280, October 1981.

[115] G. Hachtel, T.R. Scott and R.P. Zug, "An Interactive Linear Programming Approach to Model Parameter Fitting and Worst Case Circuit Design," *IEEE Trans. on Circuits and Systems*, vol. CAS-27, pp. 871–881, October 1980.

[116] G.D. Hachtel, R.K. Brayton and F.G. Gustavson, "The Sparse Tableau Approach to Network Analysis and Design," *IEEE Trans. on Circuit Theory*, vol. CT-18, pp. 101–113, January 1971.

[117] G.D. Hachtel, M.R. Lightner and H.J. Kelly, "Application of the Optimization Program AOP to the Design of Memory Circuits," *IEEE Trans. on Circuits and Systems*, vol. CAS-22, pp. 496–503, June 1975.

[118] G.D. Hachtel and A. Sangiovanni-Vincentelli, "Third-Generation Simulation Techniques," *Proc. of the IEEE*, vol. 69, n. 10, pp. 1264, October 1981.

[119] J.M. Hammersley and D.C. Handscomb, *Monte Carlo Methods*, London, Methuen & Co Ltd., 1964.

[120] I. Harada, H. Kitazawa and T. Kaneko, "A Routing System for Mixed A/D Standard Cell LSIs," in *Proc. IEEE International Conference on Computer-Aided Design*, pp. 378–381, November 1990.

[121] I. Harada, H. Kitazawa and T. Kaneko, "A Layout System for Mixed A/D Standard Cell LSIs," *Institute of Electronics and Communication Engineers Trans. on Electronics*, vol. E75-C, n. 3, pp. 322–332, March 1992.

[122] R.H. Hardin and N.J.A. Sloane, "A New Approach to the Construction of Optimal Designs," *Journal of Statistical Planning and Inference*, vol. 37, pp. 339–369, 1993.

[123] R. Harjani, R.A. Rutenbar and L.R. Carley, "OASYS: A Framework for Analog Circuit Synthesis," *IEEE Trans. on Computer-Aided Design of Integrated Circuits and Systems*, vol. CAD-8, n. 12, pp. 1247–1266, December 1989.

[124] R. Harjani, et al., "A Prototype Framework for Knowledge-Based Analog Circuit Synthesis," in *Proc. IEEE/ACM Design Automation Conference*, pp. 42–49, 1987.

[125] D. Harrison, P. Moore, R.L. Spickelmier and A.R. Newton, "Data Management and Graphics Editing in the Berkeley Design Environment," in *Proc. IEEE International Conference on Computer-Aided Design*, pp. 24–27, November 1986.

[126] D.S. Harrison, A.R. Newton, R.L. Spickelmier and T.J. Barnes, "Electronic CAD Frameworks," *Proc. of the IEEE*, vol. 78, n. 2, pp. 393–417, February 1990.

[127] J.P. Harvey, M.I. Elmasry and B. Leung, "STAIC: An Interactive Framework for Synthesizing CMOS and BiCMOS Analog Circuits," *IEEE Trans. on Computer-Aided Design of Integrated Circuits and Systems*, vol. CAD-11, n. 11, pp. 1402–1417, November 1992.

[128] W.J. Helms and B.E. Byrkett, "Compiler Generation of A to D Converters," in *Proc. IEEE Custom Integrated Circuits Conference*, pp. 161–164, May 1987.

[129] G. Hemink, B. Meijer and H. Kerkhoff, "Testability Analysis of Analog Systems," *IEEE Trans. on Computer-Aided Design of Integrated Circuits and Systems*, vol. CAD-9, pp. 573–583, June 1990.

[130] D.A. Hocevar, P. Yang, T.N. Trick and B.D. Epler, "Transient Sensitivity Computation for MOSFET Circuits," *IEEE Trans. on Computer-Aided Design of Integrated Circuits and Systems*, vol. CAD-4, n. 4, pp. 609–620, October 1985.

[131] A.S. Hodel and S.T. Hung, "Solution and Applications of the Lyapunov Equation for Control Systems," *IEEE Trans. on Industrial Electronics*, vol. 39, n. 3, pp. 194, June 1992.

[132] S.K. Hong and P.E. Allen, "Performance Driven Analog Layout Compiler," in *Proc. IEEE International Symposium on Circuits and Systems*, pp. 835–838, 1990.

[133] R. Hooke and T.A. Jeeves, "'Direct Search' Solution of Numerical and Statistical Problems," *Journal of ACM*, vol. 8, pp. 212–229, 1961.

[134] B.J. Hosticka, "Performance Comparison of Analog and Digital Circuits," *Proc. of the IEEE*, vol. 73, n. 1, pp. 25–29, 1985.

[135] C.D. Hull, *Analysis and Optimization of Monolithic RF Down Conversion Receivers*, Ph.D. thesis, University of California, Berkeley, 1992.

[136] C.D. Hull and R.G. Meyer, "A Systematic Approach to the Analysis of Noise in Mixers," *IEEE Trans. on Circuits and Systems-1: Fundamental Theory and Applications*, vol. 40, n. 12, pp. 909, December 1993.

[137] T.E. Idleman, F.S. Jenkins, W.J. McCalla and D.O. Pederson, "SLIC: A Simulator for Linear Integrated Circuits," *IEEE Journal of Solid-State Circuits*, vol. SC-6, pp. 188–204, August 1971.

[138] *IEEE PAR 1076.1 (VHDL-A) Design Objective Document.*

[139] M. Itoh and H. Mori, "ALE: a Layout Generating and Editing System for Analog LSIs," in *Proc. IEEE International Symposium on Circuits and Systems*, vol. 2, pp. 843–846, May 1990.

[140] D. Jeong, G. Borrielo, D. Hodges and R. Katz, "Design of PLL-Based Clock Generation Circuits," *IEEE Journal of Solid-State Circuits*, vol. SC-22, n. 2, pp. 255–261, February 1987.

[141] B. Johnson, T. Quarles, A.R. Newton, D.O. Pederson and A. Sangiovanni-Vincentelli, *SPICE3 Version 3e User's Manual*, University of California, Berkeley, April 1991.

[142] E.D. Johnson, C.T. Kleiner, L.R. McMurray, E.L. Steele and F.A. Vassallo, *Transient Radiation Analysis by Computer Program (TRAC)*, Technical report, Rockwell Corp., Anaheim, CA, June 1968.

[143] T.A. Johnson, R.W. Knepper, V. Marcello and W. Wang, "Chip Substrate Resistance Modeling Technique for Integrated Circuit Design," *IEEE Trans. on Computer-Aided Design of Integrated Circuits and Systems*, vol. CAD-3, pp. 126–134, 1984.

[144] A. Jordan and N. Jordan, "Theory of Noise in Metal Oxide Semiconductor Devices," *IEEE Trans. on Electron Devices*, pp. 148–156, March 1965.

[145] Y-C. Ju, V.B. Rao and R.A. Saleh, "Consistency Checking and Optimization of Macromodels," *IEEE Trans. on Computer-Aided Design of Integrated Circuits and Systems*, vol. 10, n. 8, pp. 957, August 1991.

[146] G. Jusuf, P.R. Gray and A.L. Sangiovanni-Vincentelli, "CADICS—Cyclic Analog-To-Digital Converter Synthesis," in *Proc. IEEE International Conference on Computer-Aided Design*, pp. 286–289, November 1990.

[147] H. Kahn, "Use of Different Monte Carlo Sampling Techniques," in *Symposium on Monte Carlo Methods*, pp. 146–190. New York, Wiley, 1956.

[148] M. Kamon, M. Tsuk, C. Smithhisler and J. White, "Efficient Techniques for Inductance Extraction of Complex 3-D Geometries," in *Proc. IEEE International Conference on Computer-Aided Design*, pp. 438–442, November 1992.

[149] M. Kamon, M.T. Tsuk and J. White, "FastHenry: A Multiple-Accelerated 3-D Inductance Extraction Program," in *Proc. IEEE/ACM Design Automation Conference*, pp. 678–683, June 1993.

[150] M. Kayal, S. Piguet, M. Declercq and B. Hochet, "SALIM: A Layout Generator Tool for Analog ICs," in *Proc. IEEE Custom Integrated Circuits Conference*, pp. 751–754, May 1988.

[151] B.W. Kernighan and S. Lin, "An Efficient Heuristics Procedure for Partitioning Graphs," *Bell System Technical Journal*, vol. 49, n. 2, pp. 291–307, 1970.

[152] M.S. Keshner, "$1/f$ Noise," *Proc. of the IEEE*, vol. 70, n. 3, pp. 212, March 1982.

[153] J. Kiefer, *Collected Papers III: Design of Experiments*, Springer-Verlag, New York, 1985.

[154] B. Kim, D. Helman and P.R. Gray, "A 30-MHz Hybrid Analog/Digital Clock Recovery Circuit in 2-μm CMOS," *IEEE Journal of Solid-State Circuits*, vol. SC-25, n. 3, pp. 1385–1394, December 1993.

[155] B. Kim, T.C. Weigandt and P.R. Gray, "PLL/DLL System Noise Analysis for Low Jitter Clock Synthesizer Design," in *Proc. IEEE International Symposium on Circuits and Systems*, vol. 4, pp. 31–34, May 1994.

[156] C.D. Kimble, A.E. Dunlop, G.F. Gross, V.L. Hein, M.Y. Luong, K.J. Stern and E.J. Swanson, "Autorouted Analog VLSI," in *Proc. IEEE Custom Integrated Circuits Conference*, pp. 72–78, May 1985.

[157] S. Kirkpatrick, C. Gelatt and M. Vecchi, "Optimization by Simulated Annealing," *Science*, vol. 220, n. 4598, pp. 671–680, May 1983.

[158] D.E. Knuth, "Fundamental Algorithms," in *The Art of Computer Programming*, vol. 1. Addison Wesley, Reading, MA, 1973.

[159] H.Y. Koh, C.H. Séquin and P.R. Gray, "OPASYN: A Compiler for CMOS Operational Amplifiers," *IEEE Trans. on Computer-Aided Design of Integrated Circuits and Systems*, vol. CAD-9, n. 2, pp. 113–125, February 1990.

[160] T. Koskinen and P.Y.K. Cheung, "Modeling Behaviour and Tolerances in Analogue Cells," in *Proc. IEEE Custom Integrated Circuits Conference*, pp. 871–874, May 1991.

[161] T. Kozawa, et al., "Combine and Top Down Block Placement Algorithm for Hierarchical Logic VLSI Layout," in *Proc. IEEE/ACM Design Automation Conference*, June 1984.

[162] J. Kuhn, "Analog Module Generators for Silicon Compilation," *VLSI Systems Design*, vol. 8, n. 5, pp. 74–80, May 1987.

[163] K.S. Kundert, private communication.

[164] K.S. Kundert, "Sparse Matrix Techniques," in A.E. Ruehli, editor, *Circuit Analysis, Simulation and Design*, pp. 281–324. Elsevier Science Publishers B.V. (North-Holland), 1986.

[165] K.S. Kundert, "Accurate Fourier Analysis for Circuit Simulators," in *Proc. IEEE Custom Integrated Circuits Conference*, May 1994.

[166] K.S. Kundert, J.K. White and A. Sangiovanni-Vincentelli, *Steady-State Methods for Simulating Analog and Microwave Circuits*, Kluwer Academic Publishers, Boston, MA, 1990.

[167] C.A. Laber, C.F. Rahim, S.F. Dreyer, G.T. Uehara, P.T. Kwok and P.R. Gray, "A Family of Differential NMOS Analog Circuits for a PCM Codec Filter Chip," *IEEE Journal of Solid-State Circuits*, vol. SC-22, n. 2, pp. 181–189, April 1987.

[168] K.R. Lakshmikumar, R.A. Hadaway and M.A. Copeland, "Characterization and Modeling of Mismatch in MOS Transistors for Precision Analog Design," *IEEE Journal of Solid-State Circuits*, vol. SC-21, n. 6, pp. 1057–1066, December 1986.

[169] K.R. Lakshmikumar, R.A. Hadaway and M.A. Copeland, "Charaterization and Modeling of Mismatch in MOS Transistors for Precision Analog Design," *IEEE Journal of Solid-State Circuits*, vol. SC-21, n. 6, pp. 1057–1066, December 1986.

[170] U. Lauther, "A Min-Cut Placement Algorithm for General Cell Assemblies Based on a Graph Representation," in *Proc. IEEE/ACM Design Automation Conference*, pp. 1–10, 1979.

[171] T.H. Lee and J.F. Bulzacchelli, "A 155 MHz Clock Recovery Delay- and Phase-Locked Loop," *IEEE Journal of Solid-State Circuits*, vol. 27, n. 12, December 1992.

[172] T. Lengauer, *Combinatorial Algorithms for Integrated Circuit Layout*, Applicable Theory in Computer Science. J. Wiley & Sons, New York, 1990.

[173] J. Leonard, N. Weste, L. Bodony, S. Harston and R. Meaney, "A 66-MHz DSP Augmented RAMDAC for Smooth-Shaded Graphic Applications," *IEEE Journal of Solid-State Circuits*, vol. SC-26, n. 3, pp. 989–996, March 1991.

[174] E.T. Lewis, "An Analysis of Interconnect Line Capacitance and Coupling for VLSI Circuits," *Solid-State Electronics*, vol. 27, pp. 741–749, 1984.

[175] Z.-M. Lin, "DAVE: an Automatic Mixed Analog/Digital IC Layout Compiler," in *Proc. IEEE Custom Integrated Circuits Conference*, pp. 541–544, May 1991.

[176] E. Liu, *Analog Behavioral Simulation and Modeling*, Ph.D. thesis, University of California, Berkeley, 1993.

[177] E. Liu, H.C. Chang and A.L. Sangiovanni-Vincentelli, "Analog System Verification in the Presence of Parasitics using Behavioral Simulation," in *Proc. IEEE/ACM Design Automation Conference*, pp. 159–163, June 1993.

[178] E. Liu, G. Gielen, H. Chang and A. Sangiovanni-Vincentelli, "Behavioral Modeling and Simulation of Data Converters," in *Proc. IEEE International Symposium on Circuits and Systems*, pp. 2144–2147, May 1992.

[179] E. Liu, W. Kao, E. Felt and A.L. Sangiovanni-Vincentelli, "Analog Testability Analysis and Fault Diagnosis using Behavioral Modeling," in *Proc. IEEE Custom Integrated Circuits Conference*, pp. 413–416, May 1994.

[180] E. Liu and A. Sangiovanni-Vincentelli, "Behavioral Representation for VCO and Detectors in Phase-Lock Systems," in *Proc. IEEE Custom Integrated Circuits Conference*, pp. 1231–1234, May 1992.

[181] E. Liu and A. Sangiovanni-Vincentelli, "Behavioral Simulation for Noise in Mixed-Mode Sampled-Data Systems," in *Proc. IEEE International Conference on Computer-Aided Design*, pp. 322–326, Nov 1992.

[182] E. Liu, A. Sangiovanni-Vincentelli, G. Gielen and P. Gray, "A Behavioral Representation for Nyquist Rate A/D Converters," in *Proc. IEEE International Conference on Computer-Aided Design*, pp. 386–389, November 1991.

[183] E. Liu and A.L. Sangiovanni-Vincentelli, "Nyquist Data Converter Testing and Yield Analysis Using Behavioral Simulation," in *Proc. IEEE International Conference on Computer-Aided Design*, pp. 341–348, November 1993.

[184] B. Lokanathan, "MINimum ITerations," Department of Electrical Engineering, University of Rochester, adapted from R. Salazar and S. Sen, "MINIT algorithm for Linear Programming," *Collected Algorithms from CACM*, algorithm #333, 1968.

[185] D. Luenberger, *Linear and Nonlinear Programming*, Addison-Wesley Publishing Company, 2nd Ed., 1984.

[186] V.M. Ma, J. Singh and R. Saleh, "Modeling, Simulation and Optimization of Analog Macromodels," in *Proc. IEEE Custom Integrated Circuits Conference*, May 1992.

[187] D.G. Maeding, et al., "Combining Analog and Digital Standard Cells," in *Proc. IEEE Custom Integrated Circuits Conference*, pp. 491–494, May 1985.

[188] C. Makris and C. Toumazou, "Qualitative Reasoning in Analog IC Design Automation," in *Proc. IEEE Custom Integrated Circuits Conference*, pp. 831–834, May 1992.

[189] C.A. Makris and C. Toumazou, "Analog IC Design: Part II—Automated Circuit Correction by Qualitative Reasoning," *IEEE Trans. on Computer-Aided Design of Integrated Circuits and Systems*, vol. CAD-14, n. 2, pp. 239–254, February 1995.

[190] E. Malavasi, E. Charbon, E. Felt and A.L. Sangiovanni-Vincentelli, "Automation of IC Layout with Analog Constraints," *TCAD*, vol. 15, n. 8, *accepted for publication* 1996.

[191] E. Malavasi, U. Choudhury and A.L. Sangiovanni-Vincentelli, "A Routing Methodology for Analog Integrated Circuits," in *Proc. IEEE International Conference on Computer-Aided Design*, pp. 202–205, November 1990.

[192] E. Malavasi, E. Felt, E. Charbon and A.L. Sangiovanni-Vincentelli, "Symbolic Compaction with Analog Constraints," *International Journal of Circuit Theory and Applications, Special Issue on "Analog Tools for Circuit Design," John Wiley & Sons*, vol. 23, n. 4, pp. 433–452, July-August 1995.

[193] E. Malavasi and D. Pandini, "Optimum CMOS Stack Generation with Analog Constraints," *IEEE Trans. on Computer-Aided Design of Integrated Circuits and Systems*, vol. CAD-14, n. 1, pp. 107–122, January 1995.

[194] E. Malavasi and A.L. Sangiovanni-Vincentelli, "Area Routing for Analog Layout," *IEEE Trans. on Computer-Aided Design of Integrated Circuits and Systems*, vol. CAD-12, n. 8, pp. 1186–1197, August 1993.

[195] A.F. Malmberg, F.L. Cornell and F.N. Hofer, *NET1 Network Analysis Program*, Technical Report LA-3119, 7090, Los Alamos Scientific Lab, Los Alamos, NM, August 1964.

[196] H.A. Mantooth and M. Vlach, "Beyond SPICE with SABER and MAST," in *Proc. IEEE International Symposium on Circuits and Systems*, vol. 1, p. 77, May 1992.

[197] H.W. Mathers, S.R. Sedore and J.R. Seuts, *Automated Digital Computer Program for Determining Responses of Electronic Circuits to Transient Nuclear Radiation (SCEPTRE)*, IBM File 66-928-611, IBM Space Guidance Center, Oswego, NY, February 1967.

[198] C.A. Maulik, R.A. Rutenbar and L.R. Carley, "Integer Programming Based Topology Selection of Cell-Level Analog Circuits," *IEEE Trans. on Computer-Aided Design of Integrated Circuits and Systems*, vol. CAD-14, n. 4, pp. 401–412, April 1995.

[199] W.J. McCalla, *Computer-Aided-Design of Integrated Bandpass Amplifiers*, Ph.D. thesis, University of California, Berkeley, June 1972.

[200] W.J. McCalla and Jr. W.G. Howard, "BIAS-3: A Program for the non-Linear DC Analysis of Bypolar Transistor Circuits," *IEEE Journal of Solid-State Circuits*, vol. SC-6, n. 1, pp. 14–19, February 1971.

[201] R. McCharles, *Charge Circuits for Analog LSI*, Ph.D. thesis, University of California, Berkeley, 1980.

[202] J.A. McNeill, *Jitter in Ring-Oscillators*, Ph.D. thesis, Boston University, 1994.

[203] S.W. Mehranfar, "STAT: A Schematic to Artwork Translator for Custom Analog Cells," in *Proc. IEEE Custom Integrated Circuits Conference*, pp. 3021–3024, May 1990.

[204] S.W. Mehranfar, "A Technology-Independent Approach to Custom Analog Cell Generation," *IEEE Journal of Solid-State Circuits*, vol. SC-26, n. 3, pp. 386–393, March 1991.

[205] R. Melville, P. Feldmann and J. Roychowdhury, "Efficient Multi-tone Distortion Analysis of Analog Integrated Circuits," in *Proc. IEEE Custom Integrated Circuits Conference*, May 1995.

[206] R.G. Meyer, L. Nagel and S.K. Lui, "Computer Simulation of $1/f$ Noise Performance of Electronic Circuits," *IEEE Journal of Solid-State Circuits*, p. 237, June 1973.

[207] *MHDL Language Reference Manual*, Intermetrics, Inc., February 1995.

[208] C. Michael and M. Ismail, "Statistical Modeling of Device Mismatch for Analog MOS Integrated Circuits," *IEEE Journal of Solid-State Circuits*, vol. 27 no. 2, pp. 154–166, Feb 1992.

[209] P. Miliozzi, L. Carloni, E. Charbon and A.L. Sangiovanni-Vincentelli, "SUB-WAVE: a Methodology for Modeling Digital Substrate Noise Injection in Mixed-Signal ICs," in *Proc. IEEE Custom Integrated Circuits Conference*, pp. 385–388, May 1996.

[210] L. Milor and A.L. Sangiovanni-Vincentelli, "Optimal Test Set Design for Analog Circuits," in *Proc. IEEE International Conference on Computer-Aided Design*, pp. 294–297, November 1990.

[211] L. Milor and A.L. Sangiovanni-Vincentelli, "Minimizing Production Test Time to Detect Faults in Analog Circuits," *IEEE Trans. on Computer-Aided Design of Integrated Circuits and Systems*, vol. 13, n. 6, pp. 796–813, June 1994.

[212] L. Milor and V. Visvanathan, "Detection of Catastrophic Faults in Analog Integrated Circuits," *IEEE Trans. on Computer-Aided Design of Integrated Circuits and Systems*, vol. CAD-8, n. 2, pp. 114–130, February 1989.

[213] S. Mitra, S. Nag, R.A. Rutenbar and L.R. Carley, "System-level Routing of Mixed-Signal ASICs in WREN," in *Proc. IEEE International Conference on Computer-Aided Design*, pp. 394–399, November 1992.

[214] "Modular In-core Nonlinear Optimization System (MINOS 5.3)," Systems Optimization Laboratory, Department of Operations Research, Stanford University.

[215] M. Mogaki, N. Katoh, Y. Chikami, N. Yamada and Y. Kobayashi, "LADIES: An Automatic Layout System for Analog LSIs," in *Proc. IEEE International Conference on Computer-Aided Design*, pp. 450–453, November 1989.

[216] B.A. Murtagh and M.A. Saunders, *MINOS 5.1 User's Guide*, Technical Report Rep. SOL 83-20R, Department of Operations Research, Stanford University, Stanford, CA, January 1987.

[217] L. Nagel, *SPICE2: A computer Program to Simulate Semiconductor Circuits*, Ph.D. thesis, University of California, Berkeley, May 1975.

[218] L. Nagel and R.A. Rohrer, "Computer Analysis of Nonlinear Circuits Excluding Radiation (CANCER)," *IEEE Journal of Solid-State Circuits*, vol. SC-6, pp. 166–182, August 1971.

[219] Y. Nakamura, T. Miki, A. Maeda, H. Kondoh and N. Yazawa, "A 10-b 70-MS/s CMOS D/A Converter," *IEEE Journal of Solid-State Circuits*, vol. 26, n. 4, pp. 637–642, April 1991.

[220] R. Neff, P. Gray and A.L. Sangiovanni-Vincentelli, "A Module Generator for High Speed CMOS Current Output Digital/Analog Converters," in *Proc. IEEE Custom Integrated Circuits Conference*, pp. 481–484, May 1995.

[221] M. Negahban and D. Gaiski, "Silicon Compilation of Switched-Capacitor Networks," in *Proc. European Design Automation Conference*, pp. 164–168, March 1990.

[222] A.R. Newton and A. Sangiovanni-Vincentelli, "Relaxation-Based Electrical Simulation," *IEEE Trans. on Electron Devices*, vol. ED-30, n. 9, pp. 1184, September 1983.

[223] A.R. Newton and A.L. Sangiovanni-Vincentelli, "Relaxation-Based Electrical Simulation," *IEEE Trans. on Computer-Aided Design of Integrated Circuits and Systems*, vol. CAD-3, n. 4, pp. 308–330, October 1984.

[224] N.K. Nguyen and A.J. Miller, "A Review of Some Exchange Algorithms for Constructing Discrete D-Optimal Designs," *Computational Statistics and Data Analysis*, vol. 14, pp. 489–498, 1992.

[225] T.M. Niebauer, R. Schilling, K. Danzmann, A. Rudiger and W. Winkler, "Nonstationary Shot Noise and Its Effects on the Sensitivity of Interferometers," *Physical Review A*, vol. 43, n. 9, pp. 5022, May 1991.

[226] Z.-Q. Ning, T. Mouthaan and H. Wallinga, "SEAS: A Simulated Evolution Approach for Analog Circuit Synthesis," in *Proc. IEEE Custom Integrated Circuits Conference*, pp. 521–524, May 1991.

[227] K.A. Nishimura, *Optimum Partitioning of Analog and Digital Circuitry in Mixed-Signal Circuits for Signal Processing*, Technical report, Ph.D. thesis, University of California, Berkeley, July 1993.

[228] W. Nye, D.C. Riley, A.L. Sangiovanni-Vincentelli and A.L. Tits, "DE-LIGHT.SPICE: An Optimization-Based System for the Design of Integrated Circuits," *IEEE Trans. on Computer-Aided Design of Integrated Circuits and Systems*, vol. CAD-7, n. 4, pp. 501–519, April 1988.

[229] W.T. Nye, *DELIGHT: An Interactive System for Optimization-Based Engineering Design*, Ph.D. thesis, University of California, Berkeley, May 1983.

[230] E. Ochotta, L.R. Carley and R.A. Rutenbar, "Analog Circuit Synthesis for Large, Realistic Cells: Designing a Pipelined A/D Converter with ASTRX/OBLX," in *Proc. IEEE Custom Integrated Circuits Conference*, pp. 365–368, May 1994.

[231] T. Ohtsuka, H. Kunieda and M. Kaneko, "LIBRA: Automatic Performance-Driven Layout for Analog LSIs," *Institute of Electronics and Communication Engineers Trans. on Electronics*, vol. E75-C, n. 3, pp. 312–321, March 1992.

[232] T. Ohtsuki, *Layout Design and Verification*, T. Ohtsuki Ed., North Holland, 1986.

[233] R. Okuda, T. Sato, H. Onodera and K. Tamaru, "An Efficient Algorithm for Layout Compaction Problem with Symmetry Constraints," in *Proc. IEEE International Conference on Computer-Aided Design*, pp. 148–151, November 1989.

[234] M. Okumura, H. Tanimoto, T. Itakura and T. Sugawara, "Numerical Noise Analysis for Nonlinear Circuits with a Periodic Large Signal Excitation Including Cyclostationary Noise Sources," *IEEE Trans. on Circuits and Systems-1: Fundamental Theory and Applications*, vol. 40, n. 9, pp. 581, September 1993.

[235] H. Onodera, H. Kanbara and K. Tamaru, "Operational-Amplifier Compilation with Performance Optimization," *IEEE Journal of Solid-State Circuits*, vol. SC-25, n. 2, pp. 466–473, April 1990.

[236] R.H.J.M. Otten, "Automatic Floorplan Design," in *Proc. IEEE/ACM Design Automation Conference*, pp. 261–267, June 1982.

[237] D.O. Pederson, "A Historical Review of Circuit Simulation," *IEEE Trans. on Circuits and Systems*, vol. CAS-31, pp. 103–111, January 1984.

[238] M. Pelgrom, A. Duinmaijer and A. Welbers, "Matching Properties of MOS Transistors," *IEEE Journal of Solid-State Circuits*, vol. SC-24, n. 5, October 1989.

[239] M.J.M. Pelgrom, "A 10-b 50-Mhz CMOS D/A Converter with 75-Ω Buffer," *IEEE Journal of Solid-State Circuits*, vol. SC-25, pp. 1347–1352, December 1990.

[240] M.J.M. Pelgrom, A.C.J. Duinmaijer and A.P.G. Welbers, "Matching Properties of MOS Transistors," *IEEE Journal of Solid-State Circuits*, vol. SC-24, pp. 1433–1440, October 1989.

[241] B. Pellegrini, R. Saletti, B. Neri and P. Terreni, "$1/f^v$ Noise Generators," in *Noise in Physical Systems and $1/f$ Noise*, p. 425. North-Holland, New York, 1985.

[242] S. Piguet, F. Rahali, M. Kayal, E. Zysman and M. Declercq, "A New Routing Method for Full Custom Analog ICs," in *Proc. IEEE Custom Integrated Circuits Conference*, pp. 2771–2774, May 1990.

[243] L.T. Pillage and R.A. Rohrer, "Asymptotic Waveform Evaluation for Timing Analysis," *IEEE Trans. on Computer-Aided Design of Integrated Circuits and Systems*, vol. CAD-9, n. 4, pp. 352–366, April 1990.

[244] P.A.D. Powell and M.I. Elmasry, "The ICEWATER Language and Interpreter," in *Proc. IEEE/ACM Design Automation Conference*, pp. 98–102, June 1984.

[245] D.R. Preslar and J.F. Siwinski, "An ECL/I^2L Frequency Synthesizer for AM/FM Radio with an Alive Zone Phase Comparator," *IEEE Trans. on Consumer Electronics*, August 1981.

[246] W. Press, B. Flannery, S. Teukolsky and W. Vetterling, *Numerical Recipes in C. The Art of Scientific Computing*, Cambridge University Press, 1989.

[247] V.S. Pugachev and I.N. Sinitsyn, *Stochastic Differential Systems: Analysis and Filtering*, Wiley, Chichester, Sussex, New York, 1987.

[248] T.L. Quarles, *Analysis of Performance and Convergence Issues for Circuit Simulation*, Ph.D. thesis, University of California, Berkeley, 1989.

[249] V. Raghavan, R.A. Rohrer, L.T. Pillage, J.Y. Lee, J.E. Bracken and M.M. Alaybeyi, "AWE-inspired," in *Proc. IEEE Custom Integrated Circuits Conference*, May 1993.

[250] A.L. Ressler, *A Circuit Grammar for Operational Amplifier Design*, Ph.D. thesis, Massachusetts Institute on Technology, 1984.

[251] D. Reynolds, "A 320MHz CMOS Triple 8b DAC with On-Chip PLL and Hardware Cursor," in *Proc. IEEE International Solid-State Circuits Conference*, pp. 50–51, February 1994.

[252] J. Rijmenants, J.B. Litsios, T.R. Schwarz and M.G.R. Degrauwe, "ILAC: An Automated Layout Tool for Analog CMOS Circuits," *IEEE Journal of Solid-State Circuits*, vol. SC-24, n. 2, pp. 417–425, April 1989.

[253] R. Rohrer, L. Nagel, R.G. Meyer and L. Weber, "Computationally Efficient Electronic-Circuit Noise Calculations," *IEEE Journal of Solid-State Circuits*, vol. SC-6, n. 4, pp. 204, August 1971.

[254] J.S. Roychowdhury, "SPICE3 Distortion Analysis," M.S. thesis, University of California, Berkeley, 1989.

[255] G. Ruan, "A Behavioral Model of A/D Converters Using a Mixed-Mode Simulator," *IEEE Journal of Solid-State Circuits*, pp. 283–290, March 1991.

[256] A.E. Ruehli, "Survey of Computer-Aided Electrical Analysis of Integrated Circuit Interconnections," *IBM Journal of Research and Development*, vol. 23, n. 6, pp. 626–639, November 1979.

[257] A.E. Ruehli, "Circuit Analysis, Simulation and Design," in *Layout Design and Verification*. T. Ohtsuki Ed., North Holland, 1986.

[258] A.E. Ruehli and P.A. Brennan, "Efficient Capacitance Calculations for three-Dimensional Multiconductor Systems," in *IEEE Trans. on Microwave Theory and Techniques*, vol. 21, pp. 76–82, February 1973.

[259] R.A. Rutenbar, "Analog Design Automation: Where are we ? Where are we going ?," in *Proc. IEEE Custom Integrated Circuits Conference*, pp. 1311–1318, May 1993.

[260] K.A. Sakallah and S.W. Director, "An Event-Driven Approach for Mixed Gate and Circuit Level Simulation," in *Proc. IEEE International Symposium on Circuits and Systems*, pp. 1194–1197, May 1982.

[261] R. Saleh, S-J Jou and A.R. Newton, *Mixed-Mode Simulation and Analog Multi-Level Simulation*, Kluwer Academic Publishers, Boston, MA, 1994.

[262] R. Saleh, D.L. Rhodes, E. Christen and B.A.A. Antao, "Analog Hardware Description Languages," in *Proc. IEEE Custom Integrated Circuits Conference*, May 1994.

[263] A. Sangiovanni-Vincentelli, EECS 219 class notes, Electrical Engineering and Computer Sciences, University of California, Berkeley, Fall 1991.

[264] A.L. Sangiovanni-Vincentelli, "Circuit Simulation," in *Computer Design Aids for VLSI Circuits*, pp. 19–112. P. Antognetti and D.O. Pederson and H. DeMan Eds., Sijthoff & Noordhoff (The Netherlands), 1980.

[265] A.L. Sangiovanni-Vincentelli, "Circuit Simulation," in *Computer Design Aids for VLSI Circuits*. Sijthoff & Noordhoff, The Netherlands, 1980.

[266] A.L. Sangiovanni-Vincentelli, "Automatic Layout of Integrated Circuits," in *Design Systems for VLSI Circuits*, pp. 113–195. De Micheli, et al. Eds., Martinus Nijhoff, 1987.

[267] H. Schouwenaars, D. Groeneveld and H. Termeer, "A Low-Power Stereo 16-bit CMOS D/A Converter for Digital Audio," in *Proc. IEEE International Solid-State Circuits Conference*, 1988.

[268] M.H. Schultz, E. Trischler and T.M. Sarfert, "SOCRATES: A Highly Efficient Automatic Test Pattern Generation System." *IEEE Trans. on Computer-Aided Design of Integrated Circuits and Systems*, vol. 7, n. 1, pp. 126–137, 1988.

[269] C. Sechen, *VLSI Placement and Routing Using Simulated Annealing*, Kluwer Academic Publishers, Boston, MA, 1988.

[270] C. Sechen and A.L. Sangiovanni-Vincentelli, "Chip-Planning, Placement and Global Routing of Macro/Custom Cell ICs using Simulated Annealing," in *Proc. IEEE/ACM Design Automation Conference*, pp. 73–80, June 1988.

[271] S.J. Seda, M.G.R. Degrauwe and W. Fichtner, "Lazy-Expansion Symbolic Expression Approximation in SYNAP," in *Proc. IEEE International Conference on Computer-Aided Design*, pp. 310–317, November 1992.

[272] *Semiconductor Industry Technology Workshop Conclusions*, Semiconductor Industry Association, 1993.

[273] J. Shao and R. Harjani, "Macromodeling of Analog Circuits for Hierarchical Circuit Design," in *Proc. IEEE International Conference on Computer-Aided Design*, pp. 656–663, November 1994.

[274] B.J. Sheu, A.H. Fung and Y.N. Lai, "A Knowledge-Based Approach to Analog Integrated Circuit Design," *IEEE Trans. on Circuits and Systems*, vol. 35, n. 2, pp. 256–258, February 1988.

[275] H. Shichman, "Computation of DC Solutions for Bipolar Transistor Networks," *IEEE Trans. on Circuit Theory*, vol. CT-16, pp. 460–466, 1969.

[276] H. Shin and A.L. Sangiovanni-Vincentelli, "A Detailed Router Based on Incremental Routing Modifications: MIGHTY," *IEEE Trans. on Computer-Aided Design of Integrated Circuits and Systems*, vol. CAD-6, n. 6, pp. 942–955, November 1987.

[277] Y. Shiraishi, M. Kimura, K. Kobayashi, T. Hino, M. Seriuchi and M. Kusaoke, "A High-Packing Density Module Generator for Bipolar Analog LSIs," in *Proc. IEEE International Conference on Computer-Aided Design*, pp. 194–197, November 1990.

[278] Y. Shiraishi, J. Sakemi and M. Kuzuwada, "A High-Packing Density Module Generator for CMOS Logic Cells," in *Proc. IEEE/ACM Design Automation Conference*, pp. 439–445, June 1988.

[279] J.B. Shyu, G.C. Temes and F. Krummenacher, "Random Errors Effects in Matched MOS capacitors and Current Sources," *IEEE Journal of Solid-State Circuits*, vol. SC-19, pp. 948–955, 1984.

[280] J.M. Shyu, *Performance Optimization of Integrated Circuits*, Ph.D. thesis, University of California, Berkeley, November 1988.

[281] J.M. Shyu and A.L. Sangiovanni-Vincentelli, "ECSTASY: A New Environment for IC Design Optimization," in *Proc. IEEE International Conference on Computer-Aided Design*, pp. 484–487, November 1988.

[282] R.P. Sigg, A. Kaelin, A. Muralt, W.C. Black, Jr. and G.S. Moschytz, "An SC Filter Compiler: Fully Automated Filter Synthesizer and Mask Generator for a CMOS Gate-Array-Type Filter Chip," in *Proc. IEEE International Conference on Computer-Aided Design*, pp. 510–513, 1987.

[283] J. Singh and R. Saleh, "IMACSIM: A Program for Multi-level Analog Circuit Simulation," in *Proc. IEEE International Conference on Computer-Aided Design*, November 1991.

[284] M. Sitkowski, "The Macro Modeling of Phase-Locked Loops for the SPICE Simulator," *IEEE Circuits and Devices*, vol. 7, n. 2, March 1991.

[285] T. Smedes, "Substrate Resistance Extraction for Physics-Based Layout Verification," in *IEEE/PRORISC Workshop on CSSP*, pp. 101–106, March 1993.

[286] T. Smedes, N.P. van der Meijs and A.J. van Genderen, "Extraction of Circuit Models for Substrate Cross-talk," in *ICCAD*, pp. 199–206, November 1995.

[287] T. Souders and G. Stenbakken, "Modeling and Test Point Selection for Data Converter Testing," *IEEE International Test Conference*, 1985.

[288] T. Souders and G. Stenbakken, "Cutting the High Cost of Testing," *IEEE Spectrum*, pp. 48–51, March 1991.

[289] B.R. Stanisic, N.K. Verghese, D.J. Allstot, R.A. Rutenbar and L.R. Carley, "Addressing Substrate Coupling in Mixed-Mode ICs: Simulation and Power Distribution Synthesis," *IEEE Journal of Solid-State Circuits*, vol. SC-29, n. 3, pp. 226–237, March 1994.

[290] C.H. Stapper, "Modeling of Defects in Integrated Circuit Photolithographic Patterns," *IBM Journal of Research and Development*, vol. 28, n. 4, pp. 461–475, July 1984.

[291] G. Stenbakken and T. Souders, "Test-Point Selection and Testability Measures via QR Factorization of Linear Models," *IEEE Trans. on Instrumentation and Measurement*, June 1987.

[292] G.N. Stenbakken and T.M. Souders, "Linear Error Modeling of Analog and Mixed-Signal Devices," *Proc. IEEE International Test Converence*, pp. 573–581, 1991.

[293] D.K. Su, M. Loinaz, S. Masui and B. Wooley, "Experimental Results and Modeling Techniques for Substrate Noise in Mixed-Signal Integrated Circuits," *IEEE Journal of Solid-State Circuits*, vol. SC-28, n. 4, pp. 420–430, April 1993.

[294] D.K. Su, M.J. Loinaz, S. Masui and B.A. Wooley, "Experimental Results and Modeling Techniques for Substrate Noise in Mixed-Signal Integrated Circuits," *IEEE Journal of Solid-State Circuits*, April 1993.

[295] K. Swings, S. Donnay and W. Sansen, "HECTOR: A Hierarchical Topology-Construction Program for Analog Circuits based on a Declarative Approach to Circuit Modeling," in *Proc. IEEE Custom Integrated Circuits Conference*, pp. 531–534, May 1991.

[296] K. Swings, G. Gielen and W. Sansen, "An Intelligent Analog IC Design System Based on Manipulation of Design Equations," in *Proc. IEEE Custom Integrated Circuits Conference*, pp. 861–864, May 1990.

[297] K. Swings and W. Sansen, "DONALD: A Workbench for Interactive Design Space Exploration and Sizing of Analog Circuits," in *Proc. European Design Automation Conference*, pp. 475–479, February 1991.

[298] K. Swings and W. Sansen, "ARIADNE: a Constraint-Based Approach to Computer-Aided Synthesis and Modeling of Analog Integrated Circuits," *Analog Integrated Circuits and Signal Processing*, vol. 3, n. 3, pp. 197–215, May 1993.

[299] G. Szentirmai, "FILSYN—A General Purpose Filter Synthesis Program," in *Proc. of the IEEE*, pp. 1443–1458, October 1977.

[300] E. Tan, "Phase-Locked Loop Macromodels," M.S. thesis, University of California, Berkeley, August 1990.

[301] T. Tanaka, "Parsing Electronic Circuits in a Logic Grammar," *IEEE Trans. on Knowledge and Data Engineering*, vol. 5, n. 2, pp. 225–39, April 1993.

[302] R. Telichevesky, K.S. Kundert and J. White, "Efficient Steady-State Analysis Based on Matrix-Free Krylov-Subspace Methods," in *Proc. Design Automation Conference*, June 1995.

[303] L.J. Tick, "The Estimation of the Transfer Functions of Quadratic Systems," *Technometrics*, vol. 3, pp. 563–567, 1961.

[304] C. Toumazou and C.A. Makris, "Analog IC Design: Part I—Automated Circuit Generation: New Concepts and Methods," *IEEE Trans. on Computer-Aided Design of Integrated Circuits and Systems*, vol. CAD-14, n. 2, pp. 218–238, February 1995.

[305] J. Trontelj, L. Trontelj, A. Pleteršek, A. Vodopivec and G. Shenton, "Synthesis and Layout Compilation Automation of Mixed Analog-Digital ASICs," in *Proc. IEEE International Symposium on Circuits and Systems*, vol. 2, pp. 816–819, May 1990.

[306] L. Trontelj, J. Trontelj, Jr. T. Slivnik, R. Sosic and D. Lucas, "Analog Silicon Compiler for Switched Capacitor Filters," in *Proc. IEEE International Conference on Computer-Aided Design*, pp. 506–509, 1987.

[307] J. Vandewalle, H. De Man and J. Rabaey, "The adjoint switched capacitor network and its application to frequency, noise and sensitivity analysis," in *Circuit Theory and Applications*, vol. 9, pp. 77–88, 1981.

[308] N. Verghese, D. Allstot and M. Wolfe, "Fast Parasitic Extraction for Substrate Coupling in Mixed-Signal ICs," in *Proc. IEEE Custom Integrated Circuits Conference*, May 1995.

[309] N.K. Verghese, D.J. Allstot and S. Masui, "Rapid Simulation of Substrate Coupling Effects in Mixed-Mode ICs," in *Proc. IEEE Custom Integrated Circuits Conference*, pp. 1831–1834, May 1993.

[310] J. Vital, N.C. Horta, N.S. Silva and J.E. Franca, "CATALYST: A Highly Flexible CAD Tool for Architecture-Level Design and Analysis of Data Converters," in *Proc. European Design Automation Conference*, pp. 472–477, September 1993.

[311] B. Wang, J.R. Helums and C.G. Sodini, "MOSFET Thermal Noise Modeling for Analog Integrated Circuits," *IEEE Journal of Solid-State Circuits*, vol. 29, n. 7, July 1994.

[312] K.M. Ware, H-S. Lee and C.G. Sodini, "A 200 MHz CMOS Phase-Locked Loop with Dual Phase Detectors," *IEEE Journal of Solid-State Circuits*, vol. 24, n. 6, December 1989.

[313] T.C. Weigandt, B. Kim and P.R. Gray, "Analysis of Timing Jitter in CMOS Ring Oscillators," in *Proc. IEEE International Symposium on Circuits and Systems*, May 1994.

[314] D. Weiner and J. Spina, *Sinusoidal Analysis and Modeling of Weakly Nonlinear Circuits*, Van Rostrand Reinhold Company, New York, 1980.

[315] J. White, A. Sangiovanni-Vincentelli, F. Odeh and A. Ruehli, "Waveform Relaxation: Theory and Practice," *Trans. Society Computer Simulation*, vol. 2, n. 1, June 1985.

[316] B.C. Williams, "Qualitative Analysis of MOS Circuits," M.S. thesis, Massachusetts Institute on Technology, 1984.

[317] L.A. Williams and B.A. Wooley, "MIDAS—A Functional Simulator for Mixed Digital and Analog Sampled Data Systems," in *Proc. IEEE International Symposium on Circuits and Systems*, May 1992.

[318] G. Winner, A. Nguyen and C. Slemaker, "Analog Macrocell Assembler," *VLSI Systems Design*, vol. 8, n. 5, pp. 68–71, May 1987.

[319] G. Work, G. Talbot, A. Ferris, N. Henderson and P. McGuiness, "A 20 ns Color Lookup Table for Faster Scan Displays," in *Proc. IEEE International Solid-State Circuits Conference*, pp. 310–311, February 1985.

[320] H. Yaghutiel, A.L. Sangiovanni-Vincentelli and P.R. Gray, "A Methodology for Automated Layout of Switched-Capacitor Filters," in *Proc. IEEE International Conference on Computer-Aided Design*, pp. 444–447, 1986.

[321] H. Yaghutiel, S. Shen, P.R. Gray and A.L. Sangiovanni-Vincentelli, "Automatic Layout of Switched-Capacitor Filters for Custom Applications," in *Proc. IEEE International Solid-State Circuits Conference*, pp. 170–171, February 1988.

[322] I.A. Young, J.K. Greason and K.L. Wong, "A PLL Clock Generator with 5 to 110 MHz of Lock Range for Microprocessors," *JSSC*, vol. SC-27, n. 11, pp. 1599–1607, November 1992.

INDEX